T0146061

Why Mars

New Series in NASA History

Why Mars

NASA and the Politics of Space Exploration

W. Henry Lambright

Johns Hopkins University Press
Baltimore

© 2014 Johns Hopkins University Press
All rights reserved. Published 2014
Printed in the United States of America on acid-free paper
9 8 7 6 5 4 3 2 1

Johns Hopkins University Press
2715 North Charles Street
Baltimore, Maryland 21218-4363
www.press.jhu.edu

Library of Congress Cataloging-in-Publication Data

Lambright, W. Henry, 1939–
   Why Mars : NASA and the politics of space exploration / W. Henry
Lambright.
      pages cm — (New series in NASA history)
   Includes bibliographical references and index.
   ISBN-13: 978-1-4214-1279-5 (hardcover : alk. paper)
   ISBN-13: 978-1-4214-1280-1 (electronic)
   ISBN-10: 1-4214-1279-9 (hardcover : alk. paper)
   ISBN-10: 1-4214-1280-2 (electronic)
   1. Space flight to Mars. 2. Mars (Planet)—Exploration.
3. Astronautics and state—United States. 4. United States—Politics
and government. 5. United States. National Aeronautics and Space
Administration. I. Title.
   TL799.M3L36   2014
   629.43'543—dc23   2013028978

A catalog record for this book is available from the British Library.

Special discounts are available for bulk purchases of this book. For more informa-
tion, please contact Special Sales at 410-516-6936 or specialsales@press.jhu.edu.

Johns Hopkins University Press uses environmentally friendly book
materials, including recycled text paper that is composed of at least
30 percent post-consumer waste, whenever possible.

To Nancy

Contents

Preface

In 2006, a conference took place at Syracuse University's Minnowbrook Center in the Adirondack Mountains of New York. The focus of the conference was on the "great stories" of humanity, including Greek tales such as Homer's *Odyssey*. The conference organizer, Kaye Lindauer, asked me to speak about a contemporary great story, namely, space exploration. The immediate interest of those attending my talk, a cross section of professionals, was the Moon and Apollo, about which I had written. However, it was clear that most also wanted to discuss "what next," or Mars. I left the conference feeling that Mars would be my next book.

I needed resources to accomplish this work. NASA, which celebrated its fiftieth anniversary in 2008, was interested in having its history told. It opened a competition involving various topics, and I was fortunate to be an awardee. NASA provided funds, but left it to me to do the research and writing without any constraints.

I soon commenced research. In undertaking this task, I felt a responsibility not only to examine the past and convey an active present, but also to analyze what it takes to sustain a very long and difficult quest. NASA and its allies have chosen to keep at the Mars endeavor over many decades in spite of ever-shifting political winds. That the program has achieved much in spite of obstacles is testament to the persistence of scientists, engineers, managers, and the political appointees heading NASA. The Mars record has flaws to be sure, and these are chronicled in this work. But, for the most part, the Mars story is remarkable. Exploration is a struggle. Individuals and their institutions have stretched to perform deeds that are daunting. They have been motivated mainly by the lure of Mars and its association with life.

This book emphasizes what they have done to formulate missions, establish priorities, and get the funds to accomplish technical miracles. It is thus a political history of the Mars program. It is about decisions, policy, and power—the push for exploration. It is about leaders behind NASA's Mars program, and their

Washington, D.C., travails. It is primarily about the robotic program that has taken NASA, the nation, and Earth from a Mariner flyby in the mid-1960s to Mars Science Laboratory's Curiosity rover in the second decade of the twenty-first century. The robotic program is immensely valuable in itself, especially in regard to finding evidence of Martian life, present or past. It is also essential to eventual human exploration of Mars.

I have had much help in accomplishing my own Mars project. Once underway in 2008, I had the aid of many at NASA's history office: Steve Dick, NASA historian, at the outset and then Bill Barry, his successor; Steve Garber, who read early drafts of the manuscript; and archivists who helped me locate materials, including Liz Suckow, John Hargenrader, Colin Fries, and Jane Odom. Nadine Andreassen, on the staff of the history office, helped maneuver funds through NASA's bureaucratic complexities. Jens Feeley, of the Science Mission Directorate, was always available to help me set up essential interviews with busy agency officials. I wish to thank the many individuals inside and outside NASA who gave their time for interviews. There are too many to list separately. I also had the help of history staff at the Jet Propulsion Laboratory (JPL) in California, including Julie Cooper and Charlene Nichols. Erik Conway, JPL historian, aided me enormously with background information and interview arrangements. I also drew on voluminous files at the National Archives, where David Pfeiffer and his staff helped me greatly.

At Syracuse, I was assisted by a sequence of students, particularly Erin D'Loughy, Kimberly Pierce, Madison Quinn, Bindya Zachariah, and Dayana Bobko. Also helping me were staff at the Center for Environmental Policy and Administration of the Maxwell School, where I am based: Carley Parsons and Marlene Westfall Rizzo. I am grateful also for the assistance of the Johns Hopkins University Press staff, including Bob Brugger and Melissa Solarz. An anonymous Johns Hopkins reviewer provided valuable advice. Jeremy Horsefield added his editorial skills to measurably improve the book.

My sons and their wives—Dan and Sue, Nathaniel and Kristina—and my grandchildren, Ben, Katie, Bryce, and Darius, have been a source of inspiration. They will all someday, I believe, see humans on Mars. Finally, I owe a special sense of gratitude to my wife, Nancy. She got me started on this project by urging me to accept the invitation to speak at the Minnowbrook Conference. She then endured its frustrations along the way and provided the final push for its conclusion. To all who helped, directly or indirectly, I hope the final product is worth the time you gave me. Any errors are my responsibility.

Why Mars

Introduction

At 1:25 a.m. (EDT) on August 6, 2012, NASA's Mars Science Laboratory (MSL), encased in a larger spacecraft and protected by a heat shield, hit the atmosphere of Mars. After a journey of eight months and 352 million miles, MSL embarked on what NASA called "seven minutes of terror." In this brief span of time, MSL would need to decrease its speed from 13,000 miles per hour to almost zero—or it would crash. Failure was unthinkable for a $2.5 billion mission at a time when NASA was under budgetary siege. The Mars atmosphere immediately caused the spacecraft to slow, but seven miles above the Mars surface the spacecraft was still flying at 900 miles per hour. At this point, the spacecraft unfurled a giant, 51-foot parachute.

As the spacecraft's rate of descent gradually diminished, MSL disconnected from the spacecraft that had been carrying it to this point. The spacecraft flew off, and retro-rockets blasted from MSL, causing it to come to a virtual hover two stories above the Mars surface. At one ton in weight, containing delicate instruments, the car-sized machine was too heavy to complete its landing with retro-rockets or airbags. Instead, for the first time, a newly invented device attached to MSL, called the sky crane, deployed, and cables carefully lowered the machine to the ground. Finally, with cables disconnected, the sky crane rocketed away from what NASA had now safely placed on the surface—the nuclear-powered MSL rover called Curiosity. All this happened automatically

154 million miles from Earth. The landing occurred at 1:32 a.m. (EDT). It took another 14 minutes for radio signals to go from Mars to Earth and reach the Jet Propulsion Laboratory (JPL) in Pasadena, California. Allen Chen, flight dynamics engineer at JPL, received the information. He announced excitedly, "Touchdown confirmed. We're safe on Mars!"[1]

Never before had a technology this complex gone to Mars. MSL, with its Curiosity rover, climaxed a multiyear program geared to "following the water." Its goal was to discover whether Mars was now or previously capable of being inhabited. MSL Curiosity would not actually find life. It aimed at locating the chemical "building blocks" of life. A later flight, or series of flights, would return a sample of Mars soil and rock to Earth's laboratories for analysis. Such a mission of far greater expense lay well in the future. But this particular mission was critical to Mars exploration—a milestone in a long-term quest that had begun over a half century earlier, building on what had gone before, enabling what might come ahead.

Getting to this point was a remarkable accomplishment, not only in science and technology, but in public policy and program implementation. Not only did NASA have to surmount severe technological barriers, but it also had to meet daunting political challenges along the way. In some ways, the political problems were greater than those that were technical. Large technical achievement, especially when dealing with government and costing billions over many years, does not happen automatically. It takes a strong push of political advocacy from inside and outside NASA to make Mars a funding priority, establish a program, and carry it out successfully. Who does what to forward Mars exploration? How? The answers are critical to the history of NASA and the Red Planet.

The intent of this book is to illuminate the role of key individuals and institutions that have constituted a moving force for policy action in Mars exploration. Its thesis is that an informal and changing coalition of advocates inside and outside NASA has sought to make NASA the institutional embodiment and lever for their quest to the Red Planet. The influence and limits of this coalition, as well as their scientific and political strategies, have shaped the course and pace of the Mars exploration program.

The study contends that over the long haul, the advocacy coalition has propelled Mars exploration forward. This has been particularly the case as it has turned individual missions into an integrated and sequential whole, beginning in the early 1990s. It has built political support for this program and sustained it in the face of changing times and opposition. The actors most critical to coalition leadership and influence affecting the Mars exploration program have been senior officials of NASA. Decisions and strategies in Washington, D.C., have powered (or frustrated) exploration on Mars.

The focus of this book is not the history of science, or advance of technology, or cultural aspects of Mars. Such subjects come up, but not as foreground. This book seeks to reveal and analyze the politics and policy behind Mars exploration.

The Mars Exploration Program

From NASA's establishment in 1958, the space agency looked to Mars as a compelling prize, the one place, beyond the Moon, where robotic and human exploration could converge. Over the years the human space venture to Mars remained a dream, on NASA's agenda, but always on a distant horizon. NASA's Mars robotic program—the focus of this study—has now been actualized, marking one of NASA's greatest achievements.

What has been the nature of NASA's Mars exploration program? How was it created and sustained? How has it adapted to scientific findings and shifting political winds? What have been the barriers to the program? How was opposition countered? Where is the program going? These and other questions have not been answered adequately in the existing literature. Most writing about Mars deals with specific missions and emphasizes the technical aspects. The people, institutions, politics, and policy behind the technical exploits get relatively little attention. NASA's role, although mentioned, is seldom addressed in depth. What is significant is that the missions form part of an ongoing government effort that has lasted over half a century and promises to extend indefinitely into the distant future. Mars is a federal program, but it is also a destination, a place and a magnet for the human imagination. For advocates of robotic and human Mars exploration—who seem often to disagree as much as they agree—it is a great quest, a difficult and noble journey into the unknown.

Mars exploration has evolved from the Mariner flybys in the 1960s, which provided the first blurred glimpses of the Red Planet, to orbiters and landers in the 1970s. Later, in the 1990s, NASA created machines capable of not only landing but also roving the planet. The Clinton administration in 1996 set as a national goal that NASA embark on "a sustained program to support a robotic presence on the surface of Mars."[2] By the early twenty-first century, NASA was building an intricate infrastructure on Mars, a technical system involving orbiters, landers, rovers, laboratories, and communications systems. NASA, moreover, had company on Mars, as other nations sent their own devices. The names of the machines have become well known not only to scientists but also to the public over the years: Mariner, Viking, Pathfinder, Mars Global Surveyor, Spirit, Opportunity, Phoenix, MSL with its Curiosity rover, and others. With modern technology, citizens on Earth can participate in an epic adventure and explore Mars through robotic machines of incredible capacity. These machines extend human senses of sight, sound, and touch across millions of miles. They have taken NASA, America, and the world to a period that John Grotzinger, chief scientist of the MSL, called "the golden era of Mars exploration," a time of "extended, overlapping, and increasingly coordinated missions."[3]

The evolution of the program has not been all positive. Nor is the future certain. There have been expensive failures amidst the successes. There have been ebbs and flows in scientific and public enthusiasm, heights of exultation, depths of despair. For Mars exploration, between Viking in 1976 and Mars Observer's launch in 1992, there was a long gap in missions, and then Mars Observer itself became what was called a $1 billion failure. But NASA maintained the quest in the 1990s and into the new millennium. That it did so was not easy. It was a test of scientific, bureaucratic, and political resilience. The key issue in understanding the Mars exploration program is one that is generic in American democracy: how to maintain a long-term, large-scale, high-risk, and expensive federal research and development (R&D) program in the face of competing scientific, bureaucratic, and public priorities and ever-changing political winds.

Big Science

Mars exploration is a striking example of big science. Big science is characterized by large organizations, multidisciplinary teams, great expense, government management, and, often, political controversy. There are various examples of big science, at NASA and other agencies, but this Mars endeavor especially illuminates "programmatic" or "distributed" big science. Much big science is concentrated in a single huge machine. But distributed big science in this case consists of missions (also called *projects*) that make up an extended, multidecadal program of exploration. Projects may be closely or loosely coupled. Some of the individual missions are "big" by any standard, in the sense of having billion-dollar costs. Others are moderate by big science standards, costing hundreds of millions. There was a time in the 1990s when Mars missions were pushed hard to be "faster, better, cheaper" (FBC), attributes that usually meant smaller. But Mars missions have subsequently grown, with MSL listed at $2.5 billion.

Moreover, when the individual missions are aggregated in a coherent program, the combination is obviously very large in scale.

While specific numbers are hard to delineate or aggregate precisely over a half century or more, it is virtually certain that NASA has spent more money on Mars than any other single planet over its existence as an agency. In recent years, approximately one-half of the planetary budget has gone to Mars, and the planet has had its own director in NASA Headquarters and at JPL. Whereas the Cassini mission to Saturn was an example of concentrated, multibillion-dollar big science, the missions to Mars are spread out, with projects of varying scales. The Mars missions are distributed in time—taking place over decades. They are also distributed in purpose—orbiters, landers, and rovers. This is not only big science but an example of a large technical system in action. When Phoenix landed on Mars in 2008, an orbiting Mars satellite photographed the event. At the same time, Spirit and Opportunity roved the terrain. Similarly, Mars Reconnaissance Orbiter "watched" as MSL descended to Mars in 2012. The national policy to create a "sustained . . . robotic presence" on Mars was realized.

Big science is important to better comprehend because it represents a high priority within an agency and within a national budget. Big science projects can be high-visibility "flagships" for an agency and often for a country. In execution, big science projects link government, university, and industry into large and diverse teams, increasingly with international partners. Big science entails extremely challenging management and political issues. It absorbs much of the agency's money and prevents smaller efforts from being undertaken. The political dilemmas—who gets what, when, and how—can be the most difficult of all to resolve in making a long-term big science program succeed.

The Call of Mars

At the most general level, Mars exploration is about understanding Earth's arguably most interesting neighbor. Mars is both like and unlike Earth. It is the one planet people on Earth can see in some detail. Its rich, red color was always a source of speculation prior to the space age. The ancient Romans named the planet Mars after their red god of war. The fourth planet from the Sun, it travels an elliptical orbit. At its closest approach to Earth, it is 48,700,000 miles (78,390,000 kilometers) away. Mars's diameter is about 4,200 miles (6,790 kilometers), which is a little over half the diameter of Earth. Mars takes about 687 Earth days to go around the Sun. While the length of a Mars year is much longer than that of Earth, the Mars day is remarkably similar. Mars rotates on its axis once every 24 hours and 37 minutes. Mars also, like Earth, has seasons. During the Martian winter, observers see polar caps that are relatively large, and which then shrink during the Martian summers. Mars has an atmosphere, but the atmosphere is much thinner than that of Earth. It consists chiefly of carbon dioxide, with small amounts of nitrogen and other gases. The atmosphere of Earth, in contrast, is heavy on nitrogen and oxygen. Because of its distance from the sun, Mars is extremely cold. The temperature ranges from –191°F to –24°F (–124°C to –31°C). It has two moons, Phobos and Deimos.[4]

The similarities and differences are striking. They have contributed to the human desire to know more about Mars. Beyond these factors, there are at least three reasons that Mars has long been a special magnet for scientists and public alike. First, there is the question of life on Mars. For many years prior to NASA's establishment, and even for some time afterward, there was a belief in many circles that there was life on the surface. When it became clear that such life would have difficulty surviving, various scientists suggested there were still possibilities in sheltered places on Mars, what they termed oases. Even scientists who today believe no life exists on the surface admit possibilities under the surface, in permafrost. No one expects that such life is highly developed. Most likely it is bacterial. But even if no life at all exists now, it may have once existed when Mars apparently had a very different climate and substantial water flowed. So the question of life on Mars, now or in the past, has always been a compelling driver for the Mars exploration program, in many ways the most significant one in terms of NASA history. For when optimism about life on Mars reigned, the program did well in obtaining resources. When pessimism took hold, the program languished.

There is a holy grail for the Mars robotics program: it is called Mars Sample Return (MSR), the retrieval of soil and rock from Mars. Samples would be brought back to Earth for examination in laboratories around the world to detect signs that will answer the big question of life. The challenge is to bring back "the right stuff," and do so in a way protective of possible Martian life as well as human life from contamination. MSR is a monumental test, perhaps the single most complex and important mission for NASA since Apollo. It lies ahead and probably will require international collaboration owing to projected cost. But it has enormous value as a goal, on which there is widespread consensus. It gives direction and sequence to missions leading to it.

The second driver is the desire to send human life to Mars. The robotic program thus becomes a precursor to human missions, much as robotic Ranger and Surveyor missions were scouts for the Apollo voyage to the Moon. This makes the human spaceflight program a potential ally of the robotic program. NASA leaders can envision useful connections between robotic activity and human Mars journeys. When President George H. W. Bush proclaimed in 1989 that America should return humans to the Moon and then go on to Mars, his declaration also carried an understanding that robotic missions and human flight were partners in exploration. The same understanding applied to his son George W. Bush's similar declaration in his 2004 "Vision for Space Exploration." President Barack Obama in 2010 decided to bypass the Moon and emphasize Mars as the ultimate destination for humanity, sometime in the 2030s. An asteroid landing would be the interim step. Again, robotic flight would pave the way.

Robots may be partners with human spaceflight over the long haul, but in the short run they often compete for limited funds and represent different cultures. The robotic science program of NASA sees the human space effort as embodying an engineering and astronaut culture that wants to go to Mars "because it is there." This is indeed a motivation, a "frontier" attitude, sometimes with manifest destiny overtones, to extend the human presence to outer space. "We go because we must." "It is in our genes," goes the refrain. There are those human spaceflight advocates who see Mars as not only a mountain to be climbed but a territory to be "terraformed" and settled. NASA's "can do" engineers see exceptional technological challenge in human Mars exploration. Astronauts see romance and adventure. In contrast, robotic-oriented scientists seek basic understanding in comparative planetology, with emphasis on Martian life issues.

The field centers that report to the human spaceflight directorate in NASA's Washington headquarters are different from those that report to the robotic science office. The human spaceflight program, as noted, has astronauts, and they are more than just another set of employees within NASA; they are in many ways the agency's public face and source of inspiration to young people. Mars is the only planet where human missions are likely in the twenty-first century. The human spaceflight institutions and their cadre recognize the precursory importance to them of the robotics program. The interests of robotic science and human exploration potentially connect in the MSR mission. NASA needs to know what environmental hazards astronauts would face on Mars and also if they can convert the Red Planet's physical resources into assets for long-term stays. Also, if NASA cannot bring rocks and soil back to Earth, it may not be able to bring human beings home. The technologies of MSR have direct relevance to human spaceflight. The short-term budget competition nevertheless can interfere with cooperation concerning Mars across NASA divisions.

The third driver is political. Space has always engendered rationales concerned with national leadership, prestige, pride, inspiration, and competition with other nations. It has been used by governmental officials to enhance cooperation also. Mars in the 1960s was a destination both the United States and the Soviet Union sought to reach first during the Cold War. It was a target of Cold War rivalry. During and after the Cold War, advocates of both human and robotic programs also called for using Mars as a political symbol for a joint enterprise uniting the rival superpowers. More recently, Mars has been seen as a vehicle for global cooperation. While much "Mars Together" rhetoric has been about human expeditions, it includes the robotic program in its collaborative embrace. NASA and the European Space Agency (ESA) have sought in particular to work together on robotic missions, but relations have been turbulent.

Finding life on Mars, sending life to Mars, and using Mars for political ends have been the key drivers behind Mars exploration for decades. At different points in history one or another of these rationales has dominated. Sometimes more than one has influenced Mars exploration. Which one has resonated has depended on innumerable factors, including the strategies of Mars advocates. The most consistent theme affecting the direction and pace, as well as ups and downs, of the robotic program has been the prospect of finding evidence of Martian life. The connection with life has made Mars special among planets in the solar system.

The Policy Process

Mars exploration has evolved through a series of programs, projects, and struggles over the years. There have been five periods in policymaking for Mars to date, with a sixth debated at the time of writing. The first, characterized by fly-bys and orbiters, was that of Mariner. It extended from 1958 to the early 1970s, when the aborted Voyager and Viking initiated a second era that culminated in twin landings in 1976. Viking turned out to be a single mission, rather than the start of a program of sequential missions. The third era was one of hiatus and agitation to regain momentum. It started in the wake of Viking and did not end until 1992, when Mars Observer went up. Mars Observer failed, giving rise to a fourth era that extended through the 1990s.

Mars missions were now part of programs called Discovery and Mars Surveyor. This fourth era, begun in the early 1990s, was characterized by the FBC mantra and saw a number of great successes, such as Pathfinder and Mars Global Surveyor. It also featured an abrupt shift in orientation due to the discovery of a Mars meteorite that was said to have fossils of microbiological Martian life. NASA sought to speed the pace toward MSR. This had been an ultimate goal of the Mars Surveyor Program but did not have a firm deadline. The shift in the mid-1990s established a harder deadline to force action. However, the failure of two Mars missions in 1999 ended this period and gave rise to a fifth era of Mars discovery. Called Mars Exploration Program, this era emphasized an incremental "follow-the-water" strategy aimed at finding habitable places. It downplayed FBC in favor of "mission success." The climax of this program came in 2012, with the landing of the MSL rover, Curiosity.

MSL is the most challenging and costly Mars mission since Viking—in some ways it is a much more sophisticated version of the Viking follow-on that was proposed but aborted in the late 1970s. It "follows the water," but also transitions to looking for carbon and other building blocks of life.[5] A smaller, specialized orbiter project, the Mars Atmosphere and Volatile Evolution Mission (MAVEN), was launched in November 2013, ending the follow-the-water programmatic era as conceived.

Finally, a sixth era is on the NASA agenda and in planning. It initially had been characterized by major international missions, starting jointly with Europe, in 2016 and in 2018. Its ultimate goal was the return of a sample of Martian soil and rock to Earth for examination in the 2020s. However, White House cuts to NASA's budget precluded the planned 2016 and 2018 missions with Europe. NASA Administrator Charles Bolden, in response, pledged to continue Mars missions, with the path ahead to be determined.[6] He stated that he wanted the new program—called Mars Next Decade—to be more integrated with human spaceflight goals.[7]

Meanwhile, to help maintain momentum, NASA proposed a mission for 2016, called InSight. This would be a relatively modestly priced venture to use seismic instruments to study Mars's interior forces. NASA also indicated that it would contribute to the European missions via certain instruments. Most importantly, NASA, in late 2012, received approval for a $1.5 billion rover for 2020 which would build on MSL's Curiosity. How this 2020 rover mission related to MSR and Mars Next Decade was not clear at the time of writing. However, it did seem to augur that there would be a sixth era of Mars exploration, arguably beginning in 2016.[8]

Each of the programs and missions passed and proposed goes through various decision stages. In theory, these stages are straightforward. In an ideal situation, the policy process for each program or mission begins with an awareness of need or opportunity that is sufficiently compelling to get this potential program on the *agenda* of NASA decision makers. *Planning* for action follows along with *formulation* of proposals. Then comes formal *adoption* of the new program or specific mission by authoritative political decision makers. Once adopted and funded, the program moves into an *implementation* stage that can last for years. At some point during execution, there is an *evaluation* stage that can lead to *reorientation*. The last stage is *completion*. Completion of the program or mission can prelude a follow-on effort.

The sequence of decisions noted above is obviously an abstraction. Reality is not so linear and is often messy. Stages overlap. Policy proposals are *rejected*. A program or specific project can be *terminated* prior to its planned end. One set of stages can recycle and lead to the beginning of another sequence. In Mars exploration, new programs can be planned before an existing one is completed. There can also be a gap between programs, especially following failure. The process can move swiftly or slowly—or stagnate. There are technical, administrative, and political barriers every step of the way. The journey is torturous. Indeed, it has its casualties in terms of careers of scientists, engineers, and administrators. Overall, however, there is progress. Mars exploration takes place, through a "program of programs."

The Role of NASA

What or who moves the decision process forward? The key governmental institution providing direction and pace for the Mars program in the United States is NASA. NASA and the Mars politics engulfing it are the orientation of this study. NASA is the engine that powers Mars exploration. Sometimes the engine motors well; other times, it sputters. NASA is involved where major decisions are made. It is often influential in policy development, always critical in implementation. NASA provides money and management to enable programs and projects to go from agenda setting to completion—or failure. But who influences NASA? NASA has *internal* and *external* constituencies that seek to command its behavior.

The *internal* core for Mars policymaking is in NASA's Washington headquarters, the Science Mission Directorate (SMD). This division has had different names over its history. NASA has almost always had a "Science Directorate," headed by an associate administrator who reported to the NASA Administrator.

Below the associate administrator usually has been a planetary director. Below the planetary director would typically be found the Mars program. Bureaucratically, Mars is not and has not been high on the agency's organization chart. But it has a visibility that far exceeds its place in the hierarchy. At times in history, it has had a visibility beyond much else that NASA did.

The associate administrator for science has always had considerable power among directorates within NASA, generally second only to the associate administrator for human spaceflight. SMD gets ample advice from the technical community, including the National Academy of Sciences. Indeed, scientists are extremely active in the agenda-setting stage of policy and produce planning documents galore.

SMD is also subject to pressures from superiors and field centers within NASA and beyond. The associate administrator of SMD is a decision maker embracing most aspects of space science in addition to Mars. He or she integrates many factors in priority choices, including his or her own personal preferences. While key decisions affecting Mars may emanate from the associate administrator, generally a senior career official, NASA's top politically appointed executive—the NASA Administrator—makes final decisions on the most costly and controversial matters. For the SMD head, Mars is one interest among many seeking decisions and resources. From the NASA Administrator's perspective, SMD is one interest among many wanting decisions and resources. The NASA leader (like the associate administrator) has to balance a multitude of pressures. There is never enough time or money to satisfy everyone.

The associate administrator for science and the NASA Administrator have influential *external* constituencies. They consist of individuals and institutions that develop, use, fund, and learn from space science and technology. They include scientists, engineers, the president, White House surrogates (especially the Office of Management and Budget [OMB]), Congress, media, industry, universities, foreign governments, and the general public. Both internal and external constituencies seek to shape decision making by the associate administrator for science and NASA Administrator in a myriad of ways. The NASA Administrator deals most often with the external political world. The associate administrator interacts primarily with the external scientific community. Both cope with the other participants in decision making within and outside NASA.

There are many individuals (such as the deputy administrator, the agency's "number two") who are important players in Mars policy. But, in headquarters, the associate administrator for science and NASA Administrator stand out in influence over the Mars endeavor by virtue of their formal positions in NASA. The associate administrator can make Mars the directorate's science priority. The Administrator can make it an agency priority.

Mars Politics

Within NASA and its overall constituency, certain individuals and organizations stand out as Mars advocates. They press to get Mars exploration high on the agency's agenda. These advocates, inside and outside the agency, constitute a loose coalition of forces, a set of program champions with shared attitudes. They begin as a minority. They are "first movers" and seek to enlist others. They labor to persuade the associate administrator for science, NASA Administrator, and those around these officials to make robotic Mars exploration a priority and convert NASA into an advocate for policy adoption to the White House and Congress. NASA thus becomes the organizational nucleus for the coalition. They work within, around, and sometimes over NASA to get the agency to forward their goals.

If NASA's political masters convey legitimacy and resources, through policy adoption, NASA becomes an implementer all the way to possible completion of a mission and consideration of successor ventures. NASA thus is critical to decision making at all stages of policy affecting Mars exploration. This does not necessarily mean NASA gets what it wants in the myriad of trade-offs that result in presidential budgets and congressional appropriations. It means that almost always NASA has to be enlisted itself as an institutional advocate in national policymaking for Mars to achieve broader support. NASA is an object of advocacy by internal and external Mars champions, and when they succeed, NASA becomes an entrepreneurial force for Mars exploration as a national and increasingly international endeavor. The aim of advocates is to create an ever-widening gyre of support and mobilize bureaucratic power behind a sustained Mars exploration program.

Mars advocates are not monolithic. Within the Mars community, there is variance. The physical-science-oriented Mars scientists want to understand the atmosphere, geology, seismicity, and other contextual features of Mars. The biological science community wants resources and instrumentation on flight projects to focus primarily on the life question. The human exploration enthusiasts emphasize the need for robotic flights to carry "their" sensors to detect radiation and other concerns relevant to astronauts. Engineers working at NASA, at its field centers, and in industry see Mars exploration as a way to extend their technological art to innovate machines never before made. They and scientists emphasize optimal performance, often over cost. Policymakers see Mars exploration as a means to advance the high-tech economy (including employment in specific congressional districts), promote national prestige, stimulate young people to go into technical professions, and advance foreign policy goals. The media and public want excitement, drama, and the vicarious adventure of exploration, even if it is with robots. All these constituencies are potentially "pro" Mars. But they view Mars through differing perspectives.

Cohesion among Mars advocates matters. This is because they face opponents. Most participants in policy "support" Mars exploration. Few are "against" Mars. But Mars may get in the way of other worthwhile interests certain participants prefer. Rivals push back. Mars exploration champions compete with adversaries favoring a range of other priorities in planetary and space research. Who wins and who loses in these contests depends on their respective influence. Influence is based on the relative skill with which advocates make their claims, as well as other resources they can bring to bear. They can use a range of arguments, depending on whom they are trying to persuade. Aside from competing scientists, there are opponents in NASA who want to build human spaceflight craft, observe planet Earth, or pursue some other mission. Outside NASA are those in OMB or Congress who oppose Mars spending to save money in general or divert it elsewhere.

Big science, because of its scale, necessitates more than scientific and/or engineering commitment. It requires organization, money, and administrative and political will. For some advocates, NASA moves too slowly; for some critics, it moves too fast, or in the wrong direction. There is a recurring tension among advocates between those who want to travel gradually and look comprehensively (the "incrementalists") and those who wish to accelerate progress to targeted goals (the "leapers"). Whatever the pace, advocacy is essential at every stage of decision making to overcome opposition or sheer bureaucratic inertia.[9] Success in advocacy leads to funds for program execution. Success in execution helps advocates make a case for continuing a program. As closure is reached in one project, birth can occur in another. The stages of policy for various Mars projects intersect. NASA is simultaneously seeking funds for a new mission while implementing an existing project. These parallel and intertwining paths reflect the essence of *programmatic* big science.

Leadership among Mars advocates has usually been shared and has shifted. There have been individual outside advocates in the Mars community over the years who have led and been famous, notably the astronomer and writer Carl Sagan. Most, however, have been unknown to the public, including scientists and engineers in universities and NASA field centers. Frequently, advocacy is organizational, embodied in a particular entity with a special interest in Mars exploration. NASA has 10 field centers, one of which, JPL, a federally funded R&D center in Pasadena managed by the California Institute of Technology (Caltech), has been dominant in robotic Mars exploration over much of NASA's history. It has fought for Mars missions as a matter of organizational ambition, pride, and survival.

JPL is often in conflict with other centers for Mars missions and roles, especially the Ames Research Center in the San Francisco area, which has carved out a niche in astrobiology. In the important case of the Viking project, JPL was secondary to the Langley Research Center in Virginia, but still strongly involved, contributing the orbiter. Similarly, there are a handful of universities that have been consistent performers of Mars research. The same can be said of certain aerospace firms as hardware builders. Advocacy is borne of self-interest, commitment, and expertise. The technical core to Mars advocacy that has persisted over the years has comprised the performers of R&D at NASA centers, especially JPL, and in the academic Mars science community, along with certain key managers in NASA Headquarters. But even with this nucleus of interest and leadership, support has waxed and waned, and opposition by those with other priorities has always been present.

Space policy decisions are made in a context of national policy. Advocates (especially NASA) try to influence national policy. But national policy affects what Mars advocates can do. Every year, NASA contests with OMB over how much is enough. OMB is only one of the contestants in the game of budget politics, but a powerful player. Scientists, industry, the White House Office of Science and Technology Policy (OSTP), other executive agencies, the president, Congress, and even foreign nations are involved in Mars politics. When Mars advocates come out on the winning side in budget struggles with rivals and those who seek spending cuts, they are fortunate, especially in hard financial times. Success in specific Mars projects—such as MSL Curiosity—becomes absolutely essential to moving forward. And even then, Mars interests may not prevail. Mars is a program within an agency within a national policy that is interdependent with events around the world which have little to do with space. Big science—particularly the kind distributed over time in programs—is a tempting target for budget cutters. It is somewhat amazing that Mars exploration has done as well as it has.

Toward Mars

NASA's long-term direction is toward Mars. Mars exploration advances best as advocates move it to a NASA priority, and as NASA leaders themselves become champions in national policymaking. For better or worse, NASA is the power locus within the U.S. government behind Mars exploration. To move Mars exploration forward, NASA integrates often conflicting scientific, technological, and political pressures to forge a program. The advantage Mars advocates have in comparison with rivals is the long-standing public interest in Mars. Among planets, Mars has the broadest public constituency. The interest of that constituency may not be deep, but it is abiding. That reality has helped keep Mars on NASA's agenda since its founding, sometimes at the forefront, sometimes on the back burner, but always there.

To continue the journey from past to present to future, external advocates and NASA must combine science, technology, organization, politics, and faith. Science provides ends, technology and organization means, and politics funding to keep going in the face of adversity. There is also a faith on the part of many that at the end of the journey will be a prize worth all the sacrifice. Without all these factors, a multigenerational quest cannot be sustained. The task of leadership in advocacy and decision making in government is to unite the many stakeholders who might want to go to the Red Planet, neutralize or conciliate those who do not, and keep the agency moving forward, "in motion, in a desired direction."[10]

The political and policy history of Mars exploration is traced in succeeding chapters, from Mariner to MSL and beyond. It has been a long and winding process. Borrowing from evolutionary theory in the natural sciences, many analysts use the concept "punctuated equilibrium" to explain stability and change in long-term decision making.[11] In times of stability, Mars policy is forged largely by a limited number of actors in a subsystem (i.e., space sector) of the larger national policy setting. Specialists in the scientific community, NASA, congressional committees, and other groups reach agreement on a course of action which sustains a program over time. Change is incremental. The agency implements within a policy consensus.

But events or particularly influential people can upset the consensus, jar the subsystem loose, and foment significant change. These triggers for change can come from sources within, such as a NASA Administrator; they can emanate from outside the space sector, such as a president; and they can arise from Mars advocates or opponents. These forces produce important shifts in the evolution of Mars exploration, creating decision points in Mars policy.

The various conflicts within the space sector or between that subsystem and macropolicy issues drive a program forward or hold it back. Dan Goldin, NASA Administrator, told a group of scientific advisors in 1996, amidst excitement about the Mars meteorite, that scientists can set the direction for a program, but politics determines its pace.[12] Mars exploration is not a dash, in the manner of Apollo; it is a marathon, and the pace can vary, with zigs and zags along the way. The consensus has to be worked and reworked as the advocacy coalition enlarges or contracts. That is why there is not one fixed program, but a sequence of programs separated by reformulations of policy.

These reformulations are significant. They represent changes in scientific and political strategy, give rise to alterations in approach, and constitute big decisions. One shifted Mars exploration from an individual or "one-up" design to an integrated sequence of FBC missions in the 1990s. Another moved Mars exploration into a follow-the-water mode for the early twenty-first century, building on the program that had gone before. Their coherence and narratives communicated purpose to the outside world. They won political support that undergirded Mars exploration for two decades, enabling the MSL Curiosity's landing and discoveries and setting the stage for what can follow.

Paradoxically, these program reformulations were triggered by failures. As failure revealed miscalculation, so responses showed rethinking and resolve. There are lessons to be drawn from the Mars exploration experience for the leadership of long-term technical programs in American democracy in general.

The saga of Mars exploration illuminates the frustrations, failures, joys, and triumphs that surround all attempts to advance on a new frontier. Mars calls, and human beings respond as best they can. They discover through science, build "exploring machines" through technology, and organize programs through agencies like NASA.[13] In doing so, they strive to turn dreams into reality. They seek answers to age-old questions, such as, are we alone? Robots have gone first to the Red Planet. Someday, human beings will follow. What ensues in succeeding chapters is a political and policy history of NASA's robotic effort, a government program whose purpose is to pioneer.

CHAPTER ONE

The Call of Mars

Long before NASA's journey to Mars began, there were Mars enthusiasts who said it was a special place. For many it was a mysterious place that possessed life—thinking beings capable of creating engineering marvels that scientists on Earth could see. For others, it was a place for humans to explore. For all, it was an object of fascination, an embodiment of the unknown. Once NASA was formed, Mars had an institutional advocate. But it was not at the top of NASA's agenda. NASA leaders said "Moon First!"

Lowell and Life *on* Mars

In the nineteenth century, Giovanni Schiaparelli, using telescopes of his time, saw features on the Mars surface he called "canali," or channels, later translated into English as "canals." Percival Lowell, an American astronomer with considerable personal wealth, became intrigued with "Martian canals" and observed them closely with his own telescopes. He hypothesized in a 1906 book, *Mars and Its Canals*, that these markings were made by Martians. "That Mars is inhabited by beings of some sort or other we may consider as certain," he wrote.[1]

In Lowell's mind, Mars was Earth's sister planet, but she was dying, drought stricken, arid. He could see through his telescopes that the poles of Mars were white. Did it not make sense that Martian engineers constructed giant engineering systems to transport water from frozen poles to arid regions elsewhere?

These explained the canals to Lowell. What Lowell believed many other scientists of his day did not. But Lowell held intensely to his views and proselytized them to willing readers and listeners.

Among those who listened were science fiction writers, who took Lowell's speculations to greater imaginative lengths. In the first half of the twentieth century, Edgar Rice Burroughs, Ray Bradbury, Robert Heinlein, and Arthur Clarke found Mars a fruitful subject of writing. Named for the Roman god of war, Mars was always for some a potential threat to Earth. Orson Welles, in 1938, made that threat amazingly real for many. Taking his cue from H. G. Wells's *War of the Worlds*, Orson Welles used the new medium of radio to make an emergency announcement on his evening broadcast that Martians had invaded the United States, with sightings in New Jersey. Terrified, many Americans "bolted their doors and prepared for the worst."[2]

Fears of Martians, whether ruthless warriors or benign canal builders, were well at rest by the 1950s. Helped by better telescopes and the observatory Lowell had established in Arizona, scientists in the pre–World War II years learned more and more about the Red Planet. World War II spawned a range of new technologies that proved useful in astronomical research, although not intended initially for this purpose. The Office of Naval Research, in the immediate post–World War II years, supported planetary research at several universities. The U.S. Army and Air Force developed technologies that could be advanced and used potentially in connection with space exploration.

In 1948, Gerald Kuiper of the University of Chicago "used infrared spectrometry to confirm the presence of carbon dioxide in Mars' atmosphere" and at the polar caps. American astronomers, as well as those from other countries, created organizations to plan research in planetary astronomy. Mars was a focus of much attention. A "Mars Committee" emerged that enlisted scientists who would "meet annually to share the results of their observations of the red planet." Growing scientific understanding of Mars, while slow and ambiguous, indicated that Mars was extremely unlikely to have life resembling earthlings. Mars appeared too frigid and too dry for Earth-like beings. But the similarities between Mars and Earth were too intriguing to ignore and rule out possibilities of life in some form.[3]

Von Braun and Life *to* Mars

In the 1950s, a former German rocket engineer, working on missiles for the U.S. Defense Department, began writing about and advocating human spaceflight to Mars. To be sure, there would be gradual steps to Mars, said Wernher von Braun in a series of articles in *Colliers*, a high-circulation magazine at the time. He painted a sequence of technical developments over years. Between 1952 and 1954, von Braun proclaimed that there would first be a spaceship, then a space station, and then a trip to the Moon. But Mars was the ultimate destiny for human spaceflight, although such a voyage might not come for many decades. Von Braun, an engineer, whose brilliance was matched by his passion for exploration, believed that advances in rocketry would someday make extending human life to Mars possible. Von Braun's stepping-stone approach to Mars was called by some "the von Braun paradigm."[4]

Von Braun subsequently expanded the articles into four books. In his 1956 book (coauthored by Willy Ley) he assumed that plant life would greet visitors from Earth. However, it was not life *on* Mars that most interested von Braun; it was developing technology to take human beings *to* Mars that he craved.[5] For scientists and engineers who wanted to know if life existed on Mars or humans could get to Mars, there was a barrier before 1957. Technology to escape Earth's gravity did not exist—or at least had not been demonstrated—prior to Sputnik. The Soviet Union in 1957 opened the space frontier in a move that shocked the United States and the world. All of a sudden, dreams about space exploration came closer to reality.

All the same, the two motivations to explore Mars proved a double-edged sword for Mars advocates. The von Braun approach was pursued initially by engineers who emphasized developing technology to take humanity into space. The intellectual descendants of Lowell were scientists interested in discovering life through robotic means. The scientists were mainly users of space technology, not developers. The relation between the robotic program and the human spaceflight advocates was problematic from the outset. The engineers behind human spaceflight wanted to go to the Moon. They wanted to go to Mars too, but the Moon came first, since getting there was their immediate interest, and the von Braun stepping-stone approach was a method to which they subscribed. Many scientists did not find the Moon particularly interesting, especially those anxious to discover extraterrestrial life. They looked to explore Mars via robotic means for quicker answers to the question, are we alone? The interests were different, and so were the priorities of the two advocacy communities.

Thus, there was a long dual legacy of fascination with Mars. But to get started in answering this call required going beyond scientists and engineers, to politicians. Exploration required a strong additional stimulus, because government would have to get involved in a substantial way and spend a great deal of money. Just to take an initial step in the Red Planet's direction would require political will, organization, and the push of a large program. What would get politicians aboard? Which agency in government would take the lead in managing space policy?

Creating NASA

When the Soviet Union put Sputnik into space in 1957, the United States and the Soviet Union were locked in a protracted Cold War struggle. The issue was which political/economic system was superior and constituted the wave of the future which other nations would follow. Technology was a symbol of national capacity to lead. It was emblematic of national power.[6] Sputnik came as a great psychological victory for the Soviets, even though President Eisenhower downplayed its military significance. But to most observers it seemed to indicate not only rocket-lifting capacity but national power generally—not only in military missiles, but also in scientific and technical education. Even fellow Republicans were angered that the Eisenhower administration had not been sufficiently vigilant and had let Sputnik happen. It grated that the Soviet Union was the first nation in space. America's pride was bent and its prestige tarnished.

Eisenhower appointed a science advisor and science advisory committee in part to help him establish America's course in space. Although Eisenhower did not want to engage in a "race," he wanted the United States to be competitive, and that would take some time. The Soviet Union followed up Sputnik with other successes, while the U.S. effort floundered. There was no existing space agency. To the extent that there was space-related activity at all, it was found in only a few places in government and was an uncertain priority in all.

One place was the National Advisory Committee for Aeronautics (NACA), an old agency that went back to World War I and housed a number of research laboratories to advance the field of aeronautics. Another place was the Department of Defense (DOD). There was a scientific group in the Navy (Naval Research Laboratory) active in space research and poised to launch an American satellite as part of a large international science undertaking at the time called the International Geophysical Year (IGY). Another body in DOD consisted of von Braun's German rocketeers, who were working for the army on missiles. Also active was the Jet Propulsion Laboratory, which served the army via management by the California Institute of Technology. The political consensus that emerged in late 1957 and 1958 was that the American space effort was too fragmented and low priority and a new agency for which space science and technology was *the* mission had to be established.

Eisenhower was strongly influenced by his science advisors, as well as his own predilections, to establish a *civilian* space agency. The scientists feared that if DOD became the de facto space agency, it would concentrate space research on strictly military missions, and secrecy and classification would be the rule. The scientists immediately saw tremendous opportunities for space research in an agency with a nonmilitary orientation. Indeed, they wanted an agency with an agenda scientists could influence. Eisenhower, feeling pressures for space-oriented weapons which eventually helped compel him to warn against a "military-industrial complex," agreed that a civilian agency was best for the country.

The NACA, with its 8,000-person civil servant staff, was selected to be the core of the new agency. NACA brought with it three major laboratories, or field centers: Ames in California, Lewis in Ohio, and Langley in Virginia. Other facilities would be grafted onto the new agency from DOD. These would include JPL, the von Braun team, and a naval science group. Von Braun and his associates would form the nucleus of the new Marshall Space Flight Center in Alabama. The navy group would be the keystone for the Goddard Space Flight Center in Maryland.

As the White House and Congress worked on enabling legislation for the new agency, they decided that the new entity would have to have a broad charter in science and technology which would give it unusual flexibility. Space was seen as a new frontier, and no one was sure what it would entail. What was clear was that everyone wanted the agency to move quickly and begin competing with the Soviets as soon as possible. There was a general feeling that NACA was sluggish and bureaucratic.

Hence, there was attention paid, directly and indirectly, to the question of bureaucratic power. This was exemplified most clearly in giving the new agency a single leader. NACA was led by a *committee* and director under the committee. The other leading technical agency of the time was the Atomic Energy *Commission*. Again, there was plural leadership. The political architects drafting legislation wanted an individual to be in charge, one clearly responsible and accountable. The original bill created a National Aeronautics and Space Agency (NASA). The word "Agency" was changed to "Administration." The head would be called not a "director," as originally written, but an "Administrator." "Administration" and "Administrator" seemed to the political founders more substantial terms for an agency that would be charged with leading the U.S. drive against the Soviets, and which would have to work with formidable bureaucratic rivals, such as DOD.[7]

There was no question that NASA was going to have a strong robotic science emphasis, even if the human spaceflight side of the agency came to be dominant. Eisenhower's science advisors and other scientists who testified during hearings leading up to the NASA bill pressed hard to have a science mission that was explicit in NASA's legislative charter. That it did, the charter saying simply that the new agency should carry out "the expansion of human knowledge of phenomena in the atmosphere and space." The legislation moved into law with relative ease, given the sense of urgency. On October 1, 1958, NASA officially opened for business. The generality of the legislation and anxiety of the country meant that the first NASA Administrator, T. Keith Glennan, would have a lot of discretion in how he went about his job and organized NASA.

Putting the Moon First

Eisenhower's appointee as NASA's first Administrator was, in the words of historian Roger Launius, "the perfect choice."[8] He was trained as an engineer and had worked in government, industry, and the university world. Aged 52, he took leave of absence from the presidency of Cleveland's Case Institute of Technology. Glennan shared Eisenhower's view that the Soviets should not determine the U.S. space agenda. He wanted NASA to develop a space program on America's own terms. He did, however, intend to position NASA to compete with the Soviet Union and ultimately achieve leadership in this competition. Like President Eisenhower, however, he wanted NASA to be a relatively small agency. He did not favor "big government." The way he wished to get started, therefore, was to consolidate governmental institutions transferred from NACA and DOD and operate mainly through contracts with industry and universities.

The government-by-contract model became the enduring NASA approach, even though the agency expanded enormously in the Apollo years. Glennan's deputy was the previous chief operating officer of NACA, Hugh Dryden, a physicist. Under these two political appointees were various associate administrators, most of whom were government career officials. The senior associate administrator, Richard Horner, served as "general manager." Others headed various NASA programs. The key associate administrator for the new mission of spaceflight was Abe Silverstein, an engineer and NACA veteran. Under him was Homer Newell, a scientist, who came to NASA from the Office of Naval

Research. The fact that science was under what was perceived as Silverstein's human spaceflight operation bothered the scientific community. Scientists wanted the status of their unit raised to equal that of an engineering-oriented operation.

As NASA gradually began to succeed in launching rockets into space—with the Soviets still substantially ahead in weights they could lift—it became increasingly obvious that the prime arena of competition would be human spaceflight. NASA established a "man-in-space" program called Mercury and began to recruit the first group of astronauts. In various ways, Glennan began to emphasize the Moon as a possible destination. If humans were going to go to the Moon, however, robotic scouts would have to go first. Hence, from the outset, human and robotic programs were competitive in some ways and linked in others.

Glennan claimed not to be a "space cadet," but he was an active and forceful proponent of space, although not as much as some in NASA would have liked or some in Congress would have preferred. He reported to Eisenhower, who was cool to any notion of a "crash" program to catch up to the Soviets. Glennan established the four program emphases that NASA would have thereafter: human spaceflight, space science, space applications (e.g., weather and communication satellites), and aeronautics.

The man responsible for creating a space science program at NASA was Newell. Age 43 at the time, Newell had a PhD in mathematics from the University of Wisconsin and had subsequently turned to physics and high-altitude research. In 1955, when the Naval Research Laboratory was assigned the task of developing the Vanguard launch vehicle for the IGY satellite program, he was named Vanguard science program coordinator. Newell joined NASA shortly after it opened in 1958.

Recalling the mood at NASA at the time, Newell later wrote that "everything seemed to be happening at once." The agency was new and had "to sell itself." The exciting mission and novelty of the agency served to attract many young scientists, engineers, and technical managers. "In the white hot light of public interest," it had to organize a staff, prepare budgets, develop a program of activity, and work out relations internally and with external constituencies. Newell had to work on several fronts at once, and space science, like the rest of NASA, showed growing pains.[9]

Building a staff, establishing internal and external relationships, and designing an unprecedented program—all these activities took enormous time and energy. Newell found two external institutions particularly important as representing scientists' views on policy relevant to space. One was the President's Science Advisory Committee (PSAC), and the other was the National Academy of Sciences (NAS). NAS set up a Space Science Board (SSB) to advise NASA. It was a strong advocate for a higher science profile in the agency and an Office of Space Science that would be independent of the Silverstein operation.

Although these terms were not specifically used, Newell's Science Office clearly had two thrusts: "little science" and "big science." Little science referred to grants and contracts to individuals and specific groups of investigators chiefly at universities. Big science referred to major projects involving science aboard space vehicles, using NASA field centers. The science "payload" and rockets together constituted technical systems that required more money, more organization, more diverse individuals and institutions, and complex management mechanisms. Newell regarded it as headquarters' role to provide policy management for "programs," with technical management for specific missions or projects at the level of field centers. For the largest projects, this division of labor was inevitably blurred. However, the notion of decentralizing technical management was clear as NASA got under way. Headquarters and field centers would have to partner in the actual management, but there were differences in what each would do. The field centers expected headquarters to get the resources from the White House and Congress and send them to the field centers, which would handle decisions day to day to get projects carried out.

Each headquarters office had certain field centers assigned to it. For space science, this meant two field centers in particular. One was Goddard, which was given responsibility for missions in near-Earth space. The other was JPL, which was charged with deep space—meaning the Moon and planets. Other field centers could contribute in terms of their expertise as agency needs so required. Early on, Ames Research Center (primarily under the Aeronautics Office) developed an interest in "exobiology." Nobel Prize–winning Stanford biologist Joshua Lederberg lobbied NASA to concern itself with possible contamination of the Moon and Mars with earthly machines. He was intrigued by the possibility of life on other planets, especially Mars. He took his case directly to Glennan. Glennan was responsive and set up a research activity concerned with extraterrestrial life and contamination issues. Ames took responsibility for this mission. It was Lederberg who coined the field's name, "exobiology," a field detractors characterized as a science without a known subject.[10]

Organizationally, it was up to headquarters to determine direction and pace of programs and projects. That meant decisions about which programs to emphasize and how fast to go. Within headquarters, Administrator Glennan was quite clear that he wanted NASA to go at a measured pace, step by step, so as to spend taxpayer dollars prudently. Such a policy required Newell to emphasize the Moon over the planetary programs when it came to big science. Newell recollected that Glennan "just did not want to talk about planetary things."[11] Mars, therefore, would have to wait in line for resources. This policy was not what the director of JPL, William Pickering, wanted to hear, and he fought for resources to give the planets, especially Mars, greater attention.

Pushing for Mars

In 1959, Pickering received orders from headquarters to give priority to a lunar impact mission, Project Ranger. Pickering, however, wanted to give priority to planetary research. His long-term goal was to have his laboratory lead a mission involving an "interplanetary vehicle, which could give an answer to the question of life on Mars."[12] This was, in his view, the big question. He wanted JPL to "leapfrog" the Moon. He also thought that Glennan's measured, step-by-step approach would not produce "leadership" in space technology against the Soviet Union.

Pickering was 48 in 1958 when NASA got started. Born in New Zealand, Pickering had a physics PhD from Caltech and long-term association with JPL. He was named director in 1954. After Sputnik, Pickering led JPL in working with von Braun's army team and the University of Iowa's James van Allen to help launch America's first successful satellite, Vanguard, on January 31, 1958. While he wanted JPL to be a part of NASA and was pleased when the redeployment from DOD to the space agency took place, he prized his institution's autonomy. He was very conscious that JPL was connected to Caltech and thus "different," more independent than NASA's other field centers, they all being civil service laboratories. No one said it on record, but JPL also perceived itself as "better" because of its university connection. From NASA's perspective, JPL was a federally funded research and development center that worked for NASA. Moreover, NASA paid Caltech a handsome overhead fee for JPL's services. It expected JPL to be a loyal member of the NASA family.

Glennan was annoyed with the lack of responsiveness on the part of JPL (and Caltech) generally to NASA policies. The relatively mild-mannered Newell was more than annoyed since he had to deal closely with JPL and was caught between Silverstein and Pickering, two men with strong personalities. For Silverstein, JPL might not be a civil service lab, but it was a contractor and as such should bend to headquarters' direction. Pickering stoutly defended JPL's independence. Pickering, Newell wrote, was "as stubborn as Silverstein was domineering."[13] Add to this tension the fact that the president of Caltech, Lee Dubridge, believed that his university was in charge of JPL, and not NASA. Throughout 1959, meetings between headquarters and JPL took place, with Silverstein pushing the Moon as a priority and Pickering emphasizing the planets. In December, the associate administrator, Richard Horner, wrote Pickering firmly insisting that while JPL had planetary missions as well as lunar work, it was the Moon that the agency wanted stressed. Silverstein followed up this guidance with additional direction as to what he expected of JPL.

When Newell and others from headquarters went to JPL the following week, they were confronted by Pickering, who again made it clear that he disagreed with the priorities of headquarters. Eventually, the two sides agreed to communicate more closely and find compromises. JPL went along with the lunar emphasis of Washington but got resources from headquarters for exploratory work on planetary programs—Venus first, because it was closest to Earth, then Mars. Pickering was not interested in "competitiveness." His word was "leadership." He was convinced that America's best chance of getting the lead in space was an all-out effort "to proceed at once to the planets."[14]

NASA supplied funds for "expanding JPL's facilities and equipment and for increasing the staffing," as Pickering requested. However, the issue of in-house versus external work divided NASA and JPL. NASA policy (Glennan) emphasized contracting, and JPL (Pickering) wanted to maximize intramural work and technical control by his laboratory. The issues of headquarters' authority and JPL's responsiveness continued to cause tensions throughout the Glennan era. They only got worse under the next administrator.[15]

Nevertheless, NASA knew it needed JPL for expertise, as JPL knew it needed NASA, its prime sponsor, for resources. Partners in conflict, they found uneasy accommodation. NASA decided priorities, and it favored the Moon. JPL managed and performed technical work, and it favored the planets. Pickering was the prime advocate for robotic Mars missions and was not shy about pushing his views on NASA decision makers, such as Newell and Glennan.

CHAPTER TWO

Beginning the Quest

NASA began its martian quest with a long-term program called Mariner. It planned a series of increasingly challenging probes. The Apollo Moon decision indirectly helped martian advocates. Apollo raised the level of NASA funding enormously. More money overall gave more money to Mariner. Mars enthusiasts had resources they needed to get started in a serious way. There was political consensus around Apollo and Mariner.

Adopting Mariner

Work on NASA's program to Mars began at the Jet Propulsion Laboratory even before JPL had official blessings from headquarters. In 1959, Pickering told his JPL colleagues to start creating a "planetary machine."[1] JPL then formulated requirements for a spacecraft that could travel enormous distances, survive a very long time, carry sophisticated and vulnerable scientific instruments, and transmit data back to Earth. The Soviets were working on such a spacecraft, but they had not succeeded in launching one as yet. Pickering assigned John Casani, a young engineer who had joined JPL in 1956, to take charge of the initial planning and design effort.

Casani formed a team of JPL specialists to tackle this exceptional challenge. They would aim at Venus and Mars. Pickering organized the planetary team so as to separate it from the lunar group. The lunar team was under much stricter deadlines. The planetary group had deadlines also, but a decision to develop hardware had yet to be made. Moreover, the planetary unit could learn lessons from its lunar colleagues.

Casani had no illusions about the difficulty. He recalled his feeling at the time: "It would take a colossal effort on the part of an enormous number of people."[2] JPL planned a *program*, successive flights over a decade or more, with each more technically ambitious than the preceding one. In early 1960, JPL briefed Glennan on its plans: "first planetary flybys, then planetary orbiters, then orbiter-landers." As before, Glennan insisted that the planetary program not get in the way of the lunar activity. Pickering was equally insistent about JPL interests. JPL would give lunar Ranger launches priority, but once this series was completed, JPL wanted "the major program activity of the laboratory" to be planetary flight.[3]

The discussions and negotiations continued, with Newell playing a mediating role. Glennan wanted to run a cost-conscious, technically sound agency, with clear priorities, and he intended to hold the reins on Mars enthusiasts. Nevertheless, on July 15, 1960, Glennan officially approved the proposed planetary program, to be called Mariner.[4] A month later, NASA Headquarters authorized JPL to move forward with development. The issue was now not whether to go to Mars, but how quickly.

Pace would depend on many factors, especially resources. Resources, however, were related not just to JPL and other NASA interests but to NASA's priority in national policy generally. That in turn depended on Cold War competition with the Soviet Union, which also had its eye on Mars. The Soviet planetary program, like that of lunar exploration, was well ahead of the U.S. program. The leader of the entire Soviet space effort, Sergey Korolev, was a brilliant rocket engineer who dreamed of sending human beings to the Red Planet. The Soviet premier, Nikita Khrushchev, understood how space exploits could advance his nation's prestige, and he provided ample resources to Korolev. As in the United States, Korolev made the Moon the immediate priority for spaceflight. However, it was Mars that fired his imagination. When he spoke of Mars, his excitement showed, and he could become almost "ecstatic" over the possibility of exploring Mars.[5] Robotic probes would precede human ventures. In 1958, he had directed his space organization to start working on robotic flight to Mars. Mars and Earth would come into proximity every two years; he targeted 1960, the nearest opportunity for the first try with a flyby.[6]

In October 1960, Khrushchev came to New York to address the United Nations. Visiting Eisenhower, he gave the president replicas of Soviet pennants Luna 2, the Soviet probe that had struck the Moon earlier, carried. Khrushchev had a replica of another, more advanced spaceship that presumably he would show if the Soviet Mars probe succeeded. It was launched also in October, but it fell back to Earth and crashed in Soviet territory.[7] That the Soviets were trying hard was now known unmistakably by the United States. The political imperative behind the Mars program gathered force.

Competition was fanned by John Kennedy, elected president in November. During the campaign, he pledged to speed up the U.S. space program so the United States could overtake the Soviet Union. He linked U.S. fortunes in space to a national security issue: the "missile gap."

Apollo's Impact

When Kennedy took office in January 1961, he learned that the missile gap did not exist. Also, he had many priorities other than space. While willing to augment NASA's budget, he did not see a particular urgency to do so immediately. Then came Yuri Gagarin's April flight in space. The first human in space was a Russian. This was another blow to U.S. pride and prestige, one on which Khrushchev fully capitalized. Soon afterward, Kennedy suffered another personal defeat when the Bay of Pigs invasion of Cuba failed ignominiously. He directed his vice president, Lyndon Johnson, to come up with a program he could back which would garner the United States a visible and dramatic victory in the Cold War contest with the Soviet Union. Johnson came back with what would be called Project Apollo. The proposal bore the imprint of Kennedy's NASA Administrator, James Webb.[8]

In May, Kennedy addressed Congress and declared his belief that the United States "should commit itself to achieving the goal, before the decade is out, of landing a man on the Moon and returning him safely to the Earth."[9] This was a race with the Soviet Union Kennedy affirmed the United States would win. What was most important about this decision for the robotic Mars exploration program—Mariner—was that it made NASA a "presidential" agency. That is, NASA was now a priority for the nation, and its program mattered greatly to the president. All of a sudden, NASA could get resources from the White House and Congress for virtually anything it wanted to do—including robotic Mars exploration.

Kennedy's decision reinforced the Moon as NASA's priority. However, it was a major punctuation point for Mars because it removed the financial reins on the planetary program. As Newell wrote, "The renewed sense of urgency that the Apollo decision bestowed on the space program made Webb's task [as NASA Administrator] one of loosening the shackles imposed by the previous administration and stepping up the pace."[10] Webb told NASA (and later President Kennedy in a memorable White House debate the two men had) that Apollo was not only an end of the space program but a means to a broader national goal: preeminence. That latter goal meant advancing all NASA programs, not just human spaceflight.

Immediately after the Apollo speech, Kennedy increased NASA's budget by 89% and by another 101% the following year.[11] Employment at NASA surged apace. The 54-year-old Webb was neither scientist nor engineer. Trained as a lawyer, he had vast experience in government and business management. He was unusually politically skilled. Moreover, he showed a strong appreciation for science. Soon after the Apollo decision, he reorganized NASA. In doing so, he established the Office of Space Science and Applications (OSSA) with a status independent of and equal to that of the newly created Office of Manned Space Flight (OMSF). He also promoted Newell to associate administrator for space science, a level identical to that of the director of OMSF. All programs became far bigger and moved far faster. With Webb's vigorous backing, NASA acted to create a new space science community, pouring money into universities for research and graduate fellowships. This community consisted of those in traditional disciplines (e.g., geology, meteorology, chemistry, biology, and others) eager to extend their work to outer space and the planets. NASA was building a constituency that would help support the agency and its goals. That constituency included a small band of scientists who called themselves exobiologists.

Achieving Success: Mariner 4

Although NASA and the Soviet Union both sent missions to Venus, they both made Mars the overarching priority. For years, Soviet scientists and their political masters "consistently targeted the planet Mars as the singular most important objective in plans to explore space." First, they would send robotic spacecraft, and then they planned to dispatch human explorers.[12] The Soviets, like the Americans, believed that the nation to be the first to not only get to Mars but find life would have a scientific and propaganda victory of historic proportion.

The Soviets had sought unsuccessfully to reach Mars in 1960. They tried again in 1962. Again they failed. The next launch window was 1964, and the Soviets were sure to try once more. This time, the United States intended to be ready to make an attempt at a flyby. NASA was fully aware of the difficulties. JPL referred to a "Great Galactic Ghoul" lurking in space between Earth and Mars, ready to devour all spaceships that sailed into its lair.[13] NASA, understanding risk, if not the ghoul, planned to send two spacecraft.

NASA and its various constituencies were unsure what the flyby would reveal of Earth's mysterious neighbor. The media attention was frenetic and reflected the substantial public and scientific interest. In October 1962, the National Academy of Sciences Space Science Board called Mars the "primary goal . . . in the exploration of the planets." One month later, NASA gave the official go-ahead for a 1964 launch. JPL was well along in its preparations. Casani continued his association with Mariner Mars. However, Pickering appointed as overall project manager a well-regarded division manager, J. N. James. He was able to attract some of the best and brightest of JPL's engineering talent to the effort.[14]

In 1964 NASA sponsored a summer study, one aim being to nurture the exobiological community. Lederberg was there. So was Carl Sagan, 30 years old, blending astronomical and biological interests with exceptional eloquence. He was already articulating an intense belief that life was possible on Mars. Among the scientists who attended was Gerald Soffen, then 38, who had joined NASA in 1961 soon after getting his PhD in biology from Princeton.

Mariner 3, launched in November 1964, failed soon after launch. A few weeks later, NASA tried with Mariner 4, this time successfully launching the spacecraft. The Soviet Union also launched in this year and again failed. The United States had an open opportunity for a first after trailing the Soviet Union in most other aspects of space.

On July 15, 1965, Mariner 4 sped past Mars, snapping photos as it went. Politicians called this event an important Cold War victory. Virtually everyone hailed Mariner 4 and congratulated NASA and JPL. Pickering, seen by most observers as the key leader behind Mariner 4, won special praise, including a profile in the *New York Times* and the cover of *Time* magazine.[15]

But from the perspective of those academic scientists, media, and NASA/JPL personnel anxious to find evidence of life, the Mars photographs were a disappointment. Where were the canals? Instead, the photos showed craters. Also, Mariner 4 sensors found no significant magnetic field, as well as an atmosphere so thin it would allow radiation to reach Mars that might well kill anything on the surface. The media pronounced Mars boring. In effect, said the media, NASA looked for Mars and found the Moon! Pickering tried his best at a press conference following the flyby to emphasize how little of the Mars surface was viewed. Mariner 4 was not intended to deal with the life issue, he said. Hence, as far as he was concerned, "the evidence of possible life on Mars . . . is still there."[16]

Still, reaching Mars first was a great accomplishment, and President Lyndon Johnson wanted to make the most of it for public relations and propaganda purposes. Johnson had taken power when Kennedy was assassinated in 1963, and he promised to continue the journey to the Moon. He also understood the importance of Mars from a political perspective. Webb, anxious to maximize NASA's advantage from Mariner 4, coached NASA/JPL officials and scientists on how to deal with the president, Congress, and media. Bruce Murray, Caltech and Mariner scientist and later JPL director, was impressed with Webb. He was "shrewd and skilled in negotiating Washington's corridors of power." He could play the informal southern politician with Congress, but in the privacy of the NASA meeting room, he was "crisp," all "Washington, DC." "He left no question as to who was in charge and what was expected of us."[17]

A few days after the encounter with Mars, NASA and JPL officials, along with various scientists associated with Mariner 4, met with Johnson. He gave the Mariner team medals for their good work and was attentive when Pickering explained what the pictures of Mars revealed. Johnson did not express unhappiness with the failure to find Lowell's canals. "As a member of the generation that Orson Welles scared out of its wits," he declared, "I must confess that I'm a little bit relieved that your photograph doesn't show more signs of life out there."[18]

The *New York Times* and other general media called Mars "dead."[19] But what incensed many scientists of the exobiology camp was the "I told you so attitude" of some other scientists. Phil Abelson, editor of the prestigious *Science* magazine, had written an editorial on February 12, 1965, predicting that searching for life on Mars was a fool's errand. "In looking for life on Mars, we could establish for ourselves the reputation of being the greatest Simple Simons of all time," he wrote.[20]

Sagan was offended and angry. He said a photo of Earth taken from 6,000 miles out (as Mariner did for Mars) might show that no life existed on Earth. Sagan could think of all kinds of possibilities for life that Mariner 4 did not address. What about life "beneath the surface," asked Sagan, "where there might be ice deposits and, in some places, even pockets of liquid water?" He suggested that there might even be underground lakes and other habitats where life could thrive which would have been absolutely undetectable to Mariner's instruments. "Sagan insisted that critics who called Mars 'dead' were making more out of the data than anyone had a right to do."[21]

Like Pickering, Sagan pointed out that Mariner 4 had taken photos of only a small portion of the Red Planet. As he later recalled, "So I took it as my responsibility, maybe a quixotic mission, to point out the possibilities [of Martian life], which were being excluded."[22] It was a remarkable personal decision. He opened himself to media interviews and used his rare communication skills to advantage on late-night television talk shows. His campaign to rouse interest in space and especially the search for life would cost him professionally with a number of his scientific colleagues. He did not get tenure at Harvard, but secured a position at Cornell. His lifelong public advocacy, beginning with Mariner, would contribute to his fame, fortune, and ridicule.

Along with other self-described "diehards," Sagan was frustrated and in some ways desperate to reframe the life debate after Mariner 4. Exobiologists chafed at the writing off of life on Mars and Abelson's negativism. They concluded that if life did not exist, then it was important to discover, "why not?" This question was scientifically important, monumentally so, they avowed. As scientists, they needed to get at the truth. The Mariner program had to continue, in their view, and produce better photographs over far more of the Mars surface, and eventually a lander had to go.[23] The academic exobiologists were at the forefront of Mars advocacy in the wake of Mariner 4. Among agency insiders, Pickering continued to play the lead role, writing, lecturing, and seeking support.

The Mariner program continued. It was indeed a "program," not a single mission, and seen as such inside and outside NASA. NASA and JPL—with JPL as locus of decision making most of the time—planned additional Mariner flights for later in the decade. Pickering saw Mariner 4 as a critical milestone in technological development of spacecraft. "We now know how to do it," he stated.[24]

CHAPTER THREE

Leaping Forward

In 1965, even as NASA and its Mars constituency celebrated the success of Mariner 4, they thought ahead to what would come next. "Next" meant not only next in the line of Mariner projects, but the beginning of a new program called Voyager. Not to be confused with an interplanetary Voyager, launched in the late 1970s, this Voyager—a Mars lander with an automated biological laboratory payload—was conceived in the early 1960s at the Jet Propulsion Laboratory with the encouragement and involvement of the exobiology community. It was seen as the natural successor to Mariner, which featured flybys and orbiters. From the beginning, advocates knew that it was unusually challenging, but the challenge attracted top engineers at JPL. Also, the exobiologists who wanted to look for life knew they would have to develop experiments and unprecedented life-detecting equipment. NASA leaders zealously aimed to surpass the Soviet Union in the space race, and that race encompassed robotic Mars missions. For NASA, the program was also precursor to human Mars flight. Hence, in the 1960s there was relative unity within NASA and among various interests—planetary scientists, exobiologists, engineers, and administrators—about the rationale and direction of robotic Mars exploration. All three drivers for the Red Planet operated: life on, life to, and international competition.

But NASA Administrator Webb belatedly added another reason for Mars Voyager—keeping the Saturn 5 rocket alive. This use of the giant Moon rocket raised the potential cost of Voyager tremendously, ultimately making it politically unacceptable. When Congress killed Voyager, NASA substituted Viking. NASA made Viking's purpose clear: to search for life. With its budget declining, NASA felt that it had to recapture public support. It decided to put most of its planetary program energy and money behind Viking. Tom Paine, Webb's successor, raised the stakes by augmenting the complexity and costs of Viking. While planning and selling Viking, NASA landed on the Moon. Surely, if NASA could succeed with Apollo, it could succeed with Viking! Technological optimism reigned. The political consensus in the space sector and between space policy and national policy which had operated in the Apollo/Mariner era gave way to discord over Mars Voyager. What happened to the robotic program depended in large part on what happened to NASA. Eventually, the president and Congress settled the question of NASA's future via the shuttle decision. Meanwhile, a new consensus and equilibrium among NASA, scientists, the White House, and Congress were forged around Viking.

Adopting Voyager

In December 1964, following preliminary studies by NASA, JPL, and industry, NASA's Science Directorate, the Office of Space Science and Applications, officially established Voyager as a flight program. Like Mariner, it was conceived as a program, not a single project. OSSA projected a mission to launch the first Voyager spacecraft as early as 1971, with successor flights at later two-year Mars opportunities. Webb, who could deal with LBJ on a one-on-one basis, obtained President Johnson's assent to include modest definitional start-up funds in the budget Johnson sent to Congress in early 1965. By the end of 1965, buoyed by Mariner 4's success, Congress approved Voyager.

NASA started with strong scientific support for Voyager. The National Academy of Sciences Space Science Board declared in 1965, "The biological exploration of Mars is a scientific undertaking of the greatest validity and significance. Its realization will be a milestone in the history of human achievement. Its importance and the consequences for biology justify the highest priority among all scientific objectives in space, indeed, in the space program as a whole."[1] Mariner 4 findings seemed to have made it all the more imperative for Mars advocates with an interest in finding life to have a lander program. They saw no other way to answer their questions.

However, while Congress went along with the initiation of Voyager, the schedule and longer-term prospects were uncertain. The political and funding environment of NASA began to change rapidly for the worse. NASA budgets peaked in 1965–1966. The Vietnam War and Johnson's Great Society began to place increasing burdens on the overall federal budget. NASA was clearly catching up to the Russians in the race to the Moon, and some of the urgency behind NASA was ebbing. NASA was still a national priority, but other national needs had arisen. The result was less money for "new starts" or implementation of those that were authorized.

In this shifting environment, various NASA centers looked for work in alternative areas. The Langley Research Center in Hampton, Virginia, saw opportunity in Voyager. An aeronautics center, Langley could boast expertise in the science and technology of landing. JPL did not take kindly to Langley's foray into JPL's bureaucratic turf, but Langley had support for a role in Mars activity in OSSA. Edgar Cortright, Newell's deputy, reacted positively to Langley's proposed Mars entry system at a meeting in 1965. Langley got a go-ahead to continue developing its ideas.[2]

While JPL and Langley jockeyed for roles, major decisions at the NASA Administrator's level were under way with implications for both centers. The inability to get new programs authorized or funded adequately increasingly troubled Webb. He knew that he had to sell a post-Apollo program before NASA reached the Moon to avoid a major downsizing problem for his agency in the early 1970s. He was having difficulty getting the president to focus on post-Apollo goals. Johnson kept telling Webb to wait until next year. The problem was the production line of Saturn 5 rockets (the Moon rockets). To have future uses for more Saturn 5s, NASA needed post-Apollo programs, and it had none.

Once NASA got to the Moon, what would it do? Build a Moon base? Go to Mars? Decisions needed to be made. Without decisions justifying more work on Saturn 5s, von Braun's Marshall Space Flight Center might have to start laying off rocket engineers. Webb went to Johnson and Congress and explained that it made no sense to spend so much money to create an unparalleled rocket/ spacecraft system and then not keep it going and put it to use. He received sympathy, but no decisions, and decisions had to start soon with the president and his budget given lengthy technology development times.

Webb got Voyager approved by his political masters at a time when NASA's budget was still ample. It had not been authorized as a "post-Apollo" program, but Webb sought quietly to use it in this way. He did so by choosing to launch Voyager spacecraft by Saturn 5 rockets. This move in October 1965 shocked JPL, Langley, and the scientific community, because the spacecraft they contemplated did not need so huge a booster. In fact, it would enlarge the scale and substantially raise the cost of Voyager as a program. It would also complicate roles, for the decision meant von Braun would be deeply involved in management decisions—maybe in charge. Newell tried to sell the use of Saturn 5s to the Mars scientists, however. From his standpoint, OSSA should have use of Saturn 5s and would find uses for this massive capability. He told the SSB, in seeking endorsement, "Fellows, if you don't help me, George [Mueller, associate administrator of the Office of Manned Space Flight, and bitter rival of Newell] will get all the Saturn 5s."[3] However, there were many scientists inside and especially outside NASA who worried that a Saturn-driven Voyager would take money from smaller scientific robotic programs they wanted.

The Saturn 5 decision ignited a debate within the Mars science community. The debate had many nuances, but at its heart was a question of priorities. There were scientists who were not exobiologists who envisioned a string of Mariner flights to Mars at every two-year launch opportunity. They saw robotic Mars exploration in incremental and multidimensional terms, leading *gradually* to Voyager's landing. Murray of Caltech was most articulate in expressing these concerns.[4] He was an avid Mars advocate, although a skeptic about finding life on the Red Planet. He and his allies wanted a more comprehensive Mariner program that would systematically study geologic, meteorological, and numerous other disciplinary questions in addition to biology. Murray was himself a planetary geologist, and he believed that understanding the Mars physical environment came first and was intrinsic to detecting life on Mars—if there was life on Mars. Exobiologists did not necessarily disagree with this gradualist, comprehensive approach, but they were anxious to get moving as fast as possible toward Voyager. After all, they reasoned, finding life was the big prize, and why not go for it while they could?

The real pressure for more direct flight to Mars came not from scientists but from NASA leadership, and the issue was use of the Saturn 5. Once Webb made that decision, it was obvious that not science but post-Apollo needs were his reasons for the Voyager priority. Moreover, cost considerations in a steady-state NASA budget might mean eliminating possible intervening Mars Mariner flights. Doing so did not sit well with scientists generally or with JPL. But JPL found its own influence in NASA decision making slipping. In the first half of the 1960s, when headquarters was overwhelmingly preoccupied with the Moon, JPL was where the most important technical decisions affecting Mariner were made. In the second half of the decade, headquarters began pulling decisions upward as it thought about the future, and NASA funding became constrained. Plans called for managing Voyager in an Apollo mode, with a strong headquarters director making use of multiple NASA centers, industry, and universities.[5] The "incrementalists" and Saturn 5 "leaper" camps were both represented in OSSA, but OSSA was not making the Saturn 5 decision.

Pickering, seeing competition from Langley for Voyager, tried to be supportive of larger NASA decisions. He said he wanted to move toward Voyager as soon as possible but did not want to eliminate Mariner flights. Some headquarters officials described the JPL attitude as "schizophrenic."[6] As Koppes, in his history of JPL, wrote, "The ambivalence about, and outright opposition to, Voyager derived from the fundamental question of what the laboratory should be. . . . Voyager would entail a huge expansion of JPL . . . the sheer size of the project would divert the laboratory from the in-house tasks that Pickering and the senior staff considered vital to its élan and substitute extensive monitoring of industrial contracts. JPL staff were 'doers' rather than 'managers,' and Mariner-type projects allowed them to do what they had come to the laboratory to do."[7]

Voyager's high-level proponents in headquarters were aware of the resistance to Saturn 5–Voyager within the scientific community and at JPL. JPL's attitudes, and traditional independence in general, did not help its cause with NASA Headquarters in decisions about roles in the Voyager program which JPL might play vis-à-vis Langley. Nor did JPL's use of the California congressional delegation to get its way go over well with Webb.

President Johnson postponed post-Apollo decision making as long as he could. At the end of 1966, he acquiesced to Webb's importunings. As January 1967 began, Johnson sent a budget to Congress that provided $71.5 million to begin developing Voyager hardware using a Saturn 5 rocket. The proposed program would send two large orbiters and landers to Mars in 1973 (a slip from the previously projected 1971 launch) and then do so again in 1975. It was implicit that those two missions were the beginning of a major robotic exploration program that would extend further in time and destination. Mars would come first, but NASA would develop a capability to explore the solar system.

The budget also included for the first time funds to start an Apollo Applications Program (AAP) that would also use Saturn 5s in near-Earth orbit. The Skylab "space station" effort would evolve from this activity. The point of both AAP and Voyager from Webb's perspective was to sustain institutional infrastructure and technological capability in space after the Moon landing, pending the nation's readiness to make a national policy decision akin to Apollo. The only decision that could be like Apollo in size and dramatic challenge would be one about human flight to Mars. Webb was thus buying time for his agency in a deteriorating political environment. Voyager would be justified publicly in its own right, on the basis of science, but it was also a means to an unstated end—keeping Saturn rockets, von Braun's center, and human space exploration going.

Webb also wanted to link von Braun to Voyager not only technically but politically. The famed rocket engineer and Marshall Space Flight Center director had dreamed of going to Mars for years. Webb believed that von Braun could help him sell Voyager in the difficult budget climate. Telling von Braun he could build not only the Saturn 5s for Voyager but also "the main vehicle that would stay in orbit around Mars," Webb "wanted to link the Voyager to Dr. von Braun's name and to a proven management team." He even asked von Braun to move to Washington at least for a time to help sell the program to Congress.[8]

Terminating Voyager

Webb's reason for concern—the diminution of political support for space—was glaringly obvious, and that worry was magnified by a disastrous unforeseen event. Shortly after the budget submission in early January 1967, the Apollo fire of January 27 occurred. It killed three astronauts while they were training at Cape Canaveral. Immediately, almost everything at NASA was put on hold, while the agency coped with the disaster and its aftermath. Webb personally dealt with the president and the congressional investigation. He got Apollo through the six-month ordeal following the fire relatively unscathed and made personnel and organizational changes that strengthened the agency and contractor system for completing the Apollo project. But he himself was weakened as he drew the media and political focus of the investigation to himself and shielded the organization, thereby expending much of his political capital.[9]

Also, opponents of NASA in Congress from both the right and left used the Apollo fire to attack NASA and siphon funds from space to other areas of spending (such as the Vietnam War and social programs for the cities). Congress wanted to make substantial cuts, not in Apollo but in other space programs, including Voyager, projected to cost in the billions over time. In the summer and fall of 1967, debate raged in Congress over the NASA budget. Johnson, meanwhile, grew desperate to find money for Vietnam and domestic priorities and to deal with a soaring federal deficit. He was even proposing a tax increase. In August, he declared that the country's financial situation had changed over the months since he had submitted his budget. He had "to distinguish between the necessary and the desirable."[10] Apollo was protected, but Webb had to decide what other priorities to keep and what to let go. Johnson gave him leeway to choose, and Congress pushed the NASA Administrator to state his priorities unequivocally. Webb strongly resisted.

The NASA Administrator wanted to keep Voyager, a key to NASA's future after Apollo. But several senior academic scientists testified against it. Even more damaging, Webb was undermined by his own agency, or at least the Manned Spacecraft Center (MSC). In July, the Houston center had sent out a request for proposals for human missions to Mars and Venus. Webb was aghast, furious with the political insensitivity of MSC. In fighting to keep Voyager against congressional budget cutters, Webb had taken great pains to link it rhetorically with scientific discovery, not human spaceflight. His allies in Congress had done the same. Virtually everyone knew that the mood of Congress and the country was against Mars decisions involving human spaceflight at this point.[11]

But Houston did not get the bureaucratic strategy. Legislative opponents of NASA immediately seized on the Houston announcement as ammunition in the context of Johnson's statement about deciding "between the necessary and the desirable." They charged that Voyager was a "foot in the door" for human spaceflight to Mars.[12] Now they had what they considered the smoking gun of evidence. Support for Voyager, tenuous at best, evaporated. Saying they had to nip a covert human Mars program in the bud, legislative opponents persuaded Congress to kill Voyager in late October 1967. To make their point unmistakably clear, they also terminated a Mariner orbital flight of 1971 which NASA had proposed to help locate a place for Voyager to land. The only planetary mission remaining was a two-Mariner flyby of Mars for 1969. The Mars advocates and planetary science community in general were shocked, devastated, and, to some degree, chastened.

Substituting Viking

Webb moved immediately to counter the threat and save the Mars program and planetary science along with it. He decoupled the Saturn 5 rocket from the spacecraft, and von Braun from any semblance of leadership of the program. He also sent a strong message of dissatisfaction with OSSA to NASA and the scientific community by making changes in OSSA leadership. He moved Newell to NASA associate administrator and Newell's deputy, Cortright, to a senior position in the OMSF. He appointed John Naugle, a 46-year-old physicist and experienced manager in OSSA, to take Newell's place. He told him that all existing plans for Mars were ended and to replan the robotic program for an austere environment. Specifically, Webb directed OSSA to come up with a smaller Mars project costing much less than Voyager and using an intermediate-scale rocket.[13] While Naugle led the intense scientific reorientation, Webb went to Johnson and congressional leaders and lobbied the political front.

Working furiously, NASA came up with an alternative Mars program in two weeks in early November, in time to get it inserted into the next budget Johnson was submitting to Congress.[14] In forwarding that budget at the beginning of 1968, the president declared, "We will not abandon the field of planetary exploration. I am recommending development of a new spacecraft for launch in 1973 to orbit and land on Mars. This new Mars mission will cost much less than half the Voyager program included in last year's budget. Although the scientific result of this new mission will be less than that of Voyager, it will still provide extremely valuable data and serve as a building block for planetary exploration systems of the future."[15]

NASA's overall budget, already falling, went down again in 1968, but the Mars planetary program was saved. The Mariner 1969 flyby would be followed by a reinstated Mariner orbiter mission in 1971. Then, the replacement for Voyager, which would include both an orbiter and a lander, would come in 1973. The search for life would be its rationale, along with the continuing Soviet competition on the robotic Mars front. No one at NASA dared to say anything about a possible connection to human spaceflight. Voyager was dead. In its place was a new flight to Mars which later came to be called Viking.

Webb spent most of his energy and remaining political capital in 1968 giving a final push to the Apollo Moon landing. However, he was genuinely interested in science and wanted his legacy in that area to be positive, especially in regard to Mars. Webb, Johnson, and Congress all knew that the Soviet Union was pursuing robotic flight to Mars. Naugle, the new associate administrator for the OSSA, found in one of his first meetings with Webb that the NASA leader listened attentively to his recommendations, although Naugle might have to argue at length to defend them. Webb was cool toward Pickering and JPL, but he granted Naugle's request to provide additional funds to JPL to avoid layoffs of personnel Naugle believed critical to planetary science. However, Webb, supersensitive to appearances in the wake of the Voyager debacle, told Naugle not to apply these technical people to the new Mars mission at this point, as he was still building congressional support for its approval.[16]

Naugle was responsible for reshaping Voyager's replacement. He knew that scientists had testified against Voyager, and that he had to turn them around—or at least get them to keep quiet—for the new venture to move forward. He worked closely with Harry Hess, chairman of the SSB, to form a Lunar and Planetary Missions Board. They made sure to include critics of Voyager. Their aim was to get space scientists to sort out their priorities behind closed doors rather than in public statements to the media or Congress which NASA critics could use. Moreover, while Mars was the priority now, they wanted to assure the planetary community that other missions could take their turn later. What Naugle and Hess sought was consensus on a 10-year plan, starting with Mars. Webb generally did not like science advisory committees, as he wished maximum leeway for himself in NASA policymaking. But he wanted Naugle to move ahead in forging a relatively united scientific constituency.[17]

Naugle also took the lead in deciding which center would run the new project, a dispute that went back to Voyager. Like Voyager, the replacement was seen to encompass an orbiter, lander, and automated biological laboratory. This combination was unprecedented. JPL, eager for challenging assignments at the frontier of science and technology, especially in planetary exploration, lobbied to be in charge of the whole project now that the Saturn 5 issue was gone. So did Langley. The two centers battled within NASA in the early months of 1968, with Langely's supporters pointing out that Langley was a "real" (i.e., civil service) center while JPL was a "contractor" center.[18]

Gradually, Naugle and his colleagues at OSSA settled on a recommendation to Webb, who made the final decision on the intercenter dispute. The recommendation was that JPL develop the orbiter—a spacecraft that would be based in part on its designs for Mariner—while Langley would develop the lander and be in charge of the project generally, including the biological laboratory. This decision riled Pickering, who did not give up easily. But Naugle argued that Langley had done a good job with a particular mission for Apollo (the Lunar Orbiter Program) and had stronger management capability. The latter consideration was crucial for Naugle: "Nobody felt JPL had the [management] horsepower to run a big lander-orbiter project," he recalled. There was also an issue of headquarters control, and JPL was not easy to control.

NASA wanted strong oversight of the Mars venture. This was made crystal clear in May 1968 when NASA Headquarters sent Cortright to be director of Langley. "I was comfortable with Cortright," Naugle said. "Everybody was comfortable with him. We knew he would do a good job. He would bring the resources of Langley to bear on the project." Once Cortright took the reins of Langley, Pickering backed off.[19]

Webb left NASA in October, thereby giving his deputy, Tom Paine, a chance to show his mettle in the remainder of 1968. Webb believed if Paine did so, the next president probably would retain Paine at least through the Apollo 11 launch to the Moon in July 1969.[20] Paine was seen as apolitical, a technocrat in the best sense. Age 46, Paine had come from industry and was extremely competent and imaginative. He combined zeal for space with engineering competence and vision. However, he was in a "downsizing" period of NASA's history, and he was not a downsizing kind of person.

Augmenting Viking

Naugle and Cortright debated the design of Viking. Cortright (in contrast to Pickering) had been a big supporter of Voyager and had been burned by the cancellation. He was now cautious, wanting to keep Viking's budget (and thus political visibility) low. He was willing to eliminate the orbiter (assigned to JPL). Naugle believed the orbiter was crucial to determining the best place to land. This mission was now about finding life. The more observation of the Mars surface, the better. But Naugle was also concerned about money and whether Paine and Congress would go along with a bigger, more complicated project. After much discussion in November, Cortright came down for the minimal project. Naugle opted for a project intermediate in scale. Neither favored an option that would be even more advanced, aggressive, and expensive in terms of technological development.[21]

In December 1968 Naugle met with Paine. Naugle recalled the sense at the time that the United States was still competing with the Soviet Union, and that competition included the robotic Mars program. As he later wrote, "We had to establish a good, solid, scientific mission." If "the Russians landed [on Mars] successfully in '71 or '73, what we landed . . . had to be something that would stand up against what they had done." It was in this context that Naugle proposed an intermediate design. Paine told Naugle to go with the most technically advanced and challenging Viking project possible, however. For Paine, NASA was about bold endeavors.[22] He regarded Viking (and it was he who gave the mission this name) as NASA's most important mission outside human spaceflight.[23] Paine wanted NASA to think big—and "big science" was OK. He was willing to take risks that were both technical and political.[24] However, this decision meant that the estimated cost would rise substantially from the $384 million NASA used to sell the program under Webb.

Paine's decision catalyzed organization of Viking as a project. Paine confirmed that Naugle's OSSA would be fully in charge. This meant that Naugle would have Langley, as lead center, under his authority for this particular project. Langley, an aeronautics-oriented center, was typically under another NASA directorate, but for this project, Naugle would also have responsibility. Naugle would in addition have JPL, as was usually the case for OSSA.

Paine affirmed that JPL would design the orbiter and Langley would develop the lander and be overall technical manager. The lander would carry extraordinarily sophisticated science payloads, serving many disciplines, particularly exobiology. NASA would have to involve many scientists, from both government and universities, in decision-making roles—70 as it turned out, organized into teams. There would be NASA centers other than Langley and JPL participating as necessary, particularly Ames, with its exobiology expertise, along with a number of industrial contractors.

Paine refashioned Viking into a larger-scale technical project than contemplated previously. It was destined to be a great leap beyond Mariner. He knew history would link Webb with Apollo and the Moon. Paine envisioned Mars as his potential legacy, with Viking his first major decision in this respect. Viking would rival Apollo in some ways when it came to technical ambition. The risks were immense, the technical challenges unprecedented. To bring down risk, NASA opted for redundancy as much as possible. For example, two Vikings would go to Mars in order to lower the risk of failure. Paine wanted Viking to succeed and was willing to spend more to do so. Webb had sold Viking as a single mission—not a program. That was the best he could do. What came after, no one knew for sure, but Paine and others were hopeful it would help launch a sequence of robotic missions leading to human spaceflight to Mars.

The man to whom Paine, Naugle, and Cortright looked to integrate the various parts of Viking was James Martin, a highly experienced and hard-driving engineer and project manager at Langley. Age 48 in 1968, Martin was a tall, crew-cut, no-nonsense kind of man. Martin moved rapidly to establish a project office and begin development of Viking hardware. He was joined by Gerald ("Jerry") Soffen, who left JPL to come to Langley as chief scientist for Viking. Pickering resisted the reassignment of Soffen but lost again in the intercenter dispute.

Soffen, age 42, was one of the few exobiologists at NASA. Soffen went to Langley because he figured "that was where the action" would be in exobiology. That was because of Cortright and his power within the agency. Cortright and the "Lord were very closely allied with one another," Soffen later recalled.[25] Soffen knew he would be "the interface between [NASA] engineering and the outside [science] world of academia." The external scientists, he said, "would be the only ones able to interpret our data from Mars." He would have to manage some extremely large egos among the many scientists who would be enlisted in the quest for life at Mars, most notably science superstars Joshua Lederberg and Carl Sagan.[26]

NASA under Nixon

As Viking got under way, Nixon became president, on January 20, 1969. He retained Paine and eventually appointed him NASA Administrator, but he gave him little or no access to advocate his post-Apollo vision. Paine wanted to advance a comprehensive post-Apollo program, the central element of which would be human spaceflight to Mars. It would feature also a space station, a space shuttle, and a lunar base.[27]

In July, NASA launched Apollo 11 to the Moon and Neil Armstrong took "one small step for [a] man, one giant leap for mankind." It was a remarkable moment that brought the world's people briefly together. It was an epic milestone in human history. However, the euphoria over Apollo did not transfer to post-Apollo. When Nixon and Paine flew to meet the returning astronauts, Nixon—in one of his few conversations with Paine—stressed that he supported NASA, but money was tight given the continuance of the Vietnam War and domestic economic troubles.[28]

Paine tried hard to use Apollo 11 to generate public enthusiasm for a post-Apollo human Mars mission. But winning the race to the Moon removed much of the competitive urgency space had. Paine hoped the 1969 Mars flybys (Mariners 6 and 7) would help his cause. Instead, they actually hurt to some extent. These flybys, which went up in late July and August, provided the best view yet of Mars, but like Mariner 4, they revealed a planet hostile to life. The media praised the twin probes, but some commentators asked why Paine would want to send astronauts to such a desolate planet. Indeed, critics said that robotic flight could do Mars reconnaissance relatively cheaply, and hence human flight was not necessary.[29]

Getting Congressional Support

While Paine labored to promote the goal of human exploration of Mars, it was left mainly to Naugle to sell Viking politically. He had been working the scientific community. He negotiated with the Bureau of the Budget (BOB), renamed Office of Management and Budget (OMB) in 1970. He now looked to Congress. Paine's decision to go for the most ambitious Viking option more than doubled the cost. Project Manager Martin told Naugle in August 1969 that the cost would be over $600 million. Naugle then added $150 million as a contingency from his own reserves and went to see Rep. Joseph Karth (D-MN). Karth was chair of the House subcommittee that had authorized Viking, and he was the project's most influential champion in Congress.

When Naugle told Karth that the project originally sold for $364 million was now $750 million, the lawmaker exploded. He accused NASA of "low-balling the cost to get Viking's 'feet in the door.'" He gave Naugle "a very rough time," as the NASA official had anticipated. Naugle responded that NASA was just getting started on the project, and if Karth felt strongly, he should cancel it now, before major development costs were incurred.

Naugle had lived in Minnesota for 10 years and understood Karth's problems in justifying space expenditures to his fiscally conservative district. However, the Soviet competition for Mars loomed large for Karth. Naugle recalled that he knew that Karth would not want to be the one who cancelled Viking and let the Soviet Union be the first to soft land on Mars. Viking stayed in the NASA budget.[30] With Karth leading the charge in Congress, NASA had sufficient allies on the Hill to keep Viking going. The political environment, however, was harsh, and Paine was not doing well with his campaign for human flight to Mars with Nixon. Viking's fate could not be separated from NASA's future.

Budget Pressure on Viking

Paine brought von Braun to Washington to help him promote a large post-Apollo program to Nixon. The charismatic von Braun won some converts on the body Nixon had established to advise him, the Space Task Group (STG). On September 15, STG met with Nixon in the Oval Office and presented three post-Apollo options. The options entailed a shuttle, space station, lunar base, and a human Mars mission. They varied in the aggressiveness by which to pursue these goals, especially Mars. The most aggressive would set a date, 1983, for human Mars flight at a cost of $8 billion to $10 billion a year. The least aggressive would cost $4 billion to $6.7 billion a year and would put an astronaut on Mars by the end of the century. Nixon listened to the presentation of the three options, thanked the team for its work, but made no decision.[31]

While waiting for Nixon to say something definite about NASA's future, the annual budgetary process continued, with Paine battling the BOB. Viking, like all NASA activities, awaited determinations about how much NASA would have to spend, overall, as well as on it in particular. By Christmas 1969, NASA's prospective budget for the next fiscal year was down to $3.6 billion, a sharp drop from the previous year. The issue became not whether NASA could begin a post-Apollo buildup, but whether it could even implement existing programs, including Viking. Then, just days later, the budget director, Robert Mayo, required NASA to find additional cuts owing to a last-minute decision to close a government-wide gap in funding.

The budget director went over various options with Paine, including two options involving Viking: cancellation, or delay of launch from 1973 to 1975. Mayo said BOB favored delay and so did Nixon. Paine had little choice. He called Naugle, who was at home, and asked him to come in. It was December 31, New Year's Eve. To save Viking, Paine told Naugle, they would have to set the launch back to 1975. Naugle left the meeting feeling quite depressed, as though "two years of careful planning for Viking" had been wiped out almost in the blink of an eye.[32]

In January 1970, Paine announced that NASA would have $3.5 billion in the president's budget. This was a figure Paine had earlier told Mayo was "unacceptable." It entailed not only delay in Viking but ending Saturn 5 production and reducing the number of Apollo Moon landings. Those landings were destined now to end in late 1972, and there still was no new major program to keep NASA going to prevent the agency's continuing decline.[33]

George Low, Paine's deputy, tried to soften the blow to Naugle and the Viking team, declaring in a memo to Naugle in early February, "Viking holds the highest priority of any project or program in NASA's Planetary Program. Viking holds a high priority among all of NASA's programs."[34]

That was an important statement about Viking's priority from Low, because it indicated that NASA leaders would protect Viking, even if they let other projects go. It was not only a science priority but a NASA priority. In March, Nixon issued his first policy pronouncement on space. His message was that NASA would have to live at a far different level than it had in the 1960s. He announced that "space activities will be a part of our lives for the rest of time," and thus there was no need to plan them "as a series of separate leaps, each requiring a massive concentration of energy and will and accomplished on a crash timetable." Indeed, he said, "space expenditures must take their place within a rigorous system of national priorities."[35]

What this meant, beyond the rhetoric, was that he was not endorsing *any* of the STG options. There was no decision to build a shuttle, no space station, and certainly no human Mars mission. As a consequence, NASA drifted, its future clouded. Low's memo notwithstanding, the survival of Viking was uncertain. What was absolutely clear was that Viking could not be justified as a precursor to human flight to Mars, since there would not be anything resembling such a project. Indeed, the whole human spaceflight effort was withering away.

In July, Paine announced he was resigning, effective September. In August, von Braun, who would leave NASA in 1972, complained that NASA was "waiting for a miracle, just waiting for another man on a white horse to come and offer us another planet, like President Kennedy." It was not going to happen.[36]

Fletcher Becomes Administrator

Low maintained NASA as best he could awaiting Nixon's appointee. The budget continued to fall, and more programs were cancelled, delayed, or significantly downsized. As promised, Low preserved Viking. It was not until April 1971 that James Fletcher joined NASA as its leader.

Fletcher was 51, a PhD physicist, and president of the University of Utah. He had made a fortune in industry and was comfortable with leadership. However, he was quiet and almost dull in manner in comparison with the forceful Webb and enthusiastic Paine. But he was persistent and thoroughly convinced that the nation needed a strong space program. He decided that his first important task was to persuade Nixon to adopt the Space Shuttle as NASA's next major program. He accepted—as Paine could not—the reality of a smaller NASA. But he believed that unless he could sell the shuttle, he would not be able to sustain NASA's identity as a large, independent science and technology agency focused on human spaceflight. Along with bringing the Apollo Moon program to a safe and successful conclusion, the promotion of the Space Shuttle became his central objective. He had a big problem, however. He was not well connected with the Nixon White House and was inexperienced in the ways of Washington.

While necessarily focused on the human endeavor, increasingly on the shuttle as a top priority, Fletcher was personally interested in Viking. There were institutional reasons to concentrate on Viking as well, to be sure. It was

NASA's largest planetary project. But the personal reasons were special for him. Fletcher brought a philosophical grounding in the Mormon tradition to his role as NASA Administrator. That tradition accepted the view that God created many worlds with many inhabitants. Fletcher had himself calculated that the universe had five billion worlds capable of sustaining life. While he did not believe there was intelligent life on Mars, he held to the possibility of some form of life on the Red Planet.[37]

Fletcher therefore gave Viking as much support and attention as he could muster—an important factor in sustaining the effort and helping the spirit of those directly involved in the project. Viking's budget was now estimated to rise beyond the $750 million Naugle had cited to $800 million. OMB noticed the increase, as would Congress. Fletcher had to persuade critics that the additional money was worthwhile. The scientific community's support was also flagging, with many scientists arguing for more "balance" between Mars and other science missions.

Fletcher was a consistent and relatively effective persuader. He was careful in what he said. The argument that Viking was a precursor for human flights was off the table for Fletcher. He surely considered it as such, but he muted those views. He emphasized a different rationale in which he strongly believed: that Viking was important because of the immense significance of the search for life. He also reminded others that there was still a space race with the Soviet Union under way. NASA had won the Moon race, but the contest to land and find life on Mars was still open.

Mariner 9 as Catalyst

In May 1971 NASA launched Mariners 8 and 9. Mariner 8 failed at launch, but Mariner 9 succeeded in moving toward Mars. NASA was fortunate to have had two Mariner launches in 1971. When the Mars program was revived after the Voyager demise, Newell, as NASA associate administrator, argued for spending extra money for a second Mariner. Webb had gone along with this view, as had Paine when he became NASA Administrator.[38] Shortly after the Mariner launch, the Soviet Union sent two probes, which it called Mars 2 and 3, toward the Red Planet.[39] NASA determined that the USSR spacecraft were much heavier than U.S. probes. That could only mean that they not only were orbiters (as were Mariners 8 and 9) but carried landers, and possibly even biological detection equipment. NASA sought to establish contact with the Soviets to see if there could be some cooperation in the respective endeavors, but the Soviet Union remained secretive about the objectives of its Mars program. The goal of Mariner 9 was well known—to orbit Mars and take photos of as much of the planet as possible. The aim was to scout possible landing sites for Viking—places both safe and promising potential habitability. But if the Soviets succeeded in landing with their probes, they would leapfrog Viking by more than four years.

By November, it was clear that the United States would reach Mars first with Mariner 9. There was considerable speculation and debate among scientists about what Mariner 9 would see. A host of scientists associated with the Mariner program gathered at JPL's mission control center in Pasadena, California, as the probe approached. Sagan, optimistic as ever, was a principal investigator on the project and eagerly awaited.

On November 12, the evening before Mariner was scheduled to go into orbit, a remarkable panel discussion was held at Caltech, entitled "Mars and the Mind of Man." There were numerous media representatives there to hear the discussion and also interview Sagan and other scientists. The panel was chaired by *New York Times* science editor Walter Sullivan. In addition to Sagan, the key planetary scientist on the panel was Murray of Caltech. Science fiction writers Arthur Clarke and Ray Bradbury also were on the panel. Both had written of life on Mars.

Sagan and Murray had become prominent antagonists on the Mars life issue. Sagan pointed out that scientists (like Murray) had been too quick to reject the possibility of life-forms on Mars. "There have been excesses in both directions," argued Sagan. "And one direction was the premature conclusion that there isn't life on Mars." While data were incomplete, said Sagan, all the necessary elements of photosynthesis—the life process of plants—existed on Mars: water, sunlight, and carbon dioxide. The ice caps, which advanced and receded with the change of seasons, could leave pools of water that could contain life forms.

The science fiction writers were in Sagan's corner, but Murray dismissed such views. He charged that even good scientists (like Sagan) were deluding themselves. Life on Mars was "wishful thinking." He attributed such thinking, which made scientists misinterpret data, to a "deep-seated emotional desire" on the part of humanity to find "another Earth" somewhere. Mars was not that Earth.

Privately, Sagan and Murray would debate at length. Andrew Chaikin has written that Sagan's comments would "make Murray want to roll his eyes." Sagan was three years Murray's junior, but they appeared to be on "opposite sides of a generational divide." They respected one another and would eventually become friends. But with Mariner 9 in the offing, they simply disagreed strongly. Sagan privately once snapped at Murray, "You at Caltech live on the side of pessimism." Murray did not respond, but thought, "And you at Cornell, Carl, live on the side of optimism." Murray, on the panel, told Sagan and the others he accused of "wishful thinking" that Mariner would settle the issue, and he expected it to provide "the observational stick" to force Sagan and his allies to recognize reality.[40] For Murray, the absence of evidence proved his point. But Sagan came from a different perspective. As he later wrote, "The absence of evidence is not evidence of evidence."[41]

Everyone would have to wait a while, however. The next day, Mariner 9 successfully moved into Mars orbit. Going into Mars orbit was itself a great first. But a huge dust storm swirled about Mars, obscuring almost totally images of the planet. NASA decided to simply let Mariner 9 go around and around Mars and wait out the storm.

Meanwhile, later in November, the two Soviet probes joined Mariner 9 in circling the Red Planet. The USSR spacecraft did not have the fuel to match that of Mariner 9 and thus wait a lengthy time to see Mars. They soon released their landers. Only one made it successfully to the Mars surface. This landing was indeed a significant first. However, something went wrong and after 20 seconds the probe's communications ceased. It transmitted no images of Mars.[42]

The next round of Mars exploration thus depended on Mariner 9, once the waiting was over. In late December 1971 the dust cloud gradually cleared. Scientists at JPL saw four dark spots in the northern hemisphere that peered up at them. They thought them to be big craters. But eventually, as views improved, it became obvious that these dark spots were mountains, and in fact gigantic volcanoes. The scientists' reaction at the time was a collective "Oh, my God!"[43] The largest of these volcanoes had to be 15 miles high, three times the height of Mount Everest. The crater at its summit was the size of Rhode Island; its base that of Arizona. Soon they saw an enormous rift valley that extended at the Mars equator region 2,500 miles, as well as a canyon system that was 75 miles wide and 4 miles deep. Moreover, they eventually saw what seemed like channels (the "canals" of Martian lore) that might have been carved by running water.[44] Michael Carr, a young scientist on loan from the U.S. Geological Survey, recalled his sheer excitement when he saw the images. Mars "was a wonderland. It had unfolded before us."[45]

Humanity's image of Mars was suddenly transformed. The science fiction writers in the pre-space age had painted a Mars that was like Earth. The Mariners of the 1960s had made Mars into a replica of the Moon. Mariner 9 revealed a unique planet. For the first time, earthlings saw the real Mars, and it was spectacularly different from Earth or the Moon. Murray admitted he was totally surprised. As a geologist as well as planetary scientist, he could detect signs of an extremely active planet in the past. He had himself "been the victim of his own preconceptions about Mars, even as he'd been warning his colleagues and the public not to fall prey to their own."[46] While Murray still did not believe there was evidence of life on Mars, he was now open to the possibility of being yet surprised again. As for Sagan and the exobiology community, they became more convinced that if they looked in the right places in the right way they could find life. For skeptics and optimists alike, the importance of Viking enlarged immensely.

Mariner 9 was a strong catalyst for Viking. But Viking would not succeed if NASA did not survive.

Significance of the Shuttle Decision

As Mariner 9 brought enthusiasm to NASA's space science program, Fletcher was about to obtain what Webb and Paine could not: a post-Apollo human program that would end the steady decline in budget and sustain the agency for the long haul—and make a big science effort like Viking possible. It had been a struggle. OMB staff in the summer of 1971 had proposed to reduce NASA's budget below $3 billion, denying it the Space Shuttle, further eliminating specific Moon flights, and making cuts in a host of other programs. Viking certainly would have been affected adversely.

On August 12, 1971, however, Caspar (Cap) Weinberger, deputy director of OMB, wrote Nixon that the cuts were going too far. He pointed out that "there is real merit to the future of NASA, and to its proposed programs." To keep cutting NASA would give comfort to those "at home and abroad" who say "our best years are behind us, that we are turning inward . . . and voluntarily starting to give up our super-power status, and our desire to maintain world superiority." The United States, he argued, should "be able to afford something besides increased welfare, programs to repair our cities, or Appalachian relief and the like."[47]

Nixon wrote a note on Weinberger's memo: "I agree with Cap." Subsequently, George Shultz, now the budget director, was informed that "the president approved Mr. Weinberger's plan to find enough reductions in other programs to pay for continuing NASA at generally the 3.3–3.4 billion dollar level, or about 400 to 500 million dollars more than the present planning target."[48]

Fletcher did not know about the exchange, nor did OMB staff, and they continued to fight the remainder of the year, with Fletcher forced more and more to couch the shuttle's rationale in cost-benefit terms, striking deals with the Department of Defense as a shuttle user, and making various technical compromises to save money. But by the end of the year, it was clear that NASA was being "stabilized" in line with the Weinberger-Nixon-Shultz exchange. On January 5, 1972, Fletcher and Low flew to meet Nixon at his western White House in San Clemente, California. There, Nixon told them that "space flight is here to stay" and publicly announced that NASA was going to develop a reusable space shuttle. This was a multibillion-dollar, long-term commitment.[49]

Without question, this decision saved human spaceflight, and possibly NASA as an independent agency. The shuttle decision was "the central space choice of the 1970s." It meant that for the remainder of the decade "the United States would carry out those space missions that could be afforded within a fixed NASA budget after Shuttle development costs had been paid."[50] While essential to NASA, the shuttle decision was a mixed blessing. It secured a measure of agency stability overall, but it constrained all other programs. Also, the shuttle provided services to low-Earth orbit. It was not about exploration of deep space. The exploration missions now became the sole prerogative of the robotic science program.

The Viking project in particular emerged as the de facto flagship for NASA's exploration effort in the first half of the 1970s. To be sure, there were other missions to the planets in effect or on the drawing board. But Viking stood out in this period as an exploration priority in scale and purpose. Viking was exciting to scientists and also a magnet for the media and public because of its search for life. Thanks to Mariner 9, there was a new anticipation for the quest. And thanks to the shuttle decision, NASA had a fighting chance to get the resources to make it possible.

CHAPTER FOUR

Searching for Life

The shuttle decision provided a measure of stability for the agency. However, there were still annual budget fights between NASA and the Office of Management and Budget, and Fletcher was clear about his need to protect Viking, NASA's top priority after the shuttle.[1] Fletcher's personal interest in Viking was symbolized by a globe of Mars he featured prominently in his office.[2] He was willing to move money from lesser NASA priorities to Viking to assure success and personally looked into technical matters affecting development.

The approach NASA took with Viking was "Apollo-style." Murray, a continuing critic, called this method the "great leap forward" strategy. It was a "technology-forcing" approach, he charged. Murray preferred the steadier, gradual, "evolutionary" style of Mariner. Sagan, for one, argued that the incremental strategy was not ambitious enough. The reason was that the end—to find extraterrestrial life—justified the means. NASA's strategy could be even bolder than it was in his view.[3] For Sagan and others in the exobiology camp, Viking, not the shuttle, was NASA's most important project.

Fletcher saw Viking as the beginning of a program of robotic Mars landers, each more sophisticated than the next, leading to the return of a Martian soil and rock sample. In May 1972, Deputy Administrator Low told Naugle that Fletcher wanted Naugle to brief him on the state of NASA's thinking about Mars Sample Return. In August, following the briefing, Naugle established a study group to start a planning process.[4] Fletcher was looking ahead, with a multiproject Viking program in mind. However, there was no approved Mars mission beyond this Viking mission. It had to succeed in its search for life, or NASA's Mars program would face a crisis. This crisis would be in maintaining cohesion among Mars advocates, much less sustaining political support within the space policy sector and among national policymakers.

Forcing Technology

The problem for Fletcher was that the existing Viking project truly presented a number of unprecedented technical challenges to NASA. These issues made implementation extremely daunting as the Viking team got fully under way in the early 1970s. NASA officials were grateful for having until 1975 to resolve them. They realized that the previous 1973 deadline was too close. However, even 1975 looked demanding in 1972. No one fretted more than James Martin, the project manager. "I am worried about the fact that Viking has a fixed launch window," he said. "It opens August 11, 1975 and closes about the 20th of September. And the window will close whether we launch anything or not." If NASA missed it, the agency could not launch again to Mars for more than another two years. This narrow launch window meant that all decisions related to Viking had an urgency that could not be avoided. It caused Martin to drive the project's team of scientists, engineers, and contractors ceaselessly.[5] Naugle established his own team of engineers to help him oversee Martin. To avoid offending Martin, he had the group report to Martin first. However, members of the team had the right to go all the way up to Fletcher if they saw an issue not being addressed.[6]

Among the technology development issues, none were more important or perplexing than developing the lander and life detection hardware. Much of the hardware development pressed the state of the art. Some issues seemed to go beyond. Martin initiated a list of "Top 10 Problems" as a management tool to concentrate energy and focus. Some problems were solved quickly, but others stayed on his list for virtually the entire time before launch.[7]

The prime contractor was Martin Marietta, but there were a myriad of subcontractors. Some of these performed well—on time and within cost. Others did not. One was fired early in the project because it could not do the work it was assigned and claimed that it was not feasible. NASA found another firm that said it could meet the requirements.[8] Even among those who clearly tried, there were challenges that were wholly novel.

The ones that proved most complicated related to Viking's distinctive mission—to find life. There were two basic instruments, one being the automated biology laboratory and the other one called the Gas Chromatograph Mass Spectrometer (GCMS). Initially, the biology lab contained four experiments selected by NASA through a competitive process. In March 1972, it became necessary to eliminate one of the experiments, a difficult decision that reflected on not only the nature of the experiments but the size, weight, and cost requirements of the biology laboratory.[9] Naugle, drawing on independent scientific advice, had to make the unpopular decision. The decision proved even more unfortunate when the scientist whose research was removed, Wolf Vishniac, died in Antarctica while trying to prove that his approach would work.[10]

But Naugle had no choice. The laboratory could be "no bigger than a gallon milk carton" and had to weigh "no more than 30 pounds." It had to be this small to fit aboard the 1300-pound Viking spacecraft, itself an unprecedented technical system.[11] The biology lab would have the equivalent of "three rooms of instruments on Earth, plus the people to work them." The lab would have 40,000 parts, including 22,000 transistors. There were tiny ovens where soils would be heated and ampoules of water and nutrients which would have to be broken by remote control at just the right time to mix with Martian soils inside the box. Bottles of radioactive gases, Geiger counters, and three chromatographs would be placed in the laboratory. There would even be a xenon lamp to duplicate sunlight inside the laboratory to be used in a photosynthesis test.[12]

Failures in the biology unit required major redesign in the September–December 1973 period. Martin established a task force to evaluate progress in early 1974. The task force, concerned that the contractor responsible for the lab, TRW, could not make the launch date window, recommended simplifying the project by dropping more experiments. TRW disagreed, and NASA decided to persevere, keeping the three it had. By September, TRW was still behind schedule, but catching up.[13]

Meanwhile, the GCMS proved almost as difficult a system to engineer. It had to isolate and identify organic molecules in Martian soil down to five parts per million. Doing that required virtually inventing a new technology. Then there were also many computer issues that came up again and again.[14] Everything was complicated by the fact that what landed on Mars had to go through a severe "decontamination" process so NASA did not bring microscopic life *to* Mars.

NASA decided that the urgency of the deadline required unusual management measures. If Martin needed help in pushing contractors, higher-ups became involved. Naugle and Cortright made trips to pressure contractors. In some instances, Fletcher himself had meetings with CEOs of particular contractors, "setting off an alarm in the front office" to impress on them the seriousness of delays.[15] Fletcher was especially forceful in getting topside attention from the computer contractor, Honeywell. He insisted that Honeywell put its best talent on the Viking project. Many contractors, especially those associated with the "Top 10 Problems," had to work overtime, even seven days a week, to accelerate progress. NASA personnel associated with Viking customarily worked 60-hour weeks. Viking was by far the most complex robotic mission NASA had undertaken to date. It was in some ways proving more technically difficult than Apollo because NASA knew so much less about Mars than the Moon.

Why did NASA push so hard? A. Thomas Young, Martin's deputy, explained years later: "You don't go to Mars that often. You do push things to do it . . . people who get involved in this business are . . . pushing the envelope and trying to get a little bit more here and a little bit more there. I think that's why you're willing to invest ten years for something that might blow up. Because if it really does work, it's been extraordinary."[16]

Where to Land?

In addition to technology development questions, there was the nettlesome issue of where to land. Soffen, the chief scientist, organized a special team of researchers, including exobiologists, to determine the best places from their point of view. They had Mariner 9 information, but not much else. Throughout 1972, the siting team argued over options. NASA wanted a siting decision by the end of 1972, but the scientists could not agree.

In November 1972, Lederberg came to Fletcher's office and pounded on his desk. He complained that the polar region was not being given adequate consideration because "the engineers"—Lederberg's shorthand for Martin and his project management officials—had not done their homework on the potential of the polar region for finding life.[17] The difficulty "the engineers" had was to balance safety in landing with Lederberg's preferred site. Indeed, the scientists on the siting team were sharply divided on the polar issue. Sagan, for example, favored equatorial sites as possibly wetter and thus more amenable to life. In a letter to Martin, he pointed out that most water at the poles was frozen, and the biology package could not "detect organisms which extract their water from ice."[18]

NASA decided to delay the decision on siting to 1973. In February of that

year, Fletcher asked Naugle two questions, reminding him that it was important to emphasize the "possibility of finding life" in choosing a place to land.
Fletcher's first question was whether Lederberg and the scientists believed that
the best chances of finding life were at the 73° polar latitude, and the second
was whether liquid water had to exist "*now* [Fletcher's emphasis] or could it have
existed once, for life 'signatures' to be detected?"[19]

Naugle told Fletcher the answer to his first question was "no!" The optimal
place to find life was where there was liquid water. As for Fletcher's second
question, "signatures" of life could be found not only where there was now but
where there had been liquid water in "the distant past." The problem, however,
was that Viking might "not be able to distinguish between biological and non-
biological types" of signatures.[20]

In early April, Lederberg conceded that 73°N, his preference, was not viable,
or at least not acceptable to those emphasizing safety. That did not end the
debate, which one NASA official trying to facilitate agreement called "trau-
matic."[21] Finally, in late April 1973, Fletcher got a consensus view, which was
announced May 7. There would be two prime sites and two backups. All four
were chosen with particular emphasis on the possibility of finding water and
life. However, safety in landing took equal or even greater consideration. The
debate had lasted one year.

The first region, called Chryse, was at the northeast end of a vast 3000-mile
rift canyon near Mars's equator. The second, called Cydonia, was farther to the
north and east, but not as far north as Lederberg had originally wanted. It was
near a polar cap (44.3°N) where the scientists hoped water might be left from a
previous melting. Backup sites featured similar considerations.[22]

Evaluation: Money Issues

In 1974, technological development continued and progress was painstakingly
made. One reason for the progress was that NASA was pouring more money
into Viking. If NASA could not get funds from OMB/White House and Congress, it reprogrammed funds from other accounts. The result was that Viking's
budget escalated and Congress grew restive over what it called overruns. Under
congressional pressure, top NASA officials evaluated the project. Viking was up
to $930 million in cost. Given other expenses, Viking was well into the billion-
dollar class of major projects. NASA had to postpone other new starts to pay
for Viking.[23]

In March, Fletcher appointed Rocco Petrone, director of the Marshall Space

Flight Center, NASA associate administrator ("general manager"), the number three position in the agency. He replaced Newell, who had retired in late 1973. Petrone had a reputation for being a hard-nosed and tough manager. Fletcher appointed Naugle to be Petrone's deputy. Noel Hinners, a 38-year-old NASA geochemist who had headed lunar programs, took Naugle's position as associate administrator for space science and applications. Low, the deputy administrator, also wanted to replace Martin as project manager, not because of anything he had done wrong, but because Low felt "a fresh look" could "clean up the problems, and provide new impetus to the project." Naugle and Cortright argued vehemently against what they believed would be a major disruption in Viking. Low pulled back.[24]

Petrone announced he would enforce cost discipline in regard to Viking. NASA had little choice but to do so. In August, Congress specifically warned NASA to stop reprogramming funds from other missions to rescue Viking and scheduled hearings on Viking for November to underline its concern.[25] Nevertheless, Hinners got a call from Cortright. "I need $30 million more for Viking," he said. Hinners refused. A half hour later, he received a call from Low, who said, "Give him the money, Viking is important. [We] can't afford to have it fail. It is a priority." Hinners dipped into his reserves without taking it from other programs.[26]

With congressional hearings approaching, Petrone and Hinners established budget caps for Viking, and Petrone stated that any deviation from the caps would require his personal approval. Viking managers would have to document attempts to find economies within the project budget before asking headquarters for any additional money. Petrone himself had a reserve set aside for emergencies, but he emphasized he would keep close control of it where Viking was concerned. He required Hinners to supply him "with weekly status reports on project costs and manpower levels for Martin Marietta, JPL, TRW, and Honeywell throughout the winter of 1974." Petrone not only followed the numbers but specifically directed Martin Marietta, the prime contractor, to work harder. To save money, Petrone and Hinners killed a backup Viking 3 system that was being built as a contingency in case either of the two Vikings ran into trouble, along with various other organizational adjustments.[27]

Hence, when congressional hearings by the House subcommittee overseeing space science took place in November, Petrone and other NASA officials could claim strong action to control costs. Petrone declared he "drew the line" on Viking costs, and Langley and other managers would have to answer to him

if they wanted to cross that line. A congressional staffer told the media after
the hearings that NASA had to be careful about cost escalation. "The space
program," he said, "is sort of a national luxury—like a mink coat for your wife.
It's a nice thing to do if you can afford it."[28]

Fletcher's Hope

Senior NASA officials were intensely involved with Viking and its implementation. Mindful of the huge cost escalation, they also discussed, behind the scenes,
how Viking could help promote NASA, and vice versa. Sagan was especially
critical of NASA's public relations policies. He and Low had dinner one evening
in September 1974 to ponder what to do about NASA's need for more public
support.[29]

Fletcher continued to think beyond Viking to possible follow-ons. Viking
was one mission, one project. Should it not be the first step in a long-term
program of Mars exploration? Only Viking was approved. Given lead times
on development and windows of opportunity for Mars flights, NASA had to
begin soon to advocate follow-ons to the White House and Congress. Issues of
technology and funding were critical. What was possible? What could NASA
afford? Should NASA send a mission in 1979—the next window—or wait? In
November 1973, Fletcher had written the chairman of the Atomic Energy Commission (AEC), Dixie Lee Ray, that NASA might need the nuclear batteries
AEC supplied (called RTGs) for future Viking flights. These could include
roving vehicles. Nothing was firm of course, but he wanted to alert her of his
thinking.[30]

Fletcher kept in close touch with Martin, and the Viking project manager
proposed that while Viking 3 had been killed, the spare parts for the backup were
still there and could be assembled as part of a post-Viking venture. Cortright,
Langley's director, also lobbied Fletcher, writing him in July to propose sending
a lander-rover combination—a "Viking '79 mission."[31] Hinners, however, told
Fletcher and Low in November 1974 that a follow-on in the near future was not
likely given his office's budget and competing priorities. The National Academy
of Sciences Space Science Board had done studies for NASA about post-Viking
projects, he pointed out. The "benchmark" for Mars exploration had to be an
MSR mission, the board asserted. This was the best way to get at the life issue.
At the same time, the board preached "balance" in the planetary program in
the future. Hinners said that the way spending was going and with other large
missions waiting, he could not see how NASA could launch a sample return mis-

sion before the 1990s. Missions prior to MSR, by implication, such as the rover, might also have to wait beyond a date his office had studied, 1984.[32]

This was not what Fletcher wanted to hear. The question of an MSR mission was of such significance, the NASA Administrator said, that he would not wish to foreclose it indirectly by budget decisions that locked NASA into a particular trajectory. He said he might even take the question to the president—that is, keeping the option alive by getting more ample funding.[33] Low told Hinners not to take any "irreversible steps" precluding sending a rover. But he also directed him not "to spend any significant resources on it since it is very unlikely that this will be an early start."[34]

Richard Goody, the chair of the SSB, captured the dominant mood of the scientific community in connection with Viking and what could come next. In comments to the media in December, he held that it was unlikely NASA would find life on Mars, but if it did, the whole future of Martian exploration would be affected. He declared, "You really can't make any decisions [about follow-ons] until you see what Viking does."[35]

So, Fletcher waited on advocating a Mars program beyond Viking. The good news, however, was that, as 1974 ended, NASA knew that it had a better chance to win the Mars race with the Soviets and possibly make history by finding life using robotic techniques with its stationary landers. In 1973, in an effort to beat the United States to Mars, the Soviet Union had sent two orbiters and two landers (four separate probes). If successful, the Soviets would land in 1974, well ahead of the United States. But all failed in one way or another. One orbiter flew past Mars. Another orbiter arrived and returned limited data before failing. A lander died within seconds of reaching the surface. Another lander separated early from its mother ship and missed the planet. The Soviets, bitterly disappointed, chose not to launch in 1975.[36]

Viking Launches

In 1975 the various technology development problems affecting Viking gradually gave way to solutions. Martin's "Top 10 Problems" were narrowed and then resolved enough for NASA to schedule a launch for August. As August approached, media interest expanded, and so did the angst of all associated with Viking.

From Fletcher on down, there was heightened anticipation of what could go wrong. Fletcher warned associates about how to frame the project in media interviews. He thought NASA might have overreached in emphasizing "life" as the goal. His staff suggested that NASA speak of the launch as about "comparative planetology."[37] Hinners sought to shift the rhetoric to "find[ing] evidence of life," and to get Sagan to "tone down his rhetoric." "But I could not get him to change much, and you didn't want to quell his passion in any event," Hinners recalled. Sagan "was a tremendously effective advocate, and salesman," he stated.[38] Sagan, who, more than anyone, had framed Viking as a quest for life, worried that the lander would crash and Viking would not discover what was waiting to be found.[39] Soffen worried about the biology laboratory. He lamented the decision to drop one of the four experiments it carried. "There was no way to keep it," he confided. But he worried: what if it had "been the one to detect life?"[40]

Martin was concerned about everything, but particularly the decision to kill the backup system. What if Viking failed? After all, the four Soviet spacecraft sent to Mars had failed in 1974 either to reach Mars or to perform once there. Would Viking suffer the same fate? As late as July 1975, one month before launch, Martin badgered Robert Kraemer, Hinners's deputy for planetary missions, about needing a third Viking. It had been killed by Petrone and Hinners, and Kraemer pointed out what Martin already knew—there was no money! Moreover, to go up in 1977 (the next window) would take another launch vehicle (a Titan-Centaur), and the only way to get one would be by "stealing" it from another mission that had been waiting in line. He promised that the spare parts and other hardware of the partially built Viking 3 would be kept for a possible succeeding window.[41] Martin was not encouraged. He worried that the political window on Mars exploration might be closing: "I think we will have to find something exciting to have another mission to Mars," he complained in a media interview.[42]

Shortly before launch on August 20, 1975, a valve issue came up on the launch vehicle. Fletcher, Low, Naugle (now the senior NASA associate administrator), and Hinners all sat around a table in a teleconference with Martin, who was at Cape Canaveral, peering at drawings of the valve, trying to figure out how to fix the problem.[43] That the top officials of the agency were so engaged indicated how important all viewed Viking. Fortunately, the problem was solved. Viking 1 went up. The four-ton spacecraft sped away from its Cape Canaveral launchpad "atop a Titan-Centaur rocket, a bright orange and yellow colored flame behind it. Burning solid rocket fuel that built up to 2.4 million pounds of thrust in seconds, the Titan-Centaur and its payload of instruments were 30 miles out over the Atlantic in two minutes and moving 5000 miles an hour." After going briefly into a "parking orbit" 100 miles above Earth, the rocket engine lifted Viking out of orbit at a speed of over 25,000 miles an hour. "We're finally on our way to Mars," Martin beamed. "All systems are working fine," he said. "It's been sheer hell," commented Kraemer. "There were times I thought we'd never make it."[44]

On September 9, Viking 2 went up—successfully. NASA officials were elated. As the Vikings sped toward Mars, NASA planned political strategy. In December, Low instructed the Science Directorate to prepare a supplemental request for funds in the event of "spectacular results" emanating from Viking. He wanted it ready to go to the White House and Congress by July 1976.[45]

Reaching Mars

On June 19, 1976, Viking 1 swung into Mars orbit. "After eight years, we're finally in orbit," a relieved James Martin exclaimed.[46] This was an achievement in and of itself. But NASA knew that this was but the first step. Soon, Viking was transmitting photos of Mars's surface, including the region where NASA planned a landing for Viking 1. Various NASA officials and others gathered at Mission Control, Jet Propulsion Laboratory. Gentry Lee, JPL mission planning director, recalled how grown scientists and engineers behaved like 10-year-olds as pictures of Mars from the orbiter came in. They whooped and yelled and ran to the screen where images appeared with cries of "wow."[47]

While the images were fascinating and spectacular, they also produced anxiety. Sagan, a member of the landing site team, remarked that Viking could see the larger-scale features, and many were menacing. But what about smaller-scale features the orbiter could not see? If Viking landed the wrong way on a boulder the size of a trash can, it might be wrecked.[48] Looking at images of the previously selected landing site, Harold Mazursky, on loan to NASA from the U.S. Geological Survey (USGS), felt that the risk was acceptable.

Martin, however, was not so sure. He conferred with superiors at NASA Headquarters up to Fletcher. NASA had scheduled the first landing for July 4, aligned with the national celebration of the country's 200th birthday. President Gerald Ford was "enthralled" and eager to make the landing part of the celebration. Fletcher and other top managers told Martin not to worry about the scheduled July 4 landing date. If he believed the site in question was too dangerous, they said, he should delay the landing and look for a place that was safer. On June 28, Martin informed a vast media assemblage that had gathered for the historic event that the chosen site "had too many unknowns, and could be hazardous."[49] Fletcher, meanwhile, informed the president that the July 4 rendezvous was out. Martin "would have thrown his badge on the table if we'd taken the risk of landing on July 4," Hinners recalled.[50]

The vital importance of Viking to NASA kept Fletcher intimately involved. On July 1, Fletcher announced from JPL's Mission Control Center that NASA had found an alternative 150 miles northwest of the original place. "Mars is a lot different planet than we thought it would be," he stated. "By a combination of intuition, wise judgment, and a little bit of luck, we found a site close by that exceeded all expectations."[51] But just a few days later NASA examined radar signals of the new site from Earth, and they indicated that the orbital images could be wrong as to the smoothness of the terrain. Again, Martin and his team decided they had better keep looking, and senior NASA officials once more went along with the judgment.

This time the reconnaissance was even more thorough, using orbiter photos, radar, and expert analysis. The problem was that as NASA looked farther from the original site, it moved more distant from potentially fruitful places of scientific interest. The search for life was the prime announced purpose of the mission, and that purpose was in danger of being jeopardized. The Viking team had to find a place that was both reasonably safe and scientifically interesting, which was becoming extremely hard to do.

The meetings of scientists took place every day for long hours, amidst growing frustration about getting consensus on a place to land. There was no time for personal lives. Viking dominated all schedules. Gentry Lee worried that the landing date would coincide with the birth of his first child. Tim Mutch, a Brown University geologist in charge of the lander's camera system, tested the system again and again, so often that he became mesmerized by his routines and, at one point, confused testing with reality. He went home one evening to tell his wife how well the photos had gone only to be reminded that the actual work lay ahead.[52] Everyone was on edge and getting cranky. Minds wandered and speculations roamed amidst the nervousness and loss of sleep. Mazursky imagined great floods taking place on Mars carving giant canyons. His USGS colleague, Mike Carr, countered that the surface features were more likely caused by slow-moving streams that took eons to carve the cleavages. Observers called Mazursky "the great inundator" and Carr "the long, slow trickler."[53]

No one was more frustrated or tense than Martin. "We always had it in the back of our own minds that Mars would not cooperate, and it hasn't," he complained.[54] One day he exploded over a trivial matter, signaling to everyone the exasperation they all felt about the exigency to make a decision soon about where to land.[55] Finally, at midnight, July 14, the landing-site team reached agreement on a particular site. It was 200 miles to the northwest of the original target in the plains of Chryse, where water was believed to have flowed.[56] Announcing the decision the next morning, Martin said Viking 1 would land July 20, a date that marked the anniversary of the first Moon landing of Apollo.

Landing

July 20 came. The Soviets had landed twice, once in 1972 and then in 1974. The first lander had survived 20 seconds and the second most likely crashed, neither transmitting pictures back. Would the United States meet a similar fate? Mars was 212 million miles away as the flight controllers at JPL made the decisions that separated the lander from Viking's orbiter. Then came the slow descent, begun with a parachute, braced by retro-rockets, as the lander neared the surface. Because of the distance between Earth and Mars, Viking could land—or crash—19 minutes before anyone on Earth would know which fate had occurred. NASA had prepared two press statements: one for success, one for failure. Naugle called the wait the longest of his life.[57] It was "nail-biting time," Martin later said. Mutch looked at his shoes as he waited and silence engulfed the mission control room. Dreading failure, he composed a statement of condolence for friends standing near him.

"A muffled prayer came over the loudspeaker. 'Come on, baby,' said a voice." Finally, the waiting and agony ended: "We have touchdown."[58] When the signal came that Viking had landed safely at 5:12 a.m. (PDT), everyone at JPL gave a loud cheer, followed by hugs, laughs, and other expressions of sheer relief. Hinners cried, as did Lee (whose wife was still days away from giving birth).[59] Pictures later showed that Viking came within 10 feet of hitting a huge boulder, and almost certain failure.[60]

Politicians and the media joined in the celebration. Headlines across the United States and beyond congratulated NASA for what the *New York Times* called a "superb and triumphant achievement."[61] As Viking sent back the first color pictures of Mars, revealing a light blue sky (later determined to be an imaging error; the sky was pink) above reddish land, there was rapt attention to the mission. President Gerald Ford was among those who greeted the news and photos with awe and excitement. He personally called to congratulate Fletcher, Martin, and the NASA team.[62]

Viking had passed its first great test in landing. Now all it had to do was find life.

Searching

This was what Viking was all about, at least that was the message NASA had communicated to the public: the search for life. The core of the scientific team for exobiology consisted of six biologists. Lederberg was one. The leader was Harold "Chuck" Klein of Ames. Sagan was not officially on the biology team. His role was in the site-selection group. However, he was deeply involved with the exobiologists and was unquestionably Viking's public face. Never had a NASA robotic mission been conducted in such a fishbowl environment. The media were present in force, hanging on to every word the scientists and Martin said.

Sagan had continually fanned the flames of public expectation by his comments. His book *The Cosmic Connection* had appeared in 1973 and became a best seller. He was a regular on the late-night television program the *Tonight Show*, hosted by Johnny Carson. While other exobiologists speculated about finding microbial life, Sagan spoke of "macrobes." These would be organisms large enough to be seen by Viking's camera. "For all we know," he said, "there is a thriving population of large organisms on the planet. Nothing in our present understanding of Mars excludes this possibility."[63] Talking about Viking, Sagan's rhetoric could soar. "Viking will be remembered, if it works, the rest of human history. It deals with the deepest question that human beings have asked as long as they have been human beings."[64] He was correct, of course, but Sagan's parenthetical "if it works" tended to be de-emphasized in the translation from scientist to media to public. Hinners kept trying to control him to some degree, without success.[65]

The scientists on the biology team varied in their assessments of the prospects. When asked, Klein declared that "among the biologists on the team, the odds go all the way from one chance in ten down to maybe one chance in a million. Depends on which one of the biologists you talk to. Mine are one chance in 50, which I think are not bad [odds]." The key, he said, was the "payoff," if life were found.[66]

Klein was worried, however, about the downside of failure. Most of the prominent scientists involved in Viking were academics. He was a government scientist and could feel the pressures of nonsuccess after such a public relations buildup (and $1 billion investment). He confided that NASA was putting too much emphasis "on the question of life." He worried about what would happen if Viking failed to find life.[67] No one knew more than he that the selection of experiments represented something of a "shotgun" approach. Were these shots in the dark despite the careful efforts of NASA and distinguished scientists? He would rather have followed the incremental approach that Murray had expounded, particularly since so little was known about the surface chemistry of Mars.[68] But NASA as an institution had gone for the "great leap" strategy, and he was now fully part of that effort.

Did Viking Find Life?

Klein's task as head of the biology group was exceedingly difficult. There were three chief experimenters: Gil Levin, Norm Horowitz, and Vance Oyama. Levin was a private-sector researcher, a PhD in sanitary engineering, whose experiment was called "Labeled Release." Oyama was a NASA Ames biochemist, and his experiment was called "Gas Exchange." Horowitz was a Caltech geneticist, and his experiment was known as "Pyrolytic Release." The three men—especially Levin and Horowitz—disliked one another intensely. Horowitz was openly contemptuous of Levin's and Oyama's experiments and called them "irrelevant." Klein got the nickname "Rabbi," for his efforts to keep peace among the researchers.[69]

Once Viking had settled on the ground a few days, the experimenters' work commenced. The activity was slow and painstaking but gradually produced results that some of the biologists found extremely provocative. NASA policy was that the scientists announce results quickly, as they became available, along with statements about confidence and uncertainty. The first promising results gained conspicuous headlines. But then, after a few more days, results from Viking looked less promising, and this fact was also reported by the media, but with lesser prominence. The problem was that some of the findings were compatible with life, but they could also be interpreted as reflecting evidence of a "strange and unexpected Martian surface chemistry." Klein declared, "We have at least very preliminary evidence for a very active surface material. . . . [It looks] very much like a biological signal." On the other hand, it could be chemical data that "may mimic biological activity."[70]

Levin thought that his experiment was credible as to identifying life. It had been his test that had provided initial enthusiasm. "The cork literally popped," recalled Soffen. But then the scientists took a closer look. At this point, Soffen recalled, "No one wanted to say [publicly] 'We found it,' and then say 'sorry'— the whole credibility of science is shot to hell! So there was a lot of resistance to getting up and saying there was life."[71] What Klein did say was that "Mars is really talking to us and telling us something. The question is whether Mars is talking with a forked tongue or giving us the straight dope."[72] Was it life or bizarre chemistry that Mars was communicating? Then, on August 12, came results from the GCMS. It could not find any life-indicating organic molecules. This came as a profound shock to many Viking participants. For Soffen, the GCMS findings—no positive findings—were "a real wipe out." Informed of these results, Soffen said to himself, "That's the ball game. No organics on Mars, no life on Mars."[73]

Not everyone shared Soffen's gloom, but the results were surprising and disappointing. It was subsequently surmised that Mars's ultraviolet sunlight produced highly reactive compounds that broke down the organic molecules.[74] But who knew for sure what was going on?

On September 2, Fletcher wrote President Ford that Viking was providing significant information about Mars's geology, atmosphere, and planetary evolution. However, "the search for forms of life remains inconclusive."[75]

A lot now depended on Viking 2, the enhanced significance of which the media reported. The second Viking had been circling Mars for some time and was set to land September 3. As with Viking 1, there had been issues with where to land, and alternatives searched, but those questions had now been resolved. It would set down at a place called Utopia Plain, hundreds of miles to the north and halfway around the planet from the Viking 1 site. Whereas the first Viking landed at a Mars latitude equivalent to that on Earth of Mexico City, Viking 2 would land at the Mars latitude analogous to that of Montreal.[76]

The landing was successful, and again the president congratulated NASA. There was renewed hope. Soffen said that the discovery of even the simplest organic compound—inextricably associated with life as we know it—"would do it for us."[77] Once again, the biology experiments found indications compatible with life. But that was not strong enough proof for most Viking scientists. Hopes were pinned on the GCMS instrument once more in the quest for organics.[78] By the end of September, NASA had many enticing findings, but hard and convincing evidence for life was still not there. As Klein had feared, the promising results could have resulted from "a bizarre chemical system beyond immediate explanation."[79] The GCMS data were again negative. On October 1, the *New York Times* reported that Viking 2 had found "no organic matter," and while results were "preliminary," Viking scientists conceded that the findings "did not bode well for life-on-Mars theories."[80] Levin did not go along with the consensus. He told his fellow researchers, "We agreed at the outset that if the results came out a certain way, we'd say 'yes' to life. My experiment came out this way. I discovered life."[81] But Horowitz "overpowered Levin," and he persuaded the others that Levin had not done so.[82]

The Verdict

The first phase of the Viking mission ended November 8 when Mars passed behind the Sun, interrupting radio communications with the Red Planet. Viking would resume after a while and continue operating in an "extended mission." On November 9, however, NASA held a Viking news conference. Soffen, Klein, Sagan, and Klaus Biemann of MIT, leader of the team analyzing GCMS data, were present to answer media inquiries. Relevant data and analyses from Viking 1 and 2 were now in. The scientists stated that the evidence of life was contradictory. Klein, responding to a reporter's questions, said of the data, "I would say that on the basis of incomplete evidence, which is where we are today, we cannot say conclusively that there is life on Mars. I would say we cannot say conclusively that there is not life on Mars."[83] The media representatives pressed for a yes or no answer; instead, they heard the scientists declare an absence of what Sagan called "conclusive explanations of what we're seeing." Sagan pled for the media to have "an increased tolerance for ambiguity."[84]

The scientists tried to explain that Viking had sampled just two places on a large planet and dug just a few inches beneath the surface, but such caveats were "lost in the noise." What was perceived by the media and conveyed to Congress and the public was "stripped of nuance, laden with finality." The verdict: "no life on Mars."[85]

CHAPTER FIVE

Struggling to Restart

Viking had failed! At least that was what many critics believed. NASA knew better. There was much that Viking contributed in new knowledge about Mars. But the agency and the Mars community were deeply disappointed on the life front, the central purpose of the mission. In the wake of Viking, NASA debated intensely how Mars fit into its future. NASA was getting desperate for "new starts." Hinners, associate administrator for science, warned Congress that without starting new flight projects in the pipeline NASA's planetary program was on a "going out of business" trajectory.[1] NASA had geared much of its strategy to Mars, in hopes that Viking would yield evidence of life. NASA had believed that it had ample reason to do so. It did not have firm plans as to what it would do if it did not find life.

After internal debate, NASA decided it needed to go beyond a lander to a rover, what it called Viking 3. But that would cost over $1 billion at a time when NASA had on its agenda the Hubble Space Telescope and an outer-planet mission, Galileo. NASA deferred a Mars decision, and Mars went on the agency's back burner. Outside advocates subsequently lobbied for renewing the Mars dream, but it took eight years, from 1976 to 1984, for NASA to launch a new start to the Red Planet—Mars Observer. The design and objective of the new mission were vastly different from the Viking 3 vision. NASA lowered its ambition to what it could get its political masters to accept. The Mars advocacy coalition was weakened by Viking's results after so much talk about life. Moreover, there was ample opposition to Mars from those favoring other space priorities.

Ford: What Next for Mars?

In telephone calls congratulating NASA for Viking 1 and then Viking 2's successful landings, President Gerald Ford twice asked senior agency officials about future Mars efforts. Naugle, after Viking 2, explained that NASA had been waiting on what Viking produced before defining and proposing a follow-on project. Naugle said there were three options NASA was considering. One was a Viking with wheels that could rove rather than be stationary. Another was to go back with better instruments to unravel the surface chemistry puzzle. A third was to bring a sample of Mars soil back for analysis in Earth based-laboratories. Ford replied that he assumed he'd be hearing from Fletcher at some point on NASA's plans.[2]

There were opportunities for Mars launches in 1981 and 1984, and decisions had to be made soon to take advantage of them. Could Viking (a single project) be turned into a multimission Mars program? Should it be? Martin was actively promoting the idea of a mobile Viking, not just to superiors up to Fletcher, but externally. At a news conference preceding the Viking 2 descent early in September, Martin declared, "We believe it is possible to make a mobile lander. We believe it is possible to launch by 1981, if such a program is approved." Martin said that "we have learned very exciting things from the surface of the planet and I believe we need to now take advantage of that knowledge." When asked about Martin's comments, Naugle stated that NASA and the Office of Management and Budget were negotiating next-year budget proposals and no decisions had been made, but the agency was "looking hard" at the rover issue. Martin was putting the price of a Viking 3 between \$350 and \$450 million plus launch costs.[3] Outside NASA, Sagan argued for a rover. As he later wrote,

> I found myself unconsciously urging the [Viking] spacecraft at least to stand on its tiptoes, as if this laboratory designed for immobility, were perversely refusing to manage even a little hop. How we longed to poke that dune with the sample arm, look for life beneath that rock, see if that distant ridge was a crater rampart. And not so very far to the southeast, I knew, were the four sinuous channels of Chryse. For all the tantalizing and provocative character of the Viking results, I know a hundred places on Mars which are far more interesting than our landing sites. The ideal tool is a roving vehicle carrying on advanced experiments, particularly in imaging, chemistry, and biology.[4]

Not all Mars scientists agreed that a rover was the logical follow-on mission. Tim Mutch, the geologist from Brown University who had directed Viking lander camera activity, was not so sure this rover approach was the best next step in Mars research. Klein, NASA's chief biologist, was sure it was *not* a fruitful approach. He said he felt he had to "speak out against the rover concept." Traveling around Mars taking biological samples would not necessarily resolve the question of life there, he told the media.[5]

By mid-September, NASA had settled its internal debate in favor of a rover mission as a Viking follow-on. It did not want to repeat the existing Viking mission, but go beyond, to take the next step in exploration. The rover would make measurements, take photographs, and collect Martian soil samples that would be returned to Earth through a later Mars Sample Return mission. Naugle said that NASA was thinking in terms of an MSR mission that would launch as early as 1986.[6] The rover might well be nuclear powered to assure it longevity and range. The disappointing results from the Viking soil experiments were countered to some extent by Viking 2 orbiter findings that indicated that the permanent northern polar cap of Mars was composed entirely of frozen water. If that was indeed the case, there might be water elsewhere, maybe in the permafrost. Where there was water, there was the potential for life.[7]

Proponents of Viking 3 were calling for use of the 1981 opportunity, but this would depend on NASA's getting adequate money in its FY 1978 budget. Fletcher made it known that NASA was considering a Viking 3. "We must go with what is going to sell [to the public] in addition to what is popular with scientists,"[8] he declared. Fletcher said he might discuss Viking 3 with White House officials or at least during the budget dealings under way with the administration. A possible complication in managing a Viking 3 project had been resolved with decisions about center roles made recently.

These decisions made the Jet Propulsion Laboratory lead center for future planetary missions, assigning Langley other tasks. Pickering had recently waged a determined campaign to get this designation for JPL prior to his retirement in 1976.[9] Langley did not strongly contest the matter, as Cortright had retired in 1975 and many in the institution had found Viking the "tail that wagged the dog" of Langley's historic aeronautics emphasis.[10] Although he remained a strong advocate for Viking 3, Martin was now uncertain about his own role in view of the assignment of lead center for planetary exploration to JPL.[11]

The problem for NASA was that by October 1976 it was clear that a Viking 3 rover mission would be another $1 billion project, and NASA officials were admitting that "in the absence of the spectacular selling point of life on Mars, it would be difficult to persuade Congress to finance such a project." "If we had found life, or even a reasonable hint, we would have gone berserk," Naugle recalled. "We would have sent landers at every opportunity."[12] Even Fletcher was now expressing disappointment and a sense of lost opportunity. "If you found life," he declared, "you might be making a manned mission to Mars before too long. But we weren't that lucky."[13]

With the November 9 Viking news conference and the translation of scientists' "ambiguity" into a public perception of *failure* to find life, NASA's challenge in defining and selling a Viking 3 worsened. Its plans for making a plea for additional funds rested primarily on Viking's exobiology results, and those results had been disappointing to virtually everyone. One NASA official, Oran Nicks, wrote, "It had been a little like waiting for Christmas as a kid, only to find on Christmas morning that Santa did not come through."[14] As the projected spending plan for the succeeding year made its way through the White House budgetary process, NASA was still equivocating about Mars exploration. It did not negotiate with OMB a specific Viking follow-on mission funding in its proposed budget as of mid-November.

President Ford would be evaluating the fiscal 1978 budget soon, and NASA would have an opportunity to "appeal" beyond OMB to the president to make last-minute changes before the budget was finalized.[15] President Ford was attentive to Mars thanks to Viking, and he had virtually invited a proposal from Fletcher for a successor project. But this would be Ford's last budget, owing to his loss to Jimmy Carter in the November election. The initiative would have to come from NASA, and it had to be a strong case to get more money on appeal.

The conversations about an add-on for Mars took place in the context of a growing scientific pessimism and debate about NASA's finding life on Mars via a rover. For many scientists, the biological explanation might have won over the chemical explanation of results if the Gas Chromatograph Mass Spectrometer had found organic molecules. Even Murray, longtime doubter, now JPL director, admitted he might have changed his mind. But Sagan opined that too much was being placed on the GCMS experiment. The positive biology experiments were "a thousand times" more sensitive than the negative GCMS test. But Lederberg, Sagan's ally, was not backing him up this time. "Occam's Razor clearly points toward a chemical hypothesis," he said.[16]

Many scientists with outer-planetary and telescope interests worried that pushing too hard for Viking 3 would jeopardize non-Mars science missions.

Mars scientists themselves divided over strategy. Sagan and Gerald Wasserburg, a Caltech geophysicist, attended a NASA science advisory committee meeting, and their altercation illuminated the split among scientists who were advocates of Mars research. They both reported on work with the National Academy of Science's Space Science Board on Viking follow-on options. Wasserburg argued that NASA should de-emphasize "search for life" in future Mars missions in favor of physical and chemical science. He expressed a "horrible fear" that *all* future Mars missions would be jeopardized by continued ambiguous biology results. Sagan countered that the SSB panel displayed "differential timidity" in science priorities. The biology instruments could be improved, he argued, thereby strengthening the chances of finding life. Moreover, Sagan complained, the SSB panel was not representative of the exobiology community. Wasserburg disagreed strongly with Sagan, declaring that the SSB body had a full spectrum of views, from conservatives to "fanatics, like yourself."[17]

Decision to Wait

The policy debates continued within NASA and among NASA, its scientific advisors, OMB, and others. There were doubts that a Viking 3 could be developed in time for a 1981 launch and maybe not even a 1984 launch window. NASA kept looking for a "compelling reason" to propose Viking 3 to Ford. The Cold War competition argument no longer worked since the United States had "won" the Mars race. Media reports described NASA as being in a "pressure cooker" on Mars decision making in early December.[18] Hinners recalls his feeling at the time that he did not want to put Hubble and Galileo—which OMB had approved for new starts—at risk. He also remembered "a sense that other missions had waited on Mars, and now it was their turn." Without question, advocates for Hubble and Galileo lobbied hard, and the Mars advocacy coalition was comparatively splintered and exhausted. In the end, Hinners said, "We decided to wait and digest the knowledge coming out of Viking."[19]

The president's science advisor announced the decision on December 16 in discussing Ford's last budget (FY 1978). H. Guyford Stever said there was money for "a large, orbiting telescope and a mission to place a photographic satellite in orbit around Jupiter." But Stever stated that NASA's budget would not include additional money to begin work on another Viking that would be launched in 1981 and land on Mars. He noted, however, that a launch could be accomplished in 1984, or later, depending on future decisions to be made.[20] These would have to be made by a new president, along with a new NASA Administrator.

With the decision to pass up the 1981 opportunity, the bulk of the Viking team disbanded, starting with Martin. With no Viking 3 immediately ahead, and Langley no longer the lead center, Martin decided to leave NASA for a job as vice president for Martin Marietta, the Viking contractor. There would be an "extended mission" to analyze data from Viking, NASA said. That would keep a modest portion of the Viking team busy for a while. But Martin was going, and in a bittersweet farewell visit to Langley in mid-December he stated that Viking would be the highlight of his career. "A lot of people haven't had this experience and never will," Martin said. "It would be selfish of us to want more than one. Viking has been tough." As for the question of life: "We haven't found 'life' on Mars, but we also haven't found 'no life' on Mars. Maybe it's not like Earth—to me that is possible too." As for whether Viking was worth $1 billion, Martin was emphatic: "Absolutely."[21]

Mars in Eclipse

The year 1977 marked a transition for Viking and a punctuation point in the Mars exploration journey. It was a year in which the Viking team gained many accolades, but it was also a year in which the experiments specifically to find life ended and critical Viking participants who had not already left moved on. On April 1, "Members of the Viking Team" received the prestigious Goddard Memorial Trophy. President Jimmy Carter, taking over from Ford, wrote his congratulations: "The Viking mission is a striking example of how our operations in deep space have opened up vistas of new worlds. Its success is the finest tribute that can be paid to the dedication and tremendous skills of the men and women who made it possible."[22] Carter's NASA Administrator was Robert Frosch. He was a 49-year-old physicist with a background in government and industry management. However, he was not particularly associated with space policy. He added his congratulations to Viking participants in writing an introduction to a special issue on Viking of the *Journal of Geophysical Research* and calling for a "continuing, highly productive planetary exploration program."[23]

Viking was fading into NASA's history. In late May, NASA ended all communication with the automated biology laboratory. Other components of the Viking equipment functioned for a while, but eventually, one by one, they expired. NASA terminated the project in 1983. NASA provided funds for ongoing data analysis. The debate over life dimmed. Even Sagan muted his position to some extent. Levin strongly held out from the dominant view. He argued that he did detect life. He would stoutly defend that position for years. He said in 1977,

"We're in the unexpected position of explaining away results that we would declare on Earth as unequivocal evidence of life."[24]

The fate of Soffen, the Viking chief scientist, symbolized the poignant reality of Viking's aftermath. The failure to find life, he recalled later, "clobbered exobiology. Absolutely laid it to waste." With planetary work now centered at JPL, he tried to transfer back to Pasadena. He wanted to continue in exobiology, maybe work on exobiology issues beyond Mars. The director of JPL, Murray, said Soffen could come back to JPL, but he couldn't work on exobiology. "I'm sorry," said Murray. "You can have a job here, but you're not going to work on exobiology." And Soffen could not work on exobiology beyond Mars. Said Murray, "Hey, if there's nothing on Mars, there's nothing anywhere." That, complained Soffen, was a "geologist's point of view."[25]

Soffen decided to stay at Langley for a while and then took an adjunct professorship at Harvard, eventually returning to NASA and settling at the Goddard Space Flight Center in Maryland where he turned to applications of space satellites to monitor planet Earth. Lederberg, Sagan, and other academics closely associated with Viking went back to their universities. Lederberg pursued research other than exobiology, as did most of the Viking scientists. Sagan, who had lived for two years in Pasadena during Viking's apex as a project, returned to Cornell. More a celebrity-scientist than ever through his popular writing and a television series he hosted called *Cosmos*, he tried to maintain his advocacy for Mars exploration and the search for life generally, but he had few allies. The advocacy coalition that had initiated and sustained Viking atrophied.

While dubious of exobiology, Murray remained a Mars champion. He encouraged JPL to pursue follow-on work, which fell under a classification of top priorities for the lab he called "purple pigeons."[26] These were areas of research with both intrinsic scientific *and* public interest. The JPL research emphasis for Mars was the concept of a Mars rover, what had once been called Viking 3. The problem for JPL and NASA was that there was minimal discretionary money for exploratory research for prototype development, and there were "purple pigeons" JPL wanted to pursue other than follow-on Mars work, some that Murray put ahead of the rover.

In 1977, a high-visibility project, Voyager, was launched. This mission had nothing to do with the earlier "Voyager" to Mars that was aborted prior to Viking. This project was conceived years before NASA's budget plummeted to take advantage of a fortuitous alignment of the outer planets. It took on the name "Voyager" because of its extensive journey in the solar system. Voyager consisted of two probes aimed at flybys of Jupiter and Saturn, with trajectories to Uranus and Neptune. Voyagers went up in August and September, carrying golden phonograph records with a message from Earth, just in case, eons in the future, Voyager probes passed into another solar system with intelligent life. President Carter's words were on this record with a message of peace.[27]

In 1978, NASA sent a mission to Venus. Like Voyager, this Venus mission had been approved before Viking landed in 1976. To the outside world, this sequence of missions—Viking, Voyager, Venus—made it appear that NASA's planetary program was healthy and robust. Within NASA, the view was decidedly different. There were scientific advocates inside NASA who wanted to work on Mars, but they had no approved flight project. In 1977 Hinners went to the Soviet Union. "I asked my Soviet colleagues," he recalled: "What are you doing about Mars?" He said that Russians replied, "We've stopped it! You found no life."[28] The Soviet Union had reoriented its program to other planets, especially Venus.

A main problem for those inside and outside NASA who wanted to elevate Mars on the agency's agenda was the Space Shuttle and its expense. This project, begun under Fletcher, continued to have top agency priority in the era of Frosch. Human spaceflight trumped space science. Gentry Lee recalled meeting Fletcher when he came to JPL and demanded support for the shuttle. "Young man," he said to Lee, "line up behind the shuttle."[29] Then, there were the other space science ventures with claims on NASA funding. Ford had approved both Hubble and Galileo; Carter inherited these projects and maintained them.

But when Hubble and Galileo were considered by Congress in 1977, they ran into a roadblock. The chairman of the House subcommittee considering NASA's budget, Edward Boland (D-MA), pushed NASA to decide between Hubble and Galileo for funding. To emphasize his point, he tried to kill Hubble and, when opposition from a united astronomy community surfaced, switched to an attack on Galileo. NASA and the scientists favoring Galileo pushed back and, after a legislative struggle, prevailed. But the result of the encounter was to point up the challenges all planetary exploration—and especially Mars—faced in the post-Viking era.[30]

This legislative battle—which was novel in its intensity over space science—made NASA leaders extremely wary of proposing any highly expensive Mars options. The SSB supported the rover concept, but NASA saw difficulty getting a Viking 3–type project through the Carter White House / OMB, much less Congress. But if not a rover, then what?

The Enigma of Carter

The enigma that was Jimmy Carter in relation to Mars policy was seen vividly in 1978. Carter, a Naval Academy graduate, was unusually technically astute for a politician. He had nuclear engineering training and expressed interest in space. In 1978, Sagan won a Pulitzer Prize for his book *The Dragons of Eden*. Carter invited Sagan in December to give a talk on space at the U.S. Naval Observatory, a talk he and his family, as well as the vice president and his family, would attend. The writer Hugh Sidey wrote that, while being briefed by Sagan about planetary exploration, Carter was fascinated. "Eyes bright with the sense of adventure, [Carter] urged that any new missions to Mars seek out mountains and valleys and old volcanoes instead of staying on the more level or gently rolling surfaces."[31]

Sagan biographer William Poundstone also commented on the Sagan-Carter exchange. He wrote that Carter was a space buff whose technical background allowed him an understanding rare among politicians. Carter seemed enchanted with Mars and the possibility of life on Mars. Were scientists certain Viking had not found life on Mars? Carter asked Sagan. Sagan pointed out the ambiguities, and that no life was believed to have been found at the two sites at which Viking landed. Then, Carter asked, why had the two Vikings landed in such boring places? Hadn't the Viking team heard the old saying, "Nothing ventured, nothing gained?" "You know," Carter told Sagan, "You ought to write a few books to really get people interested in planetary exploration. Then we could do some really exciting missions." "But Mr. President," Sagan responded, "You only need to write your name at the bottom of a single sheet of paper and we could have a rover mission to Mars." The president, according to Poundstone, just smiled.[32]

The Fall of Mars

The problem was that Carter's budget, released at the beginning of 1979, had no money for *any* new planetary missions. Although holding back on Mars—about which there was still division as to what to propose in the near term—NASA had hoped to begin work on a mission to Venus, as well as one to Halley's Comet. The Halley's Comet mission was seen as an opportunity to take advantage of Halley's once-in-a-generation return to Earth's vicinity. When Carter eliminated any new start in planetary science, NASA had to accommodate Venus, Halley's Comet, and Mars within a budgetary category allowing discretionary spending for exploring possible future work. That category now was taxed to the full and then some. Hinners and Murray had to decide on priorities, and Mars gave way to Venus and Halley's Comet.[33]

Abruptly, almost all work was cancelled at JPL on Mars. "We are not cutting back on Mars work because of diminishing interest in Mars or in the follow-up [to the Viking] mission," said Geoffrey Briggs, who had taken over as acting director of the Office of Space Science and Applications's Planetary Division. "Although the planning and design of a new Mars mission is being reduced to a fairly low level, we will be maintaining to a much higher level the analysis of Viking Mars data we now have."[34]

Steve Squyres, a recent Cornell PhD, who had been inspired by Sagan and other Viking researchers, ran into contrary advice from various senior planetary scientists. "Don't focus on Mars," he was told. It was a career "risk," they said.[35] The atmosphere at JPL was such that many scientists were discouraged from pursuing Mars work. John Beckman, planetary program manager at JPL, declared, "Our knowledge of Mars now vastly outreaches our knowledge of Venus and other planets, and it's time to even up our knowledge and balance our approach." He said he doubted NASA would get back to Mars until the 1990s.[36] Daniel McCleese, a young scientist at JPL, recalls that when he talked up Mars, "I ran into a very, very strong pushback from NASA, along the lines of 'We've gone to Mars with Viking. It is a dead planet, no longer very interesting to NASA, and we are headed for the outer planets.'"[37]

At the end of 1979, Carter's OMB called for no new starts in planetary science for a second year in a row. This was a decision that had implications not only for Mars but extending to other robotic science programs as well. Frosch appealed to Carter, who forced Frosch to choose among various projects that might get a go-ahead. Frosch was not about to sacrifice an astronomy mission, the Gamma Ray Observatory, for a Halley's Comet project, the priority for Murray and other solar system researchers. The president said he also favored the Gamma Ray Observatory because of his interest in black holes, about which he had been reading of late.[38] Planetary exploration was put in a position secondary to astronomy missions aimed at the cosmos. The mission to Halley's Comet was not going to happen, at least one led by the United States. There was a sharp disconnect between Carter's avowed personal excitement about space exploration and his spending decisions.

Whatever potential NASA had for a Viking follow-on was lost in the transition from Ford to Carter and NASA's continuing ambivalence about what to do next. The ambivalence arose from the fact that most Mars advocates wanted to take the next steps beyond Viking—a rover and then MSR. But NASA leaders also knew they could not possibly sell such an ambitious mission in the political environment the agency faced.

Keeping the Dream Alive

While NASA pondered options and postponed advocacy to the White House and Congress, a diminished but persistent band of supporters inside and outside government struggled to keep the dream of Mars exploration alive. Rather than being depressed by the consensus view of Viking's failure to find life, a cluster of planetary science graduate students at the University of Colorado Boulder were electrified by what Viking did discover. Carol Stoker, one of their leaders, read everything she could about Viking and concluded that there had been a "rush to judgment" about the absence of life on Mars.[39] Also, if life perhaps was not on Mars, then human life should be brought to Mars, she and her peers believed. Another of the leaders of the group—one who, like Stoker, would later work for NASA—was Chris McKay.[40] Meeting in halls, over meals, and eventually organizing a course, this group in the late 1970s began studying what it would take to revive robotic and human Mars exploration in the post-Viking period.

The students were given legitimacy by Charles Barth, the director of the Laboratory for Atmospheric and Space Physics at the University of Colorado Boulder. However, they acted largely on their own. Another person who energized them was Ben Clark, a former Viking team member from Martin-Marietta, the prime contractor for Viking. Clark was based in Denver and had written a paper titled "The Viking Project—the Case for Men on Mars." He extolled the possibilities and communicated with the students. The students were enthused and let their imaginations soar, considering the potential of "terraforming" Mars, engineering the planet so as to make it habitable.

As the group reached out, they gained new members, developed their "Mars Study Project," and took on a catchy and enticing name, the "Mars Underground."[41] Members wore red buttons they concealed and could flash to nonbelievers and others. The buttons conveyed the impression that the group constituted a secret society. Members discovered many other "closet Martians" willing to join their cause. Stoker recalled that the group seemed to be filling a need, attracting adherents in the manner of a "social movement." A journalist the students met suggested they hold a national workshop on Mars. He thought there would be considerable interest. The students, even as many looked to graduate, began planning such a conference for 1981. It took place, and 100 scientists, engineers, and others came. Those from NASA were primarily from JPL and Ames. Participants gave papers and networked. This ranged from serious Mars researchers to what McKay called "the lunatic fringe" of the Mars community. In contrast to the majority view that there were no intelligent Martians, this group, known as the "Face on Mars" advocates, believed that Viking had revealed a structure on Mars that resembled a face. Moreover, they charged that NASA had conspired to hide this reality from the public. They were a distinct minority at the "Case for Mars" conference, as it was called, but they were welcome. The Mars Underground was open to all Mars adherents.[42]

What the Boulder meeting did for those who attended went well beyond technical matters that consumed most of the formal presentations. "The first conference was magic," Stoker said. "People walked out of there feeling like they'd been freed from prison . . . we broke the taboo." It was possible to think about Mars and all its possibilities for those who came.[43] By the time this meeting took place, there was another advocacy group also asserting its claims. The Mars Underground was ad hoc, calling itself a "closely knit but loosely woven network of individuals, representing government, private industry, and individuals."[44] In contrast to the Mars Underground, this other body was organized, well funded, and led by established and influential scientists.

The dynamo was Carl Sagan, who refused to give up on his quest. Sagan was convinced that Mars needed grassroots advocacy. Perhaps more than any other scientist of his time, Sagan sensed the public pulse and what it would take to mobilize opinion behind space exploration. He also was frustrated with NASA and various fellow Mars scientists. He sensed a "disreputability about looking for life on another planet."[45] He wanted to take his case more forcefully to the public and political establishment. The Mars Underground and like-minded interests in the public, he felt, needed an organized interest group working on their behalf to persuade politicians to spend more money on space exploration. He passionately wanted NASA to go back to Mars and search for life.

His former nemesis, Murray, the JPL director, felt similarly about the need to generate support to study Mars and other planets. Once adversaries and still disagreeing on the life issue, Sagan and Murray nevertheless sought common ground in organized advocacy lest the nation lose sight of a noble dream. Energized by what they perceived as NASA indecision and the budget travails of the Carter years, they discussed the concept of an organized interest group with a third individual, Louis Friedman.

Friedman was an advanced-program manager at JPL with an activist temperament, broad perspective, and enthusiasm for Mars. In 1979, he was just back from a year as a staff member on a congressional science committee. He recalled how Murray had summoned him "to his office on the top floor of the administration building at the Jet Propulsion Laboratory." Murray explained that he and Sagan believed that planetary exploration was "threatened" and to save it they had "to form an organization with tens of thousands of members to demonstrate that people wanted space exploration to continue." "Would you lead it?" Murray asked. Friedman readily agreed.[46]

The next year, 1980, Sagan and Murray established an entity, the Planetary Society, with Friedman as its executive director; Sagan, president; and Murray, vice president. They gathered a board of prominent individuals. The Planetary Society was announced to be "dedicated to planetary exploration and the search for life." With Sagan actively publicizing the organization, it would eventually grow to 125,000 members, publish a newsletter, get support from foundations and wealthy individuals, and become a force in influencing federal space policy.[47] From the beginning, the Planetary Society "focused much of its activity on the exploration of Mars."[48] Sagan announced the formation of the Planetary Society in 1980 during one of his many appearances on Johnny Carson's *Tonight Show*.

The Mars Underground, the Planetary Society, and a few other external interest groups gave evidence that public support for Mars exploration remained in the wake of Viking. They were complemented to some extent by certain advocates within NASA, particularly in the field centers. One field center especially nurtured exobiology.

Perhaps the greatest casualty of Viking disappointment had been exobiology. Detractors had derided the field as being a discipline without a known subject to study. They now said, "We told you so!" Most Mars scientists were down on exobiology, as was NASA in general. What helped keep the field going, albeit with fewer researchers, was the maintenance of an institutional base at the research center, Ames. There, Harold Klein refused to give up after Viking, and he encouraged other researchers to persist, especially Sherwood Chang. They were able to mine Viking images and data. Even at Ames there were occasions when Klein "went to the mat" to protect exobiology from being shut down by an Ames center director.[49]

It helped enormously, in the view of David Des Marais, an Ames veteran, that Ames was a civil service laboratory.[50] The researchers had more leeway and security in what they did than scientists and engineers at JPL, who had to seek contracts on more fashionable and/or fundable topics. Many JPL employees had to work for agencies other than NASA, especially the Department of Defense, to survive in the post-Viking years.

There was precious little money for exobiology research at Ames. What there was came from intrepid program officers committed to the field. Des Marais estimated that the funds Ames had averaged $6 million a year.[51] Don DeVincenzi ran a modest exobiology program at NASA Headquarters and provided funds to Ames. John Rummel, NASA's planetary protection officer, also funneled some money to Ames for exobiology. Beginning with Apollo and other lunar journeys, NASA had been attentive to contamination risks from Earth and also back to Earth. Lederberg and Sagan had helped get planetary protection inserted into NASA policy. Sterilization procedures for Viking were "so vigorous that the mission's launches may have had fewer terrestrial microbes aboard than any other craft yet launched," according to Michael Meltzer, whose book on planetary protection was published in 2011.[52]

At a time when exobiology was decidedly unpopular in the Mars scientific community, the low funding levels may have been helpful in its survival. Des Marais remembered that it was "too small a budget item to attract much attention," but there was a payoff. Over time, researchers at Ames played an important role in helping to reframe exobiology from a direct search for life to a search for "habitable environments." A habitable environment could be a subject of study.[53] Moreover, it brought exobiology closer to nonlife scientists, who had been arguing for more attention to the physical setting of Mars.

Downsizing Mars Missions

In 1979, Hinners left the leadership of space science at NASA to become director of the National Air and Space Museum. Tim Mutch, inspired by his own experience on Viking, took leave of absence from Brown University to take Hinners's place. He came to Washington determined to save planetary exploration and revive the Mars program. He wanted NASA to appoint a blue-ribbon panel of planetary scientists to help him develop a rational sequence of missions, including those for Mars.[54] Although initially skeptical, he had, like others, come to believe that the next step in the Mars program should be a rover.[55]

He never got the chance to lead a return to Mars. Mutch was an experienced mountain climber. In October 1980, he took time off from NASA to go on an expedition to India. He and some companions scaled the 24,000-foot Mount Nun in the Himalayas. On the way down, Mutch disappeared and was presumed dead. The accident removed a potentially influential advocate for Mars, one the program sorely needed. Mutch had been at NASA barely a year.[56] The Viking 1 Lander site on Mars was subsequently named for him.

The planetary program at headquarters now drifted. Mutch's deputy, Andy Stofan, was an engineer, primarily interested in launch vehicles. There appeared to outside Mars advocates to be a leadership vacuum within NASA. John Naugle stepped into the breach. He had recently left NASA for industry. Seeing the precarious situation of planetary science at NASA, he persuaded Frosch to form a Solar System Exploration Committee (SSEC) as an ad hoc panel of the top-level NASA Advisory Council (NAC), with himself as chair. It was painfully obvious that the "big science" and "great leap" approaches NASA, JPL, and Sagan favored were not helping their cause.

Frosch charged the SSEC not only to develop a planetary program that would be scientifically sound but also "to define new ways to reduce costs."[57] More than ever before, it was clear to Naugle and top NASA officials that what many Mars activists might most want—i.e., a Viking 3 rover leading to MSR—was not likely to be funded for some time. Moreover, planetary scientists were again debating respective priorities publicly, thereby weakening their clout with NASA, much less with the White House and Congress.

Naugle had forged a measure of unity among NASA scientific advisors after the Mars Voyager had been killed, and it had helped NASA strengthen the science voice in support for Viking. He saw such a need once more to help not just Mars but planetary science generally. The Mars community needed a strategy for the future and had to tie that strategy to a cost political officials would be willing to pay. In 1981, Naugle returned to NASA as chief scientist, a new position, and relinquished the chair of SSEC to Hinners. It was largely up to them to get Mars higher on the NASA agenda as a new start flight project. The political context in which they worked was continuing to deteriorate, however.

President Ronald Reagan came to office in 1981 determined to lead a conservative revolution that would augment defense spending and minimize government's domestic role. NASA, like other civilian agencies, was adversely affected by this approach. The Space Shuttle, because it was linked with defense, a White House priority, was relatively protected from budget cuts by Reagan's political staff. Not so the robotic programs, which were left to the policies of OMB.

Frosch left NASA in January, and it was not until June that a NASA Administrator, James Beggs, was confirmed. Age 55, Beggs was a Naval Academy graduate who had served at NASA in a high administrative position under Webb and subsequently been an executive in aerospace industry and undersecretary of transportation. At his confirmation hearings, Beggs was quite precise about his goals—to transition the shuttle from development to routine operation and guide NASA to its "next logical step" in space, the building of a space station. Beggs was deeply interested in NASA's going to Mars with humans, but he knew that day was many years away. The space station, in his view, would help bring that day closer, however. While supportive of the robotic program, he was unswerving in his priorities.[58]

His deputy was Hans Mark, recently secretary of the Air Force under Carter. A physicist, Mark had in the early 1970s been director of NASA's Ames Laboratory. He shared Beggs's priorities in terms of the shuttle and space station. He had questions about the value of the robotic planetary program. Burt Edelson, an engineer, became associate administrator for OSSA. He emphasized applications, especially an initiative to use satellites to monitor planet Earth. The principal champion for Mars at headquarters was Geoffrey Briggs, a physicist who had come from JPL and worked on Viking. Briggs was first deputy, then acting, then director of planetary programs for OSSA.

By the time Beggs was confirmed and able to move into the NASA Administrator's office, David Stockman, Reagan's budget director, had made his move to cut NASA's budget. The axe fell particularly hard on the planetary program. A Venus mission, authorized in Carter's last budget and which Reagan inherited, was rejected. Galileo, in development, was saved by sacrificing other space science efforts. The issue for Beggs was not one of new starts, but of saving the planetary program from termination. Mark seemed to side with detractors of the planetary program. In October he circulated a memo recommending "a de-emphasis" of planetary exploration until the futures of the shuttle and space station programs were clear. Reagan had yet to decide on the space station. Advocates of planetary science, such as Murray and Briggs, saw Mark as an adversary.[59]

Outside NASA, the president's science advisor, George Keyworth, backed the astronomy program as a more fruitful investment than the planetary program at NASA, giving the former higher grades for "showbiz." Murray sought aggressively to change Keyworth's mind.[60] But the major problem was OMB, which wanted to cut NASA further in Reagan's second year. In the fall of 1981, Beggs and Stockman began a protracted contest of "chicken" in negotiating the upcoming budget. Murray, director of JPL, found himself and his organization very much a pawn in this negotiation.

As Beggs and Stockman dueled in the latter months of 1981, Stockman proposed a budgetary cut for NASA that was draconian. Beggs warned that if Stock-man persisted, Beggs would have no choice but to eliminate an entire program and that would be planetary exploration. Beggs reminded Stockman that be-hind the program was JPL, and, as Beggs put it, "JPL would become surplus to NASA's needs." Stockman refused to budge on the cuts he proposed. As Murray saw it, "NASA was playing hardball with Stockman, and JPL was the ball."[61] Beggs took his case to Stockman's political superiors in the White House.

Caltech, which ran JPL, reacted to the threat to the laboratory. Its president and influential members of its board of trustees now joined the fray. Caltech president Marvin Goldberg met with senators interested in the space program in December 1981. In particular, he convinced Senate Majority Leader Howard Baker (R-TN) to express his support for planetary exploration in a letter to President Reagan. The White House decided to preserve the planetary pro-gram and thus JPL.[62]

The planetary program was salvaged, but the only major planetary flight project NASA had for the future, as 1982 ended, was Galileo. Mars was not even a subject for debate at the policy level. It had been a "miserable" year at NASA, Naugle recalled.[63] He retired not long after. Likewise, in early 1982, Murray announced he was leaving JPL, effective July. The planetary program had "survived," he pointed out, along with JPL, and he took "satisfaction" in that. "It might have been worse," he said.[64]

While various political machinations took place involving NASA, JPL, Caltech, Congress, and the White House, SSEC labored to devise a new plan-etary strategy, including one for Mars. The fact that Hinners was its chair in 1981–1982, and Briggs its executive director in these years and subsequently, helped assure a tight coupling between NASA and SSEC. Everyone knew that the stakes were huge and NASA and planetary scientists had to reach consensus on a sellable strategy. That strategy, it was agreed, had to be based on low-cost missions.

There was much debate within the SSEC over priorities. Some wanted to reconstitute the Mars emphasis and eventually link robotic exploration to human exploration of the Red Planet. Briggs argued for a "Mars-focused pro-gram within the overall 20-year program that SSEC was contemplating." But there was a good deal of anti-Mars resentment in the group, Briggs recalled, in the sense that Mars had gotten too much emphasis in the past.[65] Most of the panel wanted a "broad scientific exploration of the planetary system." They also understood that the cost of missions had to come down. The entire planetary program had plummeted to 20% of its peak funding in the Viking era. SSEC sought to design a "core program" that would fit within the constraints, which it accomplished in 1982.[66]

SSEC formulated recommendations for a core program that would consist of "planetary missions designed to be conducted within a roughly constant budget of some $300 million per year." It called for projects to inner planets, outer planets, comets, and asteroids. For the inner planets, SSEC gave top priority to a Venus radar mission, because Venus represented a large gap in knowledge. It had also been cancelled in the 1981 budget turmoil. The original mission was now stripped down and given a new name, Venus Radar Mapper (later called Magellan).

However, SSEC closely followed this inner-planet priority with Mars. There was no question about SSEC's long-term goal for Mars—"the return of a surface sample to Earth." The cost precluded a mission of that kind under the budgetary requirements NASA had laid down. Similarly, a Viking 3 rover mission that would lead to sample return also would cost too much. Hence, SSEC recommended "early initiation of a Mars Orbiter, emphasizing investigation of geology and climate of the planet." It would be called a Mars Geosciences/Climatology Orbiter and represent the first in a line of low-cost inner-planetary missions, using off-the-shelf hardware. The Mars mission could be launched in 1990. It would determine the global surface composition of the planet and the role of water in shaping its climate. Estimated to cost $250 million, its name was eventually changed to Mars Observer.[67]

Among the behind-the-scenes proponents of the new Mars mission was Daniel McCleese. McCleese at JPL pushed from within the NASA organizational family to return to Mars "for specific narrow investigations that were opened by Viking and not settled." He was particularly interested in the water issue and the hydrological cycle on Mars. McCleese found an ally in Charles Barth of the Laboratory for Atmospheric and Space Physics at the University of Colorado Boulder (the same individual who had mentored the Mars Underground).[68] Barth was strategically located as a member of SSEC.

NASA was able to use the SSEC proposals to advantage. It got the Venus mission into the president's budget being formulated in late 1982, and Mars advocates hoped Mars would make it into the next year's budget. Prospects were improving. As the year ended, Keyworth and Mark (if not Stockman) were shifting their stance. "There was never any intention of cancelling the planetary program," Keyworth stated in November 1982. Mark extolled planetary exploration as important in NASA's overall program. He praised SSEC for finding less expensive ways to conduct missions. "I think it had done a terrific job in understanding the problem and formulating the solution," he said. The fight in late 1981 had produced "a lot of bad vibes and a lot of good dialogue." *Science* magazine quoted another observer of what had transpired as saying, "It was a catharsis that forced people to take a hard look at what they were doing and how much it would cost."[69]

The Space Station and Mars

The planetary program, and thus the robotic Mars segment, had a reprieve. However, the big question remained: What about NASA as a whole? The future of Mars could not be separated from that of NASA as an agency. That depended most on the space station. Beggs had been trying to persuade Reagan to back a space station virtually since he became NASA's leader. When the moment of decision arrived, Beggs played his Mars card, among other arguments.

In late 1983, Beggs gave the president many reasons to adopt the station. For example, a space station could open up manufacturing new materials in weightlessness. It could provide scientific research and display leadership vis-à-vis the Soviet Union, which already had a small space station in orbit. The Cold War context was a strong selling point, arguably the key one. But Reagan asked, "Why aren't you going all the way to Mars?" "That is exactly where we are going, Mr. President," replied Beggs.[70]

What Beggs meant was that the space station could serve as a staging base for expeditions to the planets and provide experience for astronauts in long-duration stays. Beggs knew he could not make Mars the major station selling point generally, but it was important as an enticement for Reagan. According to Keyworth, Reagan truly loved the space program.[71] Reagan not only agreed but also announced his decision January 25, 1984, in his State of the Union address. Like Nixon with the shuttle decision in 1972, Reagan authorized thereby an engineering development program big enough and long-term enough to keep NASA viable as an independent agency. As before, human spaceflight would set boundaries for what was possible budgetarily for other NASA programs, including potential robotic Mars missions. But it was a much wider boundary with a station than without it.

However indirectly and amorphously, the decisions for human spaceflight and robotic Mars exploration science were linked. Once again, the robotic Mars exploration program could be sold indirectly as a precursor to human exploration, even though the space station would be in low-Earth orbit and not justified publicly in any significant degree with Mars. Keyworth, in straining to explain why he was now backing the space station when he had vehemently opposed it previously, said in early 1984 that the space station was not an end in itself. It was a way to achieve broader ends, and these included Mars exploration. Keyworth said he agreed with those larger goals.[72]

In the budget announced in February 1984, NASA had a new mission, the space station, and the go-ahead for a robotic return to Mars—a low-cost or-biter called the Mars Observer. The Mars community, especially the geologists, rallied around Mars Observer.[73] However, the "program line" NASA and SSEC proposed for a series of relatively modest observer missions had not been approved. Mars Observer was a single project. There was no guarantee there would be a successor.

Mars advocates did not get much of what they wanted, but they got a first step in the revival of robotic Mars exploration. On April 9, 1983, NASA pro-cessed the last Red Planet image from the only component of the Viking system still functioning, one lander. The last "extended mission" of Viking was finally petering out. In May, JPL scientists and engineers gathered around a screen that received Mars data and celebrated the project with a poignant final farewell. "There's a lot of feeling. It's like losing a close friend," said George Gianopulos, Viking's last project manager.[74]

Viking was officially over. The Viking 3 rover was abandoned as a proposal for the time being. NASA downsized its ambition as a scientific and political strategy for recovery. But Mars Observer provided hope for a Mars future.

CHAPTER SIX

Moving Up the Agenda

The Mars Observer project kept Mars exploration alive in the mid- and late 1980s, but not in a manner that Mars advocates desired. Within NASA's Science Directorate, Mars policy fell to Briggs, the planetary director, as superiors concentrated on other matters. Briggs found little support for going beyond Observer and raising Mars exploration's status. He "made peace with the Mars program as it was," not as he wished it to be. "I was not chaffing to get its enlargement. My goal was to implement the SSEC core program, particularly the inner planets and the Mars Observer."[1] It was clear to external advocates that they had to press NASA harder if they were to get the kind of Mars program they wanted.

Advocates like Carl Sagan hoped that the Soviet Union, as a competitor or ally, could help revitalize Mars exploration in the United States. His was an end run around NASA and the space policy subsystem, which he saw as weighted against Mars as a priority. He looked for national and international policy allies. But to succeed in this macropolitical strategy, the Soviets would have to be successful technically.

Sagan looked to the White House to resurrect the Mars exploration program—and NASA generally. He and like-minded proponents did so before and after the Space Shuttle Challenger disaster of 1986. Challenger adversely affected the agency, but it did bring NASA to presidential attention. Would presidential interest make a difference for Mars? It did not do so in the case of Ronald Reagan. President George H. W. Bush, however, made a human space-flight decision in 1989 to go back to the Moon and on to Mars. The political environment for Mars policy thus wound up better at the beginning of the 1990s than it had been 10 years before. It was a long and torturous haul, but outside and inside advocates succeeded in moving Mars higher on the NASA agenda over the course of a decade.

Sagan Seeks an Upgrade

What the Mars program needed, the Planetary Society urged, was to translate Mars Observer into a robust set of continuing activities. And to help reach that goal, the program required a new and compelling rationale. Geophysical observation, the stated purpose of Mars Observer, had little public appeal. Sagan, the society's president, for whom the search for life was the prime rationale, understood that his views were not widely shared in the scientific community. "Life" was out as a motivator for most Mars researchers. But what other rationale would work to rekindle widespread enthusiasm for Mars and elevate robotic Mars exploration on the NASA and national agenda? It lay with the Soviet Union, the Society decided.

The Soviet Union had competed vigorously in the 1960s and early 1970s to explore Mars and to discover life first. It had lost in Mars exploration and abandoned the Red Planet. But just as the United States was now planning to go back to Mars, with a 1990 rendezvous, so also was the Soviet Union. The difference was that the Soviet Union was first targeting Phobos, one of the two Mars moons. Rivalry was a motivation for the United States. The Planetary Society also saw opportunity for cooperation. Here was a possible novel rationale for Mars exploration: partnership. The strategy that Sagan and his associates evolved was "To Mars . . . Together." Cooperation embraced both the robotic and human programs and gave space a political rationale, making it an instrument of foreign policy.

In 1984, the Planetary Society brought a number of scientists from the United States, Soviet Union, and Europe together in Graz, Austria, to discuss common interests in space exploration, with Mars as focus. It also commissioned a technical analysis of what it would take to go to Mars in human spaceflight. This activity was taking place at a time when President Reagan was using heated rhetoric about the Soviet Union as an "evil empire" and promoting his "Star Wars" antimissile system. NASA was not in a position actively to market cooperative programs with the Soviet Union, but the Planetary Society, as a nongovernmental organization, was able to do so.

Sagan called space a unifying force in the world. The Society's vice president, Bruce Murray, joined Sagan in this refrain. The former Jet Propulsion Laboratory director had returned from a stint in the private business sector to the Caltech faculty in 1984. He now saw Mars as did Sagan, as the flagship for reviving the entire planetary program.

What helped spur them was an ally in the Soviet Union—Roald Sagdeev. Born in 1932, Sagdeev had distinguished himself as both a physicist and technical manager in Soviet government. Elected to the Soviet Academy of Sciences at 36, he was appointed to run the Institute of Space Research at 40. The robotic space program in the Soviet Union had divided responsibility, with the Babakin Center responsible for the spacecraft, another organization the launch vehicles, and the Institute of Space Research the scientific payloads. Sagdeev was a very influential man in USSR science and had "somehow managed to put a limited form of *glasnost* [openness] into practice years before [Mikhail] Gorbachev [the Soviet leader] came along." The Planetary Society made him a member of its board of advisors, "a bold affiliation for a Soviet official" at that time.[2]

The Soviets were planning a return to Mars in 1988. The primary target was Mars's moon Phobos. Phobos was the darkest body in the solar system known at that time. It was believed to possibly contain "fossil" chemicals dating back to the solar system's beginning. The mission's intent was to drop instruments on Phobos and take new observations of Mars.[3] Since NASA's Mars Observer would not launch until 1990, the United States could learn from cooperating with the Soviet Union. The Planetary Society began actively using the expression "To Mars . . . Together," touting the robotic program as prelude to human exploration.

Gorbachev, who came to power as head of the Soviet government in 1985, established himself as a reformer, one who wanted more openness and international goodwill. The Planetary Society, meanwhile, was linking Mars with internationalism, taking advantage of Sagdeev's influence and an emerging thaw in the Cold War. It sponsored a meeting in Washington in mid-1985, "Steps to Mars." Various scientists, astronauts, and federal officials attended. The conference dealt with detailed engineering and scientific matters and gave equal attention to robotic and human roles. At the meeting, Sagan challenged the United States and the Soviet Union to rise to the emerging opportunity and connect in going to Mars. The first American woman in space, astronaut Sally Ride, spoke in support of the concept of a U.S.-Soviet human expedition. NASA Administrator Beggs attended and indicated that it was becoming possible to think about new opportunities of this kind. President Reagan had meanwhile asked Tom Paine, the former NASA Administrator, to head a National Commission on Space (NCOS) to consider space exploration in the future. The vision of "To Mars . . . Together" gained momentum as a political rationale.[4] On January 7, 1986, the *New York Times* endorsed the concept.

Then, on January 28, 73 seconds after takeoff, the Space Shuttle Challenger exploded. The explosion was seen on television by millions. All seven aboard were killed, including the first teacher in space. The immediate cause was later determined to be a seal that malfunctioned in one of the shuttle's two solid rocket boosters. The underlying causes were management errors, including a false belief in the shuttle's capacity to launch many missions over a brief period of time and failure of NASA to observe its own safety procedures. The fact that NASA included a teacher on the mission revealed the degree to which the agency had grown overly optimistic about shuttle risks as it shifted attention to developing a space station.

President Reagan gave NASA strong rhetorical support. However, he turned to an independent body to investigate the accident, which became known as the Rogers Commission after its chair, William Rogers, a former secretary of state. Unluckily for NASA, James Beggs, the Administrator, was on leave fighting a charge of illegal activity while in industry, before coming to head NASA. His deputy, William Graham, was new to the job and unprepared to defend the agency as the Rogers Commission conducted its inquiry and media launched a blistering attack on NASA. The charges against Beggs were found ultimately to be bogus, but Beggs had had to resign to fight them. In May, Reagan asked James Fletcher to return to lead NASA, moving Graham to the White House and out of the line of fire. Fletcher inherited a wounded agency. For Mars advocates, Challenger was a huge setback.

Delaying Mars Observer

The shuttle accident brought much of NASA to a halt—certainly the human spaceflight program. But to a severe degree various robotic programs were also affected, since many were dependent on the shuttle for launch. Fletcher stated that his highest priority was to "get the shuttle flying again." It was not obvious when that would be, however. He now spoke of a "mixed fleet," involving shuttle and expendable rockets. However, he had been father to the shuttle during his first tour as NASA Administrator, devoutly believed in its worth, and wanted to maintain its role as America's prime launch system as much as possible.[5]

In August, Fletcher decided that the shuttle could not return to flight early enough for all its launch assignments, including the 1990 Mars Observer mission. He set back the date for Observer's launch to 1992. Those who had banked on Observer to revive the Mars program were shocked and angry. Sagan, Murray, and Friedman knew that an expendable rocket could serve just as well for Mars Observer and that one could be made available by the Air Force. The Air Force, which had been restricted in using expendable rockets under President Carter's policy to maximize use of the shuttle, had extricated itself from this policy. President Reagan changed the Carter policy and was making expendable rockets usable for national security and commercial launches. Why not Mars Observer? As far as Planetary Society leaders could tell, Fletcher had not even considered the expendable rocket option when he made his Observer postponement decision. Moreover, they discovered he planned to move money that would now not be needed in the next two years for Mars Observer to other priorities.[6]

The Planetary Society decided to fight the Fletcher decision. "We raised a cry in the news media and elsewhere," Murray recalled. At first, it appeared that the Society had persuaded the NASA Administrator to change his mind. In September, Fletcher stated, "We have decided . . . to continue working on the Mars Observer on schedule to provide launch readiness for the 1990 opportunity."[7] It looked as though Mars Observer would go forward as planned, although it was not settled whether it would go on a shuttle or expendable rocket.

On January 2, 1987, Fletcher informed Congress of another change in NASA's plans. Instead of launching Mars Observer in 1990, NASA would delay launch to 1992. The reason he cited was the Challenger accident. With the shuttle down, there was a logjam in flights scheduled to use the shuttle. In spite of his earlier statement about a 1990 launch, he was now holding to the 1992 schedule. Something had to give, he said, and he had decided (again) that Mars Observer could be postponed two years.[8]

The new decision came as a second shock to the Mars community. In many ways, the affected scientists felt betrayed. One reason was that the Solar System Exploration Committee had carefully developed plans to reconstruct the planetary program and had made Mars Observer not only an inner-planet priority but a key to a sequence of missions demonstrating low-cost planetary exploration. In its 1986 report on an "augmented" program, SSEC had proposed that

Observer be followed by long-coveted rover and sample return missions. But it was more than a matter of scientific planning for Sagan and the Planetary Society. For them, Mars stood as a beacon of NASA's commitment to exploration in general after Challenger.

As Fletcher was postponing Mars Observer, he brought back Sally Ride, a member of the Rogers Commission, to chart NASA's post-Challenger future. To the Mars advocates, Mars Observer, whatever its limits, was central to that future.[9] As Friedman put it, Mars Observer had become "enormously symbolic" to the Mars community, as well as space enthusiasts in the general public. The reasons, Friedman declared, were threefold. First, "it was *Mars* and therefore an object that we believe should represent the focus and long-range goal of space exploration. Mars Observer was a step toward that goal. So, by putting a low priority on it, NASA was putting a low priority on the one goal that we thought was the most widely accepted and most important to the agency." Second, he said, "As the first of the Observer-class missions, Mars Observer represents the minimum planetary mission, devised to reconstruct the [Mars] program." Third, he argued, the Mars Observer represented "the first chance to cooperate with the Soviets in Mars exploration. How could the United States cooperate with the Soviets, which had announced an ambitious program, if the U.S. policy was to delay our only approved program?"[10]

Sagan, Murray, and Friedman decided to launch a multifront fight and asked the then-100,000-plus Planetary Society members to write letters to Fletcher and Congress protesting the decision. In addition, Murray took direct action by calling Dale Myers, Fletcher's deputy administrator, whom he knew, to persuade him to change Fletcher's direction. Murray found Myers not particularly interested in speaking to him when he placed his call. The former JPL director had a "solution" to propose to Myers. "Why not use the Titan expendable launch vehicle rather than the shuttle?" Murray asked. Myers shouted back at him, "Where am I going to get the $150 million" to pay for the launch?[11]

Indeed, there was a problem about money. Using a Titan would be expensive. But so would a two-year delay for a shuttle. Fletcher, feeling the heat from the Mars community, discussed with Congress the possibility of using a Titan and getting the money to pay for it, but he ran into resistance from the House Appropriations Committee chairman responsible for NASA's budget. The congressional authorization committees were supportive, but they did not have final word on funding.[12] On March 13, NASA confirmed that its spending plan for the year did not include Mars Observer and that money already approved by Congress for the project would be reprogrammed in the next two years to higher priorities.

Sagan and his allies at the Planetary Society reacted sharply. They took their case to the media, Congress, and the general public. Sagan went to the Mars Underground's third conference in Boulder and gave the keynote address, rousing the 400 attendees to the Mars cause. In speaking to the media, Sagan, on March 15, called NASA's decision "a great mistake and an example of a consistent lack of vision that NASA has had since the middle 1970s."[13]

NASA's leaders sought to head off criticism, especially with Congress. Myers wrote Congressman Bill Nelson (D-FL), chairman of the House Subcommittee on Space Science and Applications, explaining the agency's position—and that Challenger had caused a backlog of shuttle missions for the 1990 period and that NASA did not have the money to use an Air Force expendable Titan rocket.[14] He also wrote Murray, saying the obstacles to launching in 1990 were "insurmountable."[15]

But Sagan and his allies were not quieted. They condemned the decision and pointed out that delaying the launch would cost almost as much as using an expendable Titan. The real reason, the Planetary Society leaders said, was that not using a shuttle would undermine NASA's claims about the shuttle's necessity as a launch system. They specifically chastised Fletcher, who was "besieged by personal phone calls from members of Congress, some of whom were key to his important budget items like the Shuttle and Space Station." More than 10,000 Planetary Society members sent off 25,000 letters to Congress.[16] On April 27, the three Planetary Society leaders testified before Congress. NASA's public affairs director called the press release preceding the testimony an "unconscionable attack on both Dr. Fletcher and the agency in general." She recommended that NASA and JPL not cooperate with the Society in using JPL facilities for meetings involving possible joint U.S.-USSR activities the Society was promoting.[17]

She did not get her way. The next month, in May, the first International Conference on Solar System Exploration was held in Pasadena. The Soviet Union sent delegates who unveiled grand plans for exploring Mars, starting with the Phobos mission. The Soviets said they intended to sponsor a Mars rover and sample return project by the end of the century. David Morrison, head of the SSEC, spoke. He said that the United States had "fine plans. We have great inherent capability. But we have been very slow to turn those plans into actual programs."[18]

On May 14, Murray paid a personal visit to Myers in hope of mending fences and getting him to help in reversing Fletcher's Mars Observer decision. What he found was that Myers "was really steamed up." He and Fletcher were furious that the Planetary Society had created such a big public relations problem for them. Murray wanted to see whether NASA and the Society could work together and he could rebuild the personal relationship he had with Myers. "How can we help, Dale?" Murray asked, hoping he would use that opening to outline an area of future cooperation. "We're 100,000 dues-paying members, a real Mars constituency." Murray reminded Myers that Fletcher had recently brought astronaut Sally Ride to lead a task force studying long-term options for NASA's future. Ride's report was not public as yet, but there was ample reason to assume that her options would include Mars robotic and human missions.

"We may not choose the Mars options," Myers replied in a manner Murray called "gruffly." "If we do, we'll contact you," he declared. That was that! Murray had found no common ground, and Myers signaled that Murray's time was up. As Murray interpreted the meeting, Myers considered the Planetary Society just another pressure group, one giving the agency headaches rather than support.[19] The Planetary Society could not get NASA to budge on the decision to delay Observer.

The man responsible for directing science at NASA was now Lennard Fisk, 43, who had succeeded Edelson as associate administrator for space science and applications in April 1987. Fisk had been vice president for research and financial affairs at the University of New Hampshire. He was an astrophysicist who had a number of projects other than Mars on his agenda when he came to NASA. Moreover, he became increasingly interested in a new activity within NASA that would be called "Mission to Planet Earth" (MTPE).[20] Started by Edelson, this effort was rapidly developing in NASA, and the earth sciences along with it. NASA had played the lead role in the mid-1980s for determining the causes of the ozone hole over Antarctica. Some of its scientists provided technical advice contributing to the Montreal Protocol on ozone-depleting chemicals in 1987.

The issue of climate change was rising on the national and international agenda, and Fisk, like Edelson, wanted to position the agency for leading in a growing and important mission. Moreover, the NASA Administrator, Fletcher, personally cared about NASA's environmental role. In his earlier tenure as Administrator he had termed NASA "an environmental agency." His strong Mormon beliefs, which had mattered in his Viking advocacy, also made him extremely attentive to the concept of stewardship of the planet.[21]

In short, Fisk had good reasons to push the new Earth priority for the Office of Space Science and Applications. He supported, but did not stress, the Mars program. Like Edelson, he let his subordinate, Briggs, largely direct Mars activity. In addition, he took seriously the SSEC emphasis on broad solar system exploration as opposed to Mars-centered exploration. The next new start in planetary science he wanted to sponsor, pursuant to recent SSEC recommendations, was Saturn, with an expensive mission that would be called Cassini. For a while, Cassini was connected in planning with another project, the Comet Rendezvous Asteroid Flyby (CRAF). These projects were aimed at the outer solar system.

But the most important reason Fisk did not give Mars exploration the priority its advocates wanted was that he was spending a great deal of money from his program on missions that were well past Observer's development stage and were literally sitting, waiting in storage, to be launched on a shuttle. There was little Fisk could do about Fletcher's intent to use the shuttle for major science launches. That being the case, he recollected his challenge as follows:

> Everything was stacked up. We were spending $5 million a month just to watch the Hubble Space Telescope. We had Galileo [Jupiter] at the Cape and shipped it back to JPL. We had Ulysses [a project to study the Sun] at the Cape. We came at it [decision making] this way: What's the most cost/effective way to get rid of this backlog? It cost us $2 billion to stand down science in the post-Challenger period. We had to get Hubble off the ground. The most expensive science missions went first. Mars had to wait. Hubble especially had to launch. Our most expensive science mission. Tremendous hype about Hubble.[22]

Selling Mars

Although preoccupied with immediate issues, Fletcher and Fisk were aware that the Soviet Union was presenting an opportunity to direct policymakers' attention to Mars and space exploration generally. Whereas the Planetary Society stressed cooperation, NASA used the competition card.

NASA looked beyond Mars Observer to propose what it and its scientific advisors wanted—a rover and Mars Sample Return mission. Once the Soviets went to Phobos, they would likely launch such a mission to the Red Planet. NASA argued that the United States should compete. In late March, Briggs and the directors of JPL and Johnson Space Center (JSC) proposed the "Mars Rover and Sample Return" mission at a meeting with 200 representatives from industry in Houston, Texas. "This is a mission that has to happen," said Briggs. "NASA is not exactly at the zenith of its activities at the moment," he pointed out. "Nor is the planetary program." "We either do it now or after the Soviets do it," he continued. "NASA has got to get moving again."

It was made clear that this particular mission had strong support from the scientific community and NASA in general. Aaron Cohen, the JSC director, declared, "If we want to be able to carry out tasks [involving astronauts] on the Moon or Mars, we must have a firm base in science. Lew Allen [the JPL Director] and I agree that by working together, we can learn much from this mission."

NASA gave a similar briefing to contractors a few days later in Pasadena, home of JPL. JPL would be lead center in the project, with JSC assisting. Briggs again emphasized the competitiveness driver. He said that the Soviets were engaged in a similar mission and "that it is my responsibility to make sure the United States is in a position to do a mission with the same kind of complexity in the same time frame. So that really is the trigger on the timing."[23]

NASA—and Mars advocates generally—seemed to be searching for any rationale that would sell to get the agency moving again, after Challenger. Whether it was cooperation or competition did not matter, as long as the strategy was effective. The problem was not in the packaging. It was whether policymakers were in a buying mood for the product.

In August 1987, Ride's report was published. Cooperation and competition gave way to "leadership." Entitled *Leadership and America's Future in Space*, Ride listed four possible long-range initiatives that would restore the United States to a position of leadership in space. Without stating priority among the four, Ride listed them as (1) Mission to Planet Earth—Earth satellites to monitor the global environment; (2) Exploration of the Solar System—essentially the robotic program recommended in 1983 by SSEC, as augmented in 1986, to include more ambitious endeavors; (3) Building an Outpost on the Moon; and (4) Humans to Mars. This last initiative overlapped with the second, as robotic flights to Mars were essential precursors. Ride wrote, "Robotic exploration of the planet would be the first phase [of humans to Mars] and would include the return of samples of Martian rock and soil."

The problem, as Ride pointed out, was that while many space enthusiasts wanted to go beyond the shuttle and space station to bolder ventures, existing programs would continue to dominate agency spending for the foreseeable future. Given realities, she noted, many observers believed that NASA could not handle another major program.[24] These ambitions and financial realities had to be reconciled somehow, she argued, for NASA needed strategic goals and direction.[25]

Fletcher responded to the Ride study in two ways. First, he established an Office of Exploration to provide a focus for sustained thinking and advocacy for exploration in general, with emphasis on human spaceflight. With Ride leaving NASA for an academic appointment, he placed John Aron, a veteran NASA official, in charge of the new office as an assistant administrator. In addition, on August 10, he met with Vice President George H. W. Bush to brief him on NASA's plans after Challenger recovery. In this meeting, he pointed out that the Soviet Union was surpassing the United States in certain technologies and program ambition. He said that the Soviets had an advantage in both robotic and human exploration of Mars. He advised that the United States needed to choose a long-term, post–space station goal to give the agency strategic focus, and that human spaceflight to Mars was the logical one.[26]

Fletcher also wanted to discuss NASA's future with the president but was blocked in getting to Reagan. The word was that Reagan's lieutenants felt that Fletcher should get space station spending under control before trying to persuade the president to launch a Mars venture.[27]

In November–December 1987 the Planetary Society published and circulated what it called the *Mars Declaration*. Former NASA Administrators, Apollo-era officials, astronauts, Nobel laureates, actors, authors, politicians (including former president Jimmy Carter), university presidents, professors, activists, musicians, and many other notables signed on. The *Declaration* echoed Sagan's view that the two countries should go together to Mars, first robotically, and then with humans. The *Declaration* also noted that a Mars destination would give the space station greater purpose, as a step toward the Red Planet.[28]

When Reagan and Gorbachev met in a summit in December 1987, the two leaders discussed many common issues, and space collaboration was one area where partnership was deemed possible. Gorbachev spoke officially with Reagan and unofficially with others. He met with one group of Americans which included Sagan and Paul Newman, the actor, a member of the Planetary Society's advisory board. Gorbachev specifically mentioned to this group his desire that the United States and the Soviet Union collaborate on Mars exploration.[29]

While conversations took place in Washington, D.C., NASA scientists and engineers were in Moscow conferring with Soviet counterparts about how to collaborate in view of the Soviet Union's stated desire that Phobos initiate a sequence of missions to Mars. The Americans had Mars Observer approved, as well as ambitions for more missions. Beyond Phobos, to go up in 1988, the Soviets planned a mission in the mid-1990s (after the U.S. Mars Observer) which would go to Mars (rather than Phobos) and release balloons to study Mars's atmosphere. NASA entered into talks the Soviet Union initiated with the French as potential collaborators on this mission. If Mars Observer carried an antenna, it could receive data from Soviet balloons and transmit information back to Earth. The robotic programs of the two nations could thereby get collaboration on Mars started in a serious way.[30]

The Reagan budget came out in February 1988. NASA got a raise, but most of the new money went to the space station, which was suffering overruns, and a new space shuttle to replace Challenger. Money for Mars robotic flight remained tight.[31] Reagan had considered a "Kennedy-style declaration" calling for a dramatic space venture, such as a Moon base or trip to Mars, but rejected such an initiative. Instead, the White House issued in February a national space policy that called for the United States "to expand human presence and activity beyond Earth orbit into the solar system."[32] While these larger policy activities relating to NASA's future went on, NASA implemented the one Mars project it had.

Mars Observer Troubles

Mars Observer was still on target for 1992. However, it was not exactly untroubled. Its budget was mounting from the $250 million slated for the first in a series of low-cost "planetary observers" its original architects had planned.[33] The administration and Congress had approved only one Observer, not a program of closely coupled missions, and this initial "low-cost" venture was up to at least $450 million in cost in 1988, not counting launch expenses. The problems causing growth were many, but the basic reason for the cost overrun was that Observer was increasingly vital to all stakeholders: NASA, JPL, scientists, the industrial contractors, administration, Congress, and even the Soviet Union.

Seeing Observer as the first U.S. Mars spacecraft in the years since Viking, Mars scientists were desperate to get their experiments on the machine. NASA, JPL, and their political masters did not want it to fail, especially now that it embraced foreign policy purposes. NASA sought to reduce risk through various technological safeguards. Virtually all involved agreed that the delay from 1990 to 1992 made it all the more essential that the scientific payoff be substantial. Moreover, the use of a shuttle added to the pressures to make the mission worthy of the huge launch cost.

Indicative of how costs could rise was a decision in 1986 by Edelson. He had personally ordered that a sophisticated new camera developed by Michael Malin, then at JPL, be put on Observer. He thereby overruled Malin's JPL superior, who had tried to keep it off. "I'm not going to approve of any mission to Mars, or any planet that doesn't have a camera aboard," Edelson had declared.[34] The decision had merit, but so did other decisions that added expense. Briggs, as NASA official responsible for the flight at headquarters, tried hard to keep costs down, struggling with a host of stakeholders for whom technical success loomed largest in values. He did not succeed. Many headquarters officials shared the performance-oriented values of those doing the work at JPL.[35] Thus, formally and informally, the mission was redefined and grew in instruments and complexity over time.

The rising expense became so much an issue that Briggs in May 1988 asked the Space Science Board's Committee on Planetary and Lunar Exploration (COMPLEX) what instruments might be taken off Mars Observer. The committee refused to say, declaring,

> For whatever reasons, Mars Observer has now outgrown all the original
> Observer class parameters. Moreover, it is clear from the recently promulgated
> OSSA strategic plan that with the failure to establish a true Observer line, MO
> almost surely represents the only mission to Mars by this nation in the coming
> decades. COMPLEX therefore takes the position that in these circumstances
> MO cannot be judged by the criteria for science return that would apply to
> Observer-class missions as initially conceived by the Solar System Exploration
> Committee. Consequently, the potential surrender of any current mission
> capability that substantially addresses the primary science objectives established
> for the exploration of Mars is a matter of great concern to the committee.[36]

On July 19, Fisk went to JPL and met with Allen, the director of the facility. They agreed, via a "handshake," to descope the mission, removing certain instruments. At the same time, they concurred that NASA would add the Mars balloon relay, with funding from outside the Mars Observer project, to enable possible U.S.-USSR collaboration.[37]

The Soviet Phobos Shot

In July, the Phobos mission was ready to go. There would be two probes sent, one July 7, the other July 12. There was great expectation and attention paid to the flight in the United States. The Soviet Union had not tried to go to Mars since 1973, and it had failed then. Success potentially meant initiation of a more robust Mars program in both the Soviet Union and the United States, with collaboration a central, political purpose.

Science magazine spoke of "Mars Mania," highlighting the enthusiasm and expectation that were building. American dignitaries went to the Soviet Union to watch the Phobos launch, and U.S. scientists prepared to participate in research Phobos made possible. In return, NASA had agreed that Soviet scientists could participate in Mars Observer research. NASA and JPL worked to incorporate the Mars balloon relay system into Mars Observer development.[38]

The launch for Phobos was spectacular, and the various Western dignitaries who attended were impressed. Sagdeev, who had spearheaded the mission, called previous attempts with small bodies like Phobos a "quick kiss." This was going to be a prolonged dalliance, he said.[39] But relatively early into what was a seven-month flight, on August 31, problems arose on the Phobos 1 spacecraft. Flight controllers in the Soviet Union sent Phobos 1 a radio command that "lacked a single character." This error confused "its navigation system" and moved "its solar panels out of alignment with the sun." Without adequate power, Phobos 1 ran into severe difficulty.

A "forlorn" Sagdeev was photographed behind a model of the Phobos spacecraft at the U.S. ambassador's residence in Moscow. It was September 9, and he had to announce that Phobos 1 seemed doomed, tumbling out of control millions of miles from Earth. By November, it would be officially ruled a failure.[40] The Soviets and their allies still hoped for the success of Phobos 2. It reached Mars on January 29, 1989. Tass proudly announced that the planet Mars "has acquired one more satellite which will bring mankind closer to unraveling the mysteries of the planet."[41] It did return useful data on Mars and Phobos briefly. However, in late March, as it neared the long-anticipated Phobos rendezvous, it ceased to function.

The Phobos failure damaged the Soviet Mars program. The repercussions were dreadful for Mars advocates in the United States. At least for the moment, the dream shared across nations of a Mars Together initiative was on hold. An angry Sagdeev lashed out at engineer-managers who had had final control over technical design. Computer backups might have saved the mission, he complained, but were not available. "I hope that, in the future, space technology producers will have their absolute freedom restricted so that the world scientific community, as the end user of the technology, can have a say in decision making on spacecraft design." That remark brought a retort from Roald Kremnev, as ranking Phobos engineer, reminding Sagdeev that "space technology designers have to comply with a set of restrictions relating to the funds and the weight and size of the spacecraft, etc." Scientists cannot expect to have their own way, he countered.[42]

The disappointment was deep not only in the Soviet Union but also in America. The Planetary Society and U.S. scientists associated with Phobos had banked on success of at least one of the two probes. They viewed Phobos not as a Soviet mission, but as a mission that "transcended" national borders and could lead to a regeneration of Mars flights in the United States.[43] Scientists in the United States and the Soviet Union were linked in what they hoped would be a sequence of ever more challenging Mars missions. The Planetary Society saw Phobos as part of an overarching strategy to rekindle public interest in space, build bridges of peace between the United States and the Soviet Union, and pressure NASA to be more Mars oriented. Now, Mars exploration depended on what happened to the next flight in line, Mars Observer, in America. Also, Mars exploration depended on the man Fletcher had briefed on Mars policy following the Ride report, George H. W. Bush, elected president in November.

Using Bush's Moon-Mars Initiative

As President Bush took office in January 1989, he was well aware of NASA's troubles, as well as ambitions. He wanted to aid the agency's recovery from Challenger and chart a space policy that would make the United States clearly the unquestioned global leader in space. He also smarted over criticism from opponents that he was lacking in "vision." He soon signaled that he intended to give a higher priority to space, reestablishing the National Space Council (NSC). This was a top-level interagency body concerned with both military and civil space affairs which had existed under Presidents Eisenhower, Kennedy, and Johnson. It had been abolished by Nixon. He put his vice president, Dan Quayle, in charge. They agreed they had to give stronger direction to NASA, an agency that seemed directionless. At the same time, the Cold War was continuing to thaw, and Gorbachev kept talking about joint activity in space, the disappointment of Phobos notwithstanding.

In April, Bush chose Richard Truly to be NASA Administrator. Truly, age 52, was a retired admiral with substantial experience as a naval aviator and NASA astronaut. He had distinguished himself as a NASA associate administrator who guided the shuttle back to flight status in 1988. Truly's overriding interest was the Space Shuttle. He wanted not only to replace Challenger, as Reagan had authorized, but to enlarge the fleet from four to five. He also wanted to use the shuttle to build the space station. He regarded these tasks as his prime job.

The initiative for what became the president's Moon-Mars decision came not from Truly but from the NSC, which was looking for ways to rejuvenate NASA and help the president. On July 20, Bush went to the National Air and Space Museum and called for "a long-range, continuing commitment. First, for the coming decade, for the 1990s: Space Station Freedom, our critical next step in all our space endeavors. And next, for the new century: back to the Moon, back to the future. And this time, back to stay. And then a journey into tomorrow, a journey to another planet: a manned mission to Mars."

The decision, which launched what Bush called his Space Exploration Initiative (SEI), did not get the reception Bush and NSC anticipated. It seemed to come out of the blue. Very little political spadework had been done to prepare Congress, the media, or the American people. The consequence was that the decision was met by puzzlement, indifference, and, in the case of some influential Democrats in Congress, opposition. Given budget deficits and NASA's problems with the shuttle and space station, even proponents of Mars exploration, whether human or robotic, failed to take it seriously. Many in NASA were skeptical.[44] The skeptics included Fisk.[45]

The question of "how" to get to the Moon and then Mars was left to be worked out. NSC executive director Mark Albrecht asked NASA to address the "how" question and provide various options. NASA took the next three months on what was called the "90-Day Study." The result, which became known in November, was a 30-year program costing between $400 and $500 billion. When that estimate leaked, it shocked policymakers and dampened whatever enthusiasm that had been mustered. Albrecht and Vice President Quayle were furious. They had wanted options, and NASA came up with only one—the most expensive possibility in their view.

Aaron Cohen, of the JSC, who chaired the study, had a different point of view. He did not recall being asked by Albrecht, in a conversation they had, to provide different options with radically different costs. He determined that NASA's job was to devise an aggressive program that would do what the president wanted with costs that were realistic, as had been the case with Apollo.[46]

For advocates of Mars robotic programs, the 90-Day Study provided a large menu of projects, including several rover missions and two MSR missions. There would also be observers and communication satellites on Mars. The study proposed a rich network of robotic technology for Mars.[47] Hence, the Moon-

Mars decision of Bush seemed to give quite a boost to the Mars robotic science program's viability, even though few took the 90-Day Study seriously once they contemplated the costs. As far as Albrecht and Quayle were concerned, Truly had, directly or indirectly, sabotaged the president's decision. The relationship between NSC and NASA Administrator started badly and deteriorated further as time went on.

Whatever critics thought of SEI, the fact of presidential interest in space and Mars was important, especially with the Office of Management and Budget. In fact, the director of OMB, Richard Darman, was personally supportive of the space program. In early February 1990, Bush announced the budget that he was submitting to Congress. In contrast to cuts he was imposing on various other agencies, Bush asked for substantial increases for NASA. NASA would go from $12.3 billion to $15.1 billion.[48] The increase would provide enhanced funds for NASA, virtually across the board. Mars Observer was now linked, in planning at least, with the human Moon-Mars Program. It already was connected to U.S.-USSR robotic relations. At the end of March, Bush announced that he would be looking for ways the United States could cooperate with other nations, especially the Soviet Union, in SEI. The Soviet Union reacted positively.

A lot had happened since the original U.S.-Soviet efforts in space cooperation led by Sagan and Sagdeev. The Phobos mission had failed, casting doubt once again on Soviet capability to follow through on plans. Politically, the Soviet Empire was under siege. In November 1989, the Berlin Wall had been breached and the Cold War, to all intents and purposes, ended. Bush wanted to support Gorbachev, if only to prevent less reform-minded Soviet hard-liners from gaining power. Sagan was one space scientist who had steadfastly encouraged U.S.-USSR space cooperation, in spite of the Phobos debacle and Soviet political turmoil. However, many other U.S. scientists were wary of Soviet connections.[49]

They could see robotic Mars exploration rejuvenating, independent of Soviet collaboration. A National Academy of Sciences National Research Council (NRC) panel that had been evaluating the SEI plan under the 90-Day Study took it upon itself to warn publicly that the U.S.-USSR alliance could have negative results for the robotic science program. In particular, the NRC scientists looked ahead to MSR and admonished NASA and the United States to be careful in planning a joint mission of this complexity. A highly interdependent undertaking could make planetary science "a potential hostage to political events."[50]

If scientists were critical of SEI and wary of Soviet connections to the U.S. robotic exploration program, many in Congress were downright hostile and increasingly opposed to SEI, especially on financial grounds. In hearings by the House appropriations subcommittee responsible for NASA's budget, questions were asked aimed at distinguishing SEI money from Mars Observer and other ongoing space science projects. Congress wanted to move forward with Mars Observer—even though it was experiencing significant cost increases—but not human Mars ventures.[51]

Encouraged by possible increases in funding for space science, in spite of the SEI controversy, NASA officials and scientists were thinking more positively and creatively about post–Mars Observer options. Scott Hubbard, a scientist and division chief at Ames, spotted a request for proposals for precursory missions to Mars issued by NASA's Office of Exploration, responsible for SEI planning. He developed concepts for a program called Mars Environmental Survey (MESUR). The mission would place 16 robotic stations on Mars that would take various kinds of physical measurements. The first of these stations might be called "Pathfinder," Hubbard suggested.[52]

Hubbard went to Washington in April 1990 and met with Briggs and Jim Martin, the former Viking project director, serving as an advisor to NASA. "Martin gave me a hard time—tough questions," Hubbard recalled. But "afterward, he came up to me and said: 'It might work!'"[53] Some time later, Briggs noted his interest in the Hubbard project to his OSSA associate, Joseph Boyce. What if the station carried a rover vehicle? he asked. "A little car could drive out," Briggs continued. For years, since Viking, NASA had wanted to send a rover to explore. But "we've got to get to Mars' surface cheaply," Briggs added.

Briggs then called a close friend at JPL and broached the general idea, as well as his thought that it would save money if the Mars probe used a "hard landing" technique, such as airbags, instead of a "soft lander" approach, as had Viking, with retro-rockets. The JPL contact said JPL probably would not go for a hard lander. When Briggs asked JPL more formally, he got a response that did not give him what he wanted, particularly in regard to landing. "This is crap," he told Boyce. He then went back to Ames to study the overall idea.[54] Soon, JPL and Ames both were exploring the MESUR concept. In late May Briggs surfaced MESUR publicly to see if it would attract wider support.[55] Meanwhile, JPL got to work on rover concepts.

Donna Shirley, the manager at the center in charge of the research on rover concepts, recalled the period as one where scientists and engineers at JPL were "excited." She herself saw rovers as "scouts" for later human missions. The Bush decision of 1989 had encouraged JPL to work with JSC in planning for Mars

Rover Sample Return, which she called a "perfect precursor to a human mission." Also, "The scent of money in the air permeated the atmosphere," wrote Shirley.[56]

At the beginning of February 1991, Bush proposed a NASA budget of $15.7 billion, a figure that would have been over 13% higher than that which Congress appropriated the year before. Space science in particular would rise 21%, to $2.1 billion.[57] NASA would surely need the money to take care of all the programs under way and anticipated. Fisk presided over a science program that had been rapidly expanding. It included a Mars Observer that was rising in budget, along with a line of telescopes complementing Hubble. The new program scheduled for the sharpest rise in the future was the Earth Observing System (EOS), a suite of environmental satellites aimed at deciphering global change. This was the hardware embodiment of NASA's MTPE activity. It was the program many scientists believed Fisk to be pursuing most avidly.[58]

EOS had not been his initial priority for a new start. But once he realized that the Bush administration was anxious to show support for researching global climate change, he pushed EOS "with vigor."[59] All of Fisk's programs were scheduled to ramp up in funding in the future. New Mars precursory missions were part of the projected mix, led by MESUR, but just how was not clear.

Congress balked at Bush's NASA request. In 1990, Bush and Congress had agreed to reduce the federal deficit by cutting expenditures and raising taxes. Bush earlier had said "read my lips" about no new taxes, but he had been forced to change his mind. Now, in 1991, the results of that White House–Congress agreement began to hit NASA. By July, the House and Senate appropriations subcommittees considering space told NASA it could not get the money it expected. The Senate in particular directed NASA to think in terms of a 3% to 5% raise a year, at best. To underline that point, it cut the request of Bush substantially and told NASA to prioritize. A Senate report said NASA should "not envision any new starts" unless it could explain how it could make them "sustainable" within the agency's "limited funding profile."[60]

One way to expand possibilities was international collaboration, and the Soviet Union indicated eagerness to collaborate on Mars. But the Soviet Union was tottering politically and had huge internal schisms. In July, Bush and Gorbachev met at a summit and sought to move cooperation in space forward. But just two weeks later, Soviet hard-liners staged a coup d'état and briefly held Gorbachev captive. Gorbachev survived the coup attempt, but his days as Soviet leader were numbered, as were those of the Soviet Union. In December

1991, the Soviet Union disintegrated, replaced by a confederacy of independent states. It would take a while for Russia and other states emerging from the former Soviet Empire to reconstitute stable governments. Collaboration on Mars was now on hold.

Building a "Program"

Nevertheless, all these macropolitical events at the international and national policy level, as well as pressure from external advocates (especially Sagan), were making a difference for NASA's Mars science program. The priority was going up. Fisk saw MESUR as "part of a larger program of Mars Science Exploration in OSSA."[61] Such a Mars program did not exist at the moment. It would have to be designed. In late 1990 Briggs had left NASA Headquarters for another position, being succeeded by Wesley Huntress. Huntress was an astrochemist who came to Washington after 20 years of research and project management at JPL. He was activistic in temperament and intended to be a strong advocate internally. He saw the environment in which he served as favorable to innovation as long as costs were contained. He wanted programs that could appeal to both scientists and the public and came to his job with an agenda. He was supportive of expanded Mars activity, but his initial emphasis in 1991 was a solar system–wide program called Discovery. It featured low-cost missions that could launch frequently. For Huntress, Mars Observer was revealing a way not to run a planetary program, and he pushed for Discovery as the right way.

Complicating getting MESUR under way was its origin at Ames and the fact that Ames was under a NASA directorate other than OSSA. Fisk sought in 1991 to make MESUR "a leading example of successful intercenter cooperation in Mars exploration." His discussions about MESUR with the director of Ames did not go well, however, much to Hubbard's dismay. As Hubbard recalled, Dale Compton, the Ames director, came across as ambivalent. In contrast, JPL's director, now Edward Stone, clearly wanted the program. As before, JPL lobbied aggressively to run all planetary efforts.[62] On November 8 Fisk wrote JPL and Ames that he wanted to move ahead with a comprehensive and evolutionary program for the scientific exploration of Mars and that he had decided to make JPL the lead center in this endeavor.[63]

As Fisk was making the requisite scientific and institutional choices for MESUR, and Mars research generally, Huntress was moving Discovery forward. Discovery, with its emphasis on a range of low-cost missions, fit the times. It also matched the political situation Fisk faced. Fisk was getting pressure from congressional supporters of the Applied Physics Lab (APL) of Johns Hopkins University. APL, an entity somewhat similar to JPL, worked primarily for the Defense Department and wished to perform more substantially for NASA. These legislative allies indicated they would help NASA establish a "program line" in NASA's budget for Discovery if NASA would be willing to entertain a proposal from APL.[64] NASA had sought a program line for the Observer series years before and failed to get it owing to the traditional reluctance of OMB and Congress to provide long-term authorization for a particular program. Also, Huntress wanted competition for JPL and planned Discovery to be open to proposals from the scientific community beyond JPL. He decided that the first low-cost mission would go to APL—a relatively simple project called NEAR, for Near Earth Asteroid Rendezvous.[65]

However, Huntress wanted Discovery's second mission to be much more demanding so as to prove the point that difficult missions could also be low cost. As NASA and JPL had already been discussing the initial MESUR mission, one entailing both landing and roving, he decided to move that mission from MESUR to Discovery. That project would go to JPL. With the prospect of congressional backing for Discovery, OMB also went along with the concept of a program line, by which NASA could determine the sequence of missions within an established budget category without negotiating each mission separately as a "new start."

Such a line provided continuity. It was an extremely important and strategic move.[66] It gave NASA more power over its future missions and scientists greater sense of sustainment. The bargain NASA struck with its budgetary and political masters was that these missions would be low cost as well as open to various performers. It gave NASA greater flexibility and autonomy in choosing and managing projects. It enabled NASA to transfigure MESUR Pathfinder into what later became Mars Pathfinder, the combination lander-rover. MESUR Pathfinder got into the president's budget as part of the Discovery series. The president's proposed budget went to Congress in early 1992, with the way smoothed for Discovery's approval. Discovery was designed to be a general program, not a Mars-specific program—but it was a model for how to proceed, and it would get the first Mars effort after Observer under way.

CHAPTER SEVEN

Prioritizing Mars

The year 1992 began reasonably well for NASA from a White House budgetary standpoint. The president called for a 4.5% raise for the agency. Space science in particular was augmented, with a 9% increase.[1] Congress, however, was less interested in giving NASA more funds and ordered the Science Directorate to cut back on its most expensive programs. Congress made it abundantly clear it would not grant the president's request to fund his Moon-Mars initiative.

Who would lead NASA? The White House had come increasingly to believe that Richard Truly was not the Administrator it wanted at NASA's helm. On February 10, days after Bush's budget was announced, the president called Truly to the White House and forced him to resign. Mark Albrecht, National Space Council executive director, searched for a replacement, someone who would bring an enthusiasm to Bush's Moon-Mars vision which Truly had not.

The man he found was Dan Goldin, an aerospace executive from California. Goldin turned out to be a NASA Administrator for whom Mars was "the" priority. He might have to emphasize other programs for institutional reasons. The shuttle and especially the space station were utterly critical to NASA. But, in his heart and soul, Mars came first among his personal interests. He also had a personality—vision, self-assurance, drive, intensity—such that he could make a distinctive mark on the agency. Goldin was not an easy man for whom to work. But the science directors he appointed found they could realize their own goals through him. Finally, it turned out that Goldin would set a record for longevity in the Administrator role. All those factors would make a positive difference for the robotic Mars exploration program. What had been a slow, painful climb up NASA's agenda for Mars advocates after the Viking disappointment now turned into something quite different. Goldin was a dynamic advocate. Also, he wanted to use Mars exploration to showcase a managerial-technical strategy called "faster, better, cheaper" that fit the White House and congressional mood. Goldin intended to lead the agency and nation forward—to Mars. The result was the Mars Surveyor Program, the first program of sequential, integrated missions to the Red Planet since Mariner.

The advocacy coalition, for better or worse—and Goldin engendered many critics—had a powerful champion at NASA. He would strive mightily to remake the space-policy subsystem and enlist national policymakers in his quest for Mars.

Dan Goldin

Age 51 at the time, Goldin was vice president and general manager of TRW's Space and Technology Group. Born in New York City, he had received a BS in mechanical engineering from City College of New York in 1962. Fascinated with space since boyhood, he immediately came to work for NASA at its Lewis Research Center in Cleveland. NASA was going to the Moon at the time, and Goldin wanted to pursue research helping the agency take the next step, to go to Mars. When it became clear later in the 1960s that NASA would not be going to Mars any time soon and started retrenching, he grew restless and frustrated, leaving NASA for the aerospace company TRW. Based in California, Goldin advanced in the corporation over the years, spending most of his time on classified military and intelligence space programs. Coming out of this "black" world, Goldin was not well known in civil space circles, but in the classified field, he was considered a significant figure.[2]

The National Space Council, which presided over both national security and civil space endeavors, saw the Strategic Defense Initiative, known as "Star Wars," introduce innovative efficiency approaches to robotic space, which NSC called "faster, better, cheaper."[3] NASA, in contrast, seemed to NSC to be mired in the past, a bloated bureaucracy with big, expensive technical systems. Goldin was an exemplar of this new approach. He made maximum use of the latest microelectronic technology to bring down the size of space satellites. Goldin was known as a demanding, tough manager who could reshape an organization. He also was the opposite of Truly in one important respect: he was something of a visionary. Moreover, as Truly focused on the shuttle and space station, Goldin looked beyond to exploration and was passionate about Mars. It was his compelling ambition to lead NASA to the Red Planet. The opportunity to implement the Bush policy targeting Mars as a long-range goal was a significant reason that he left TRW and came to NASA, even though his tenure might be short, owing to the upcoming presidential election.

Goldin sailed easily through the confirmation process, becoming NASA Administrator on April 1. In those hearings, he indicated he would maintain all existing programs but manage them with greater efficiency. Once in office, he told his managers that large raises for NASA were not possible in the near-term future and they would have to get NASA out of the "vicious cycle" so exemplified by Mars Observer. He described the cycle as follows: "Because NASA flies relatively few missions, program officers overload each one with instruments. This makes each spacecraft expensive. Because they're expensive, they must be carefully tested before flight. This takes time and costs more money, raising the ante. In the end, so much is riding on each flight that NASA can't afford to have them fail—leading to more caution, delay, and expense."

"We've got to cut the Gordian Knot," he declared, by making spacecraft smaller, lighter, and cheaper, so that NASA can take risks and not fear making mistakes.[4] Soon aware that he could not pursue Bush's Moon-Mars program, as he would have wished, because of congressional opposition, he set his sights on remaking NASA so that it would be more capable of maintaining the programs it had and be readier for Mars if and when circumstances changed. He looked at the space station and science programs with an eye to technological innovation and management reform. He focused on the robotic program as a step toward human spaceflight to Mars.

Goldin saw himself as seeking "revolutionary" change and characterized himself as a change agent. He knew he faced opposition and spoke of his adversaries as erecting barricades to his FBC policies.[5] Among his adversaries, in his opinion, Fisk, the head of NASA's Science Directorate, stood out.

Fisk was accustomed to ample autonomy, and Goldin was not about to grant that discretion. Fisk was receptive to the notions of FBC as they applied to robotic programs, as exemplified by his support for the Discovery activity. However, he was pursuing a number of large projects at the time which he defended. He and Goldin had crossed swords earlier when Goldin had been in industry. Fisk was laboring to protect the Earth Observation System from substantial downsizing, and Goldin was seeking to sell NASA on a smaller-scale version of the system. Goldin was told by the Office of Space Science and Applications not to press his case publicly if he wanted TRW to get work with NASA, and Goldin had not forgotten what he regarded as a threat. As Administrator, he and Fisk had a tense relationship.[6]

Fisk's deputy, Huntress, the planetary division director, told Goldin that Discovery was an FBC program and was already being considered positively by Congress, and thus he won favor with the NASA Administrator at an early meeting. Goldin was looking for allies to help him with his "revolution." He saw Huntress in that category.[7]

Ironically, Huntress encountered resistance to FBC notions from the scientific community—not to Discovery in general or the MESUR effort, still in planning, but to the Pathfinder mission in particular. Huntress held that two values had to guide the planetary program: scientific worth and public interest. It was not enough to have one without the other. Pathfinder would meet his criteria and as a by-product help NASA politically as an institution.

Huntress had appointed a Science Definition Team for MESUR chaired by Cornell planetary scientist Steve Squyres, a Sagan protégé. Various scientists on the team had concerns, and Squyres passed those on to Carl Pilcher, Huntress's Advanced Studies Branch chief. On April 20, Michael Carr, a distinguished planetary geologist at the U.S. Geological Survey, whose long work on Mars included experience on Viking, wrote Huntress also detailing scientists' worries about starting a critical program (Discovery or MESUR) with so risky a project as Pathfinder. He pointed out that with Pathfinder NASA was now proposing to successfully land on the Mars surface, a feat achieved only by Viking at a much greater cost; deploy a rover, something never done before; launch in 1996 (about half the time usually taken for development); and keep costs below $150 million. He saw the likelihood of failure quite high. The chairman of the Solar System Exploration Committee, Huntress's planetary advisory body, also cautioned Huntress on Pathfinder.[8]

Huntress told Pilcher to compose a reply, because NASA had to have notable scientists like Carr aboard. The basic argument Huntress and Pilcher made to Carr and other skeptics was that the rewards made the risks worth taking. They held that it was important to demonstrate that NASA could land scientific instruments on the surface and also rove. Pathfinder was more than an engineering demonstration, they stressed. It was an enabler of science, and a way of showing "that NASA can do a quick, inexpensive, exciting, challenging project involving major departures from the way most previous planetary missions have been conducted. The positive repercussions of success could be beneficial to all NASA planetary missions including MESUR."[9] The institutional and public relations values thus were important along with the scientific and technological gains.

Huntress was very much in harmony with Goldin's reformist approach. With Goldin's active support, Congress approved Discovery, providing funding for it to begin officially in 1993. Huntress selected Tony Spear of the Jet Propulsion Laboratory to lead the Pathfinder project. Huntress regarded Spear as the kind of "out-of-the-box" manager who could make Pathfinder work. Huntress scheduled Pathfinder to launch in 1996.

Huntress wanted Pathfinder to carry a rover to Mars. Donna Shirley, JPL's rover manager, and Spear did not get along particularly well, but they were willing to cooperate to make the overall project succeed.[10] Huntress, who came from JPL, saw the NASA center as having an "old guard" that would expect Pathfinder to fail. He wanted to show them they were wrong about FBC missions, and Spear in particular was the manager to do it.[11] There were individuals at JPL who questioned whether JPL should even perform such a mission, but a senior manager at the lab, Norm Haynes, told his peers that JPL had better take it on: "If we don't do this, somebody else will."[12]

Mars on the Brain

In the remainder of the summer 1992, Goldin divided his time between visits to Congress to lobby for the NASA budget, especially funds for the space station, and trips to Russia to discuss possible future collaborative efforts with Yuri Koptev, head of the newly created Russian Federal Space Agency (RSA). Goldin also gave speeches, in which he extolled exploration of the Moon and Mars. He knew he could not sell Bush's Space Exploration Initiative but wanted to keep the idea of human exploration back to the Moon and especially to Mars in the public eye as best he could. He grew extremely close to Sagan as a key advisor. Sagan biographer William Poundstone has written that Sagan's influence on space policy was at its height when Goldin was Administrator.[13] The two men saw eye to eye on Mars as the key destination for NASA and were both passionate about the Red Planet.

Goldin reached well beyond Sagan, however, in seeking advice. One scientist he consulted was Michael Carr. Carr vividly remembers a call he received from Goldin shortly after Goldin had become NASA Administrator. He asked him to come from California to Washington. "Teach me about Mars," he requested. Carr noted, "I went. We talked a couple of hours. I told him what I know. It was clear he was smart." As far as Carr could tell, Goldin absorbed everything. He found Goldin fascinating, charming, and volatile.[14]

On September 25, NASA launched Mars Observer. For the first time since Viking landed in 1976, America had a mission to Mars actually under way. There was enormous elation at the time of the launch. "America is going back to Mars!" Huntress enthused. Goldin used the opportunity to declare that Observer was just the beginning of NASA's missions to Mars. The Russians joined in the celebration, stressing the long-term, step-by-step nature of Mars exploration, and pointing out that theirs would be the next nation to go to Mars, with a 1994 mission.[15]

In mid-October, Goldin announced a major reorganization affecting space science. The tensions between Goldin and Fisk had worsened over time. Goldin abruptly split OSSA into two major divisions: Space Science and Mission to Planet Earth. (He later added a third.) He summarily moved Fisk from a powerful associate administrator role to a staff position with no budget, that of chief scientist. Fisk left NASA not long thereafter.

Goldin called Huntress, then in Europe, and told him, "I just fired Fisk. You're the Associate Administrator. I'll see you when you get back!" Huntress was now in charge of space science. Huntress was "appalled" and also "scared."[16] He was sorry to see Fisk go. He was also worried about dealing with his unpredictable boss. But he shared Goldin's view about FBC and had a program ready to show these principles in action via Discovery. Even more importantly, he strongly agreed with Goldin about the importance of Mars, scientifically and politically, and had a project, Mars Pathfinder, that would showcase the Red Planet.

He decided he could work with Goldin. He would not let the NASA Administrator intimidate him, as Goldin did others. Nor would he challenge Goldin directly. "If you know Dan," he recalled, "You don't tell Dan, no. That's what Len [Fisk] did. You don't last that long."[17] The 51-year-old Huntress was very different from Goldin in manner, but he considered himself a change agent, like Goldin. They agreed on a basic strategy for exploration.

Mars was the centerpiece. Goldin stated in retrospect, "I made it the focus of the planetary program. The beauty of Mars is you have a launch window every 26 months. We needed a product line. If you come up with commonality in [technology], you can contain cost."[18]

As the year ended, Huntress wrote an article summing up the Goldin-Huntress strategy as it related to Mars. He proclaimed "the New Era in Mars Exploration." The long hiatus following Viking was over, he stated. NASA was indeed going back to Mars. He noted that various principles would guide the new era: NASA would go at every opportunity, meaning every 26 months, when the planets were aligned. NASA would go "together" with other nations, if possible. NASA would lead a Mars Sample Return mission to investigate the question of life. MSR, he said, was "the holy grail of planetary science." And, finally, NASA would use the robotic program to help enable human spaceflight to Mars.[19] The robotic Mars program now had a clear, succinct strategy. Whether NASA could get the new Mars strategy off to a strong start, much less sustain it, depended greatly on whether Goldin and Huntress would continue in their present, influential roles.

On January 20, 1993, Bill Clinton became president and Al Gore his vice president. In Gore, Clinton had a vice president with strong interests in science and space. Clinton abandoned Bush's NSC and assigned the space portfolio to Gore. Clinton retained Goldin, who had impressed Gore because of his emphasis on cost efficiency in space. Clinton and Gore were going to mount a "reinventing government campaign," and NASA under Goldin could be a potential example of what might be possible. However, space and Mars were not a high priority for Clinton, for whom the economy and deficit reduction were most important when he entered office. Clinton terminated Bush's SEI in one of his first space policy moves when he became president.

As always, Mars policy was influenced by NASA priorities in human spaceflight, the dominant spending category of the agency. The space station took most of Goldin's time and brought NASA considerable White House attention in Clinton's first year. The Clinton Office of Management and Budget, led by Leon Panetta, recommended at the outset of the new administration that the space station be killed. Goldin sought presidential reprieve, which Clinton granted, provided that he drastically downsize the station and reduce its costs. From February to June, this redesign process and an independent review of that process proceeded at a furious pace.

As it did so, Yuri Koptev, director of the RSA, sought a merger of his country's space station program with that of the United States. In the wake of the Soviet Union's collapse, the space program of Russia was in financial free fall, and Koptev sought help to salvage what he could. Goldin saw a space station merger as aiding NASA. He said it would save money and time because of the experience and expertise Russia would bring to the project. It would also make the space station more important to Clinton from a foreign policy perspective. Like his predecessor, the president was looking for ways to bolster Yeltsin and employ Russian scientists and engineers, so as to avoid proliferation of Russian missile (rocket) technology to U.S. enemies. Clinton wound up accepting a redesign option and forwarding the idea of a U.S.-Russian space station merger. Congress was slower to agree, as were the existing international partners. In June, the House of Representatives came within one vote of killing the space station. Nevertheless, by the end of the year, the space station program was saved and renamed—from Freedom to International Space Station (ISS). The president and congressional leadership met at the White House and agreed to stabilize the U.S. space station budget at $2.1 billion per year through completion of assembly.

This decision kept NASA going with a measure of financial stability throughout the 1990s, in large part because it kept human spaceflight viable. As had been true of the shuttle, the space station was at once a blessing and a curse for robotic science programs. Goldin spent an enormous amount of time lobbying Congress and negotiating with Koptev on the space station.[20] It was a constant headache. But ISS was a way to keep the president and vice president involved in space policy, as part of foreign policy. It was a symbol of post–Cold War U.S.-Russian cooperation for the White House. Goldin was thus important to Clinton and Gore, and that was important to his access to power generally.

In dealing with Koptev, Goldin also talked about Mars Together options for the robotic program but left it to Huntress to take the lead in this area with his Russian counterpart. He also recognized the work and contacts the Planetary Society had accomplished in its earlier Mars Together activity. Louis Friedman became involved in trying to make Mars Together work in an informal way, as an unpaid consultant, even going to Russia to facilitate negotiation.[21] However, while there was progress, the viability of Mars Together depended on success in the robotic program on the parts of *both* the United States and Russia.

Mars Observer Fails

Even though he had had to shut down the Exploration (i.e., Moon-Mars) office at NASA and move its associate administrator, Michael Griffin (a later NASA Administrator), to a new role, Goldin wanted to sustain the humans-to-Mars goal. His struggle to save the space station and zeal to reinvent NASA were in fact partly for the purpose of making the agency more capable of leading

America to Mars. With his background in robotic space technology, he saw that he could keep the Mars human spaceflight goal alive through the robotic Mars program.

Goldin had mixed views of Mars Observer. He called it an example of a Battlestar Galactica mission, the kind he eschewed, but it was too far along to change. In fact, it was heading toward Mars, and a lot rode on it. In addition to providing a mass of scientific data to the Mars research community, Observer was a stepping stone toward future robotic trips to the Red Planet. Russian scientists wished to use the maps of Mars that Observer would yield so they would know where to land missions they scheduled for 1994 and 1996. Russian and French researchers were planning to use a relay system on the U.S. Mars Observer to transmit up to 10 times the data that the Russian orbiter to be launched in 1996 could handle. Observer's results could also feed into MESUR's design.[22] And, longer term, Mars Observer was the beginning of NASA's return to Mars and eventual human journey. This was the exploration strategy of which Huntress had written in proclaiming a new Mars era.

Of the many scientists and engineers associated with Mars Observer, none waited more expectantly than Mike Malin. His camera was aboard and was central to the mission, thanks to a "command" decision made years before by Associate Administrator Edelson. So committed was Malin that he had left JPL and an academic position to set up a company so he could dedicate himself to building the camera. He thought of it as his eye. And through it, he would soon be looking at Mars.[23] Other scientists and engineers at JPL and the universities also had devoted a large chunk of their careers and professional lives to this mission.

At 6 p.m. (PDT), August 21, 1993, Mars Observer was three days away from going into orbit around the Red Planet. Suddenly, JPL lost communication with the probe. As controllers at JPL worked feverishly to reestablish contact, word of the mishap flashed throughout JPL, NASA, and the Mars community. In succeeding days, anxiety mounted, and then gloom. "It's difficult to work on something for 10 years, and expect to work on something for another 5 years and have it disappear," said Philip Christensen of Arizona State University, one of Observer's principal investigators. NASA planetary geologist James Garvin, who worked on the spacecraft's laser altimeter experiment, said, "Basically, it's my entire scientific professional career."[24]

Garvin's personal devastation was worsened by an experience soon after when he returned to NASA's Goddard Space Flight Center in Greenbelt, Maryland, where he worked. He went to the mall near Goddard with his wife. Prior to the loss of communication with Mars Observer, those associated with the mission wore "Mars Observer hats" for good luck that never came. "For some reason," Garvin recalled, he wore his hat to the mall. "Someone came up to me, a stranger. He accosted me. He said: 'How dare you wear your hat. It is a sign of failure!' "[25]

Turning Adversity into Opportunity

As it became obvious in succeeding weeks that Mars Observer was hopelessly lost, Goldin established a panel to determine what had gone wrong. Eventually, the panel held that the immediate culprit behind the demise of Observer was most likely a "massive rupture in its propulsion system."[26] More broadly, NASA was seen to have erred in thinking that the technology it used successfully in Earth orbiting would be easily adapted to Mars. The differences in requirements were much greater than anticipated.[27] Meanwhile, Goldin established another panel to chart where NASA would go next in Mars exploration. He put Charles Elachi, assistant director of JPL, in charge. He also appointed Sagan to this panel.[28]

Goldin turned to Sagan, as one true believer turned to another, for advice and encouragement. Sagan felt as did Goldin and Huntress—that NASA should go to Mars at every opportunity, meaning every 26 months. He also pressed Goldin to revive the search for life on Mars and elsewhere in the universe. Within NASA, Goldin consulted especially with Huntress. Right after the failure, they huddled over how to deal with the media and then how to persevere in their goals. "Goldin didn't beat me down," Huntress remembered. "His question was: 'What are we going to do? We need to fix this program. We can't go back on our commitment to Mars.' "[29]

There was no thought by Goldin to abandoning Mars, as had happened after Viking. Goldin told his agency and the Mars community not to be depressed. The accident could be an opportunity, he said. Goldin stressed to the media and virtually everyone else that NASA would go back to Mars, and do so as soon as possible, but in the "right" way, meaning FBC. Here was the chance to showcase what his approach could mean, and to do so on the most important and visible stage, namely, that of Mars.[30] Not long after Observer's failure, he went to Arizona State University. He addressed a group of scientists and graduate students who were depressed in feeling years of work had been obliterated in the Mars Observer demise. Don't be glum, an upbeat Goldin told the gathering. He called the loss of Observer the best thing ever to happen to the Mars program. He proclaimed that it marked the dawn of a new era in Mars research.[31]

For NASA and the Mars community, the instruments on Mars Observer were seen as vital to building on Viking and setting the stage for what came next. As advisors went through various options, Goldin and Huntress settled on replacing the proposed MESUR network of landers with a new Mars Surveyor Program. In this program, instruments that had been grouped together on Mars Observer would be separated and flown on a series of FBC—that is, smaller— spacecraft.[32] As it evolved, the planned Mars Surveyor Program "would send two low-cost spacecraft—an orbiter and a lander—to Mars every 26 months over the course of ten years." Each mission would be capped at a cost of $175 million and be limited to three years in development. It "would address science objectives centered on understanding Mars' climate, resources, and the search for water and life." The agency hoped eventually to send a robotic envoy to the surface which would be capable of collecting samples of Martian terrain and returning them to Earth.[33]

Goldin and Huntress sold this "program line" to OMB. Discovery was a model and precedent. They stressed the importance of continuity in Mars exploration. Also, Huntress commented, "We sold it to OMB on the grounds we could get more science done in a less costly way."[34] Rather than fretting over Mars Observer, NASA leaders used the accident to build momentum for a new Mars program that carried the FBC stamp.

In selling the Mars Surveyor Program, it was in their interest to emphasize the contrast with Mars Observer. A mission can be priced at different amounts, depending on what is included, ranging from a low cost for only development of the spacecraft to a high "lifetime" cost including launch and operations. As Fisk saw it, now sitting on the sidelines at the University of Michigan, Mars Observer was a $500 million mission. Including virtually everything, the cost went up to $980 million. "But it was called a $1 billion failure to help Goldin sell faster, better, cheaper," emphasized Fisk. It was, he said, "one of those rare situations when an agency uses failure and a cost-overrun to advantage, and makes the overrun seem worse than it was." The $1 billion cost, repeated continually, stuck.[35]

A New Start

In early February 1994, the White House unveiled the president's budget proposal. The request for NASA was $250 million below the previous year's budget.

This reduction was the first time since 1974 that the White House had requested a decrease in NASA funding.[36] However, within this budget, there was a new start in robotic spaceflight, the Mars Surveyor Program. The budget proposed initial funding for a new Mars orbiter, one that would be smaller and less expensive than Mars Observer. It would carry a number of the eight instruments borne by Mars Observer. It would be called Mars Global Surveyor (MGS) and be managed by JPL. Subsequent orbiters would bear the remainder of Mars Observer's instruments. In essence, the new program would do what Goldin wanted: carry out Mars Observer's missions and then some, in a "product line" of smaller spacecraft. The Mars Surveyor Program was extremely significant. For the first time since Mariner, there was political acceptance of a sustained series of spaceflights to Mars. This program would embody and be a showcase for FBC and the notion that government could be reinvented to produce more for less. It conveyed the idea that big science could be made into smaller and more affordable science through distributing missions and costs over time.

In mid-March, at a senior management meeting, Goldin indicated that finding more funds for space science was a top priority.[37] On April 1, NASA issued a Request for Proposals (RFP) to develop and build the MGS craft. Martin Marietta won the MGS task in July. The process from RFP to award was seven to eight months shorter than usual. Under Goldin's new "faster" rules, Martin Marietta would have 28 months to build MGS in order to meet the 1996 Mars launch window. Getting the RFP out this fast was new, a result of administrative streamlining. So was the award decision.[38]

Congress was meanwhile coming through with authorization of a program line for the Mars Surveyor Program, as well as the money for implementing NASA's first mission in the series. Sagan helped NASA promote the program to Congress. So did the National Academy of Sciences, which gave it endorsement. Congress listened to such supporters. So, apparently, did OMB. The country would get more science for less money. Said Huntress, "There was consensus."[39]

It was not just Goldin who was pushing FBC efficiency to the agency. Huntress was doing so also, and on April 12 he attacked NASA's "outdated technologies and attitudes." He condemned NASA's "resistance to change," saying NASA was "two decades behind the curve in engineering the way we operate our missions."[40]

In September, NASA chose a landing site for Pathfinder, scheduled to arrive at Mars in 1997. Tony Spear, project manager for Pathfinder, said that this mission would indeed reveal the new way of doing business. He declared he had "assembled a team and dumped all the formality."[41] As Huntress later commented, "Tony ran the show. He established a Skunk Works at JPL. . . . JPL didn't like this approach. It is used to certain engineering practices. JPL expected failure. Tony made it work. He recruited young engineers and scientists who did not know what had been done in the past. He pushed innovative practices. He mined their creativity and put it to work. The old guard watched with no confidence in what he was doing."[42]

Goldin's Dream

The new strategy of Goldin and Huntress also called for more international partnership. In April, Goldin initiated a study of a "Mars Together" mission with the Russians. The proposed mission would be reviewed in December by Gore, with June 1995 as the final month for a decision one way or the other. The Gore-Chernomyrdin Commission, set up to nurture U.S.-Russian collaboration in the post–Cold War era, had asked Goldin and Koptev to review possibilities of robotic space cooperation with an eye to savings in cost and time.[43]

In July 1994 Goldin told the media that while he had to give priority as Administrator to the space station, which Congress was now backing more strongly, it was Mars that was his love. He recalled that he had watched in awe when Armstrong walked on the Moon. As a young man, he had pledged to himself that he would someday be part of a Mars mission. He declared that his "deepest dream" was "that in my lifetime I will some way be responsible for a [human] mission to Mars. It would be the next noble thing we could do as a society." The first, he said, had been Apollo.[44] The Clinton administration was not interested in a human mission to Mars but was supportive of the space station, robotic Mars development, and international collaboration. As Goldin saw it, these were stepping stones to human spaceflight to Mars.

NASA was discussing Mars exploration not only with the Russians but also with the Europeans and the Japanese, who were also planning a robotic Mars mission. The Russian mission of 1994 had slipped to 1996, and that meant Russia's 1996 mission would slip to 1998. The 1998 mission would involve the French and include balloons to be released in the atmosphere of Mars. Given Russian delays and Mars Observer's demise, the United States would not be helping the Russians as previously planned via Observer, but perhaps another U.S. mission could do that. "Mars Together" was taking a more ecumenical form, as the Japanese invited the United States to include a U.S. instrument on its Mars launch, scheduled also in 1998, called "Planet B."

NASA had its Mars Science Working Group, an entity whose job was to take NAS Space Science Board recommendations and apply them more concretely to Mars planning. Now NASA and other nations established an International Mars Exploration Working Group. Its purpose was to plan a major international robotic effort to investigate Mars. In October this body convened for the first time, with Huntress chairing the first meeting. Involved were Austria, Canada, France, Germany, Great Britain, Italy, Japan, Russia, the European Space Agency, and the United States.[45]

Success internationally required success of national programs. For the United States that meant initially Mars Pathfinder and Mars Global Surveyor. In December, Spear reported that all was going well with the development of Pathfinder. Being built at JPL with a number of contractors providing components, the project was well under way. Spear said design was completed, and the spacecraft was ready to be assembled.[46]

Protecting Mars, Reviving Exobiology

In November 1994, the Republicans captured control of Congress. Led in the House by Speaker Newt Gingrich (R-GA), they brandished a "Contract for America," in which they promised to end the deficit, balance the budget, provide tax relief, and drastically cut back "Big Government." President Clinton, seeing the country shift to the right, decided that his own political future required him to get ahead of the Republicans and recapture initiative. At the beginning of 1995, he promised his own form of tax cut, deficit reduction, and government downsizing. For NASA, this meant more budget and personnel shrinkage and even greater emphasis on FBC science missions.[47]

On January 12, 1995, Goldin got a letter from OMB. It said NASA would have to reduce its projected budget over the ensuing five years by $5 billion.[48] Goldin had thought he would have some stability in budget; instead, he faced even more stress. "My immediate reaction," he later recalled, "was one of sadness and frustration . . . it took me two weeks. First you have shock, then you have denial. I went through all those emotions. Within a day, I said, 'I support the president.' Then, weeks later, I was enthusiastic. My position was, we're not going to look at this as a budget cut. This is an opportunity to take the final step in reinvention. This is an opportunity to get NASA to do what it finally had to do."[49]

The consequence was that Goldin led a drive to cut "infrastructure," while maintaining all existing programs. The aim was to do even more with even less. The Mars Surveyor Program, for example, went forward as planned, but in a context of personnel reduction and reorganization. Many employees feared for their jobs, especially at headquarters. The Science Directorate, along with other program offices, took a hit. There was much talk of a draconian personnel reduction, perhaps 50% at headquarters alone, with additional cuts throughout the centers.[50] The contractor workforce would be even more decimated. Goldin vowed revolutionary management reform, and this change included moving more control over programs from headquarters to the centers. Decentralization also meant shifting control over funding. Project managers at JPL, for example, had fewer resources with which to work, but more control over those resources. The downside was that they would have to make exceptional cases to get more money from headquarters, since headquarters had less to provide.

With his vocal enthusiasm for reform and loyalty to the president, Goldin became the poster boy for Vice President Gore and his reinventing government campaign. This campaign, under way since the advent of the Clinton administration, became more pronounced in 1995 and seemed to be especially associated with governmental downsizing. Goldin basked in favorable publicity as he did battle against "bureaucracy," but, inside NASA, he was much criticized, and his seemingly callous offhand remarks helped create fears at JPL and elsewhere. "People are terrified," said one aerospace industrial official. Goldin, in the view of his critics, equated disagreement with disloyalty. He would, it was said, kill the messengers of bad news. He called his senior managers "pale, male, and stale" and replaced half his center directors, although not the head of JPL.[51]

Planetary science suffered along with other science programs in the cutback atmosphere. However, within planetary science, Mars was relatively favored as a priority. Mars was the destination of choice, Huntress emphasized: "It is the most Earth-like planet, and it is a place where life may once have formed."[52] So, it was going to get the priority within space science generally and planetary programs particularly. As part of the Mars priority, NASA was reviving exobiology.

For years, exobiology had languished. Under Huntress, exobiology was re-invented, with Michael Meyer, one of the few relatively younger scientists at NASA retaining this interest, in the lead. In 1995, Meyer brought together a number of researchers who had persisted in conducting exobiology research despite what Sagan had called its "disreputability." Chuck Klein of Viking, now retired, was among them. They developed "an exobiological strategy for Mars exploration." The emphasis was on robotic research, but eventually there would be human exploration. Sagan was an advisor to the group. Klein and others, particularly at Ames, had worked to bridge the gap between exobiology and other fields in emphasizing "habitable environments" over life per se. Still there was an estrangement of "exobiology and the disciplines of traditional planetary science." The group called for the research communities to work together, even though that might "strike some as far-fetched, even fanciful."[53]

The Meyer role reflected not only renewed interest in exobiology but also new organizational clout for Mars within the Space Science Directorate. The same was seen at JPL. Indicating this Mars emphasis organizationally, JPL had appointed Donna Shirley to the position of manager of its Mars activities. She presided over Pathfinder and other Mars projects under the Mars Surveyor Program. In doing so, she advanced over Spear, a move that grated on Spear but did not hurt Pathfinder. She noted in May that in 1995 dollars, Viking cost $4 billion. In comparison, the spacecraft JPL were now developing to go to Mars would cost a small fraction of that amount, and they would do more science. The new Mars program would go beyond what Shirley called Viking's "veneer science" to a point where you could actually start to bore in and do some in-depth studies.[54]

Mars Pathfinder, the first of the series of Mars ventures, would be the smallest planetary spacecraft ever, at 1,500 pounds. But it would be big compared to the 1998 probe NASA was developing, which would be half Pathfinder's size. Then would come even smaller spacecraft. One JPL scientist said the laboratory's aim was "to create a 'virtual presence' for mankind" in the solar system through the various robotic devices JPL and NASA planned under the FBC regime.[55] That "virtual presence" began at Mars.

In July, as NASA celebrated the 30th anniversary of Mariner 4, the first spacecraft to successfully view Mars, there was a strong sense among a growing number of advocates inside and outside the agency that NASA was gathering momentum in respect to Mars—in spite of the funding crunch. Goldin said that the adversity was forcing the agency to be more innovative, leaner, and meaner. He spoke of NASA's long-range strategy and said that the robotic precursors would lead the way to human missions later on. "I want to go to Mars," he avowed.[56] He predicted that NASA could spend wisely in the remaining years of this century and then begin preparations for a concerted effort to send humans to Mars in the twenty-first century, with 2018 the target year.[57]

The governmental downsizing mood of the times, shared by the president and Congress, enhanced Goldin's power to get his way within NASA. But there were many who were skeptical that more with less would work.[58] Those critics inside NASA tended to keep quiet, however, fearing for their jobs. Goldin avowed he was going to do what was right, and not worry about being loved. What was right, in his opinion, was to go to Mars—by way of the FBC strategy.

CHAPTER EIGHT

Accelerating Mars Sample Return

As 1996 got under way, NASA found itself pressed between the Republican Congress and President Clinton, as they fought over the federal budget. This bitter struggle had led to the government's shutting down in 1995 because of an inability to get a budget bill passed to keep it running. That shutdown was over, and most agencies were now operating on a continuing resolution, under which Congress permitted them to spend at a level commensurate with what they spent the previous year.[1]

NASA eventually got its 1996 budget—$13.8 billion—in May, but the fiscal uncertainty continued. Clinton promised a balanced budget by the time he left office, and the Office of Management and Budget proposed cuts in NASA's budget in ensuing years. Under the OMB plan, from $13.8 billion in 1997, NASA would plummet to $11.6 billion in 2000. Making matters worse for NASA was its need to cope with Russia's torpidity in fulfilling its space station obligations, thereby helping to delay that project and adding to its expense. Under the 1993 presidential-congressional agreement, the station was getting $2.1 billion a year. With the overall budget falling, space science and robotic Mars exploration would be further squeezed. Huntress complained that NASA could not sustain its science initiatives under the projected funding.[2] Goldin, meanwhile, seemed to want more rather than fewer initiatives as he allied himself closely and vocally with Gore's "doing more with less" policy. NASA and its Mars program seemed to need a miracle to extract itself from a dire financial prospect. This miracle came—in the form of a meteorite from Mars. The agency's response was to reorient the Mars Surveyor Program from a comprehensive, gradual approach leading to Mars Sample Return to one that was targeted to achieving the sample return goal at the earliest instance. The Mars meteorite provided Goldin with what he needed—a way to raise Mars above the space subsystem level to national and even global attention. It was a punctuation point that altered the political equilibrium, at least for the moment. With support from the president, vice president, and Congress, Goldin seized the opportunity to forward his dream.

A Renewed Optimism

The politically driven financial constraints came at a time of renewed scientific optimism about the potential for life on Mars. The optimism derived mainly from discoveries on Earth and the distant cosmos. Research since Viking was turning up increasing evidence that hardy microbes and other organisms could thrive on Earth in the most unlikely and hostile places, where it was incredibly hot or cold. Indeed, these beings were accorded the name "extremophiles."

Mike Carr commented to the media, "We're in a different world. Our understanding of biology has advanced so much in the past 20 years. The probability that life could have started on Mars has greatly increased."[3] Another stimulus for renewed scientific optimism about Mars lay with discoveries far from Earth *and* the Red Planet. Scientists in the mid-1990s were beginning to discover evidence of planets around distant stars. These discoveries, which were of Jupiter-sized planets, gave hope that, sooner or later, astronomers would detect Earth-like planets.[4]

These remarkable discoveries on Earth and in the cosmos not only stimulated renewed hope about Mars but also helped spawn planning for a new initiative at NASA called "Origins" which would link planetary research with research in the cosmos around a theme of how the universe and life began. It was a theme that echoed much of Sagan's writing, and Sagan continued as a close Goldin advisor. Goldin and Huntress strained to find money to nurture the new venture, which they hoped would help reignite public interest in the space program. Michael Meyer, who headed exobiology studies under Huntress, recalled that Goldin "knew taxpayers were not interested in NASA for science's sake. The public cared about human exploration and the search for life. Goldin also was interested in biology. He thought that was the next frontier. Goldin had Huntress and me in for talks. He spoke with me alone. I would get a call from him. It was a 'heady' experience."[5]

Senator Barbara Mikulski (D-MD), a staunch NASA supporter, was alarmed about NASA's fiscal future and pressure to downsize employment, especially as it affected facilities in her state.[6] She pressured Clinton and Gore to have a "space summit" to discuss NASA's perilous future and what could be done about it.

In July 1996, the last Case for Mars conference took place at Boulder—at least under Underground auspices. Most of the original leaders of the Mars Underground had moved on. Chris McKay and Carol Stoker had gone to work for NASA. The current leaders of the Underground believed they had accomplished their objectives in keeping the dream of Mars exploration alive. Now Robert Zubrin, a firebrand engineer formerly of Martin Marietta, was poised to take a leadership role in outside advocacy, writing and speaking evangelically about Mars.[7] In two years, he would form a new organized interest group, the Mars Society. Its orientation was emphatically on the human program (in contrast to the Planetary Society, which emphasized exploration in general). He would also publish a book in 1997 called *The Case for Mars*, the title being that used by the Mars Underground for its conferences.[8]

All Mars advocates knew that the human and robotic Mars programs were potentially mutually supportive. But Zubrin held that human spaceflight should drive the robotic program and its priorities. Advocates of robotic efforts, in contrast, emphasized that science objectives were significant in themselves. In July, Ed Stone, director of the Jet Propulsion Laboratory, argued that the robotic program deserved support on its own merits. At this point in history, Stone declared, robotic explorers were essential. "At least for the questions we're smart enough to ask right now, robotic missions will suffice," he said. "Eventually, we may get to a set of questions where humans on the surface may be crucial," he declared. But not now, Stone insisted. Zubrin and those who followed him strongly disagreed.[9]

The dilemma for NASA was that there was not enough money for the desired robotic Mars program, much less human spaceflight to Mars. NASA needed cooperation among its conflicting constituencies in a time of threat. Huntress and Goldin believed that Origins conveyed an all-embracing purpose for space exploration, whether by machines or humans. Origins was not a specific program, but a theme to which various activities of NASA could contribute and which might be exciting to the public. Now the question was how to market this theme.

The Catalyst from Mars

In April 1996, Huntress was working in his office when one of his science managers came to his door and said, "You ought to see this pre-print. It makes some remarkable claims." "What is this?" Huntress asked. It was a possible *Science* magazine article about fossilized life in a Mars meteorite found in Antarctica, the manager said. It was written by a team of government and university scientists led by a NASA investigator, David McKay, based at Johnson Space Center.

While conscious that the idea of fossilized life in a Martian meteorite was directly relevant to the Origins theme he hoped to push, Huntress quickly dismissed the article. There had been other claims about life on Mars over the years. "It won't survive the peer review process at *Science*," Huntress commented. "Come back when you have a print date," he told his lieutenant. Huntress assumed that the paper would die and largely forgot the matter.[10]

Then, on July 12, the same manager appeared again at Huntress's office door, obviously agitated, and announced that the article had survived peer review. It was going to be published. The manager handed Huntress the embargoed pre-publication copy, saying, "You need to know about this!" "My eyes bugged out," Huntress recalled, as he looked at the article. It drew a number of strands of evidence together that made a compelling, if not absolutely convincing, case.[11]

Huntress immediately took the document to Goldin. "He came out of his chair," Huntress remembered. "Is this right?" Goldin asked. "Did it get past peer review? Oh my God! I've got to see the president."[12]

Goldin immediately set up a highly select (and secretive) task force to study the findings in the article and determine how to handle the public announcement. He also got an appointment with Clinton's chief of staff, Leon Panetta, the former OMB head, for July 30. Goldin decided that NASA should keep the matter "contained" until the article came out, August 16. Then, NASA would hold a press conference, with McKay and his team present, along with a scientific "skeptic" to provide balance. Goldin did not want this announcement to come across as a NASA public relations stunt.[13]

However, the potential significance of the Mars meteorite discovery for science and NASA was obvious. It would give the Origins concept credibility. The search for life would have a centerpiece, and it would be Mars. The implications for accelerating Mars exploration and returning a sample were clear. But there was enough doubt about the fossilized bacteria claims that it would be a mistake to push for an Apollo-like search.[14] Getting White House endorsement for additional Mars research and for Origins would reverse the unfavorable budget trends and counter OMB.

On July 30, Goldin and Huntress arrived at the White House. After briefing Panetta, they and Panetta went to the Oval Office and Clinton. It was 10 a.m., and Clinton initially seemed tired, but once he heard Goldin speak about the Mars rock, he became quite alert and attentive. As journalist Kathy Sawyer wrote in describing the scene, "The president's posture straightened, and his eyes opened wide. It seemed to be dawning on him that this could be big." The meeting lasted 30 minutes, and the president asked a number of questions. At the end, he told Panetta, "The vice president has got to hear this. Leon, take 'em to see the vice president."[15]

Gore was equally fascinated once he heard Goldin's story. Moreover, he "peppered" Goldin and Huntress with questions. He "asked about the age of the meteorite, when the purported Martian microbes would have been alive, and how that related to the initial bombardment of both Mars and Earth. He asked about the relative conditions that would have existed on Earth and Mars at the same time." He wanted to know "whether this life-form, if it existed, could have evolved on Mars, how the McKay group's result related to the Viking experiments of the 1970s, and what this new evidence in combination with the Viking results might mean for the possibility of current life on Mars."

Gore also asked about the implications for future policy. Sawyer wrote, "Huntress was ready with the 'Origins' pitch." That accomplished, they were ready to leave after what had been a 45-minute meeting. Before departing, Goldin pointed out that the announcement about the Mars meteorite would take place at the same time as the Republican national convention. Goldin wanted to know if that was a problem. Gore emphasized, "Follow *Science*'s process. Do not make this a political issue. Do not ever make this a political issue. This is a scientific discovery."[16]

After the meeting, Goldin called McKay, based at JSC in Texas, and directed him to come to Washington for a meeting as soon as possible. The next day, July 31, McKay and a colleague on the meteorite team, Everett Gibson, sat in Goldin's office. Goldin, flanked by Huntress, "pounded" the two with questions. The "grilling" went on "for three punishing hours." At the end of the session, Goldin asked McKay, "Can I give you a hug?" The Mars rock was exactly what

Goldin wanted—a gift of "evidence [of Martian life] that could launch a big revival of public interest in space exploration."[17]

Meanwhile, Dick Morris, Clinton's friend and political advisor in the White House, heard about the Mars meteorite. He passed the information along to his mistress, who in turn sought to sell the information about extraterrestrial life to the media. Morris also began pushing Clinton to make a Kennedy-like decision for a crash human Mars mission. The president's science advisor, Jack Gibbons, fought that possibility. "I said: 'Hell no!' " he commented in retrospect. His message to Clinton was that Morris's idea "would make us look like fools. We'd be sending a mission impossible."[18] Morris called Goldin to get his support for the crash venture. Nobody wanted more to go to Mars than Goldin, but, as Sawyer wrote, Goldin felt he had to be honest. He explained to Morris that a venture to the Red Planet was simply not technically or fundamentally feasible under current circumstances.[19]

NASA's hope to contain the news failed—the leak occurred, but not necessarily via the Morris mistress. Not only did the White House know, but also senior lawmakers concerned with space had been informed. By early August, the story was getting out in bits and pieces, and *Science* and NASA were blaming each other. With the story poised to break fully in the print and television media, and both *Science* and NASA wanting some semblance of accuracy, NASA had little choice but to act.

On August 7, NASA held a hastily arranged press conference at its Washington headquarters. Numerous officials from NASA and the administration were present, along with a McKay meteorite team. Goldin had a startling announcement to make. The research team believed that it had discovered credible evidence of fossilized Martian microbiological life in an ancient meteorite found in Antarctica.

For two years, the scientists explained, they had been examining the rock using the most sophisticated equipment available. The team found a number of factors that, taken together, pointed to fossil microbiology. They had a peer-reviewed article coming out in *Science* on August 16, explaining, in-depth, their findings.[20]

Goldin made sure he noted various caveats about the claim and the need for further confirmation by the scientific community. As planned, a skeptical scientist who had read the article was present at the conference to give dissenting views. But Goldin could not control his excitement. "We're now at the doorstep of the heavens. What a time to be alive!" he exclaimed.[21]

Goldin had invited Jerry Soffen of Viking to the conference and pointed out that he had pioneered in Mars-life research and that "we all stand today" on his shoulders.[22] Everyone present was fascinated with the announcement and anxious to know how certain the scientists were of their findings. Many congressional leaders were said to be "almost childlike in their excitement about the possibilities and very humble about the news."[23] Gore was fully engaged, and Clinton made a statement to the media that day: "If the discovery is confirmed, it will surely be one of the most stunning insights into our universe that science has ever uncovered. The implications are as far-reaching and awe-inspiring as can be imagined."[24]

Clinton said he "was determined that the American space program will put its full intellectual power and technological prowess behind the search for further evidence of life on Mars." Vice President Gore, he went on, would convene a bipartisan space summit of national political leaders to consider an appropriate response.

This "summit" was based on the one for which Senator Mikulski had earlier called. She was concerned about NASA's cuts in money and human resources. Clinton was now agreeing to the summit, with the Mars meteorite his stated reason. Goldin was extremely careful not to say anything about needing "more money" to follow up on the Mars rock. He did say that NASA would reexamine the Mars Surveyor Program strategy in view of the meteorite. He indicated that NASA might accelerate the program's timetable for MSR.[25]

Reactions

The media gave the discovery extensive coverage. As John Noble Wilford, veteran science writer for the *New York Times*, recalled of his thinking at the time, if the claims were true, "this would be the biggest story of my career—bar none."[26] His newspaper editorialized that the claims needed confirmation, but the credibility of the scientists involved meant that they had to be taken seriously. One of them, from Stanford, was Richard Zare, prominent chemist and chairman of the National Science Foundation's board. The *New York Times* said that the discovery could be "a transforming event of our time." The *Washington Post* commented that the announcement made normal politics in Washington stop "for a moment or two of wonder."[27]

Many scientists were quoted in the media, a number of whom expressed skepticism. One was Thomas Ahrens, a planetary scientist at Caltech. He called the findings "hypothetical" and said if any one of the assumptions the investigators made were false, the whole interpretation would collapse "like a house of cards." Sagan's past general statement that "extraordinary claims require extraordinary evidence" was used by critics against these particular claims. But Sagan, who had reviewed the *Science* article, emphasized the significance of the claims: "If the results are verified, it is a turning point in human history."[28] Amidst the scientific give and take, one scientist, Harry McSweet of the University of Tennessee, expressed a very personal reaction: "I don't know if this is evidence of life or not, but I want it to be."[29]

Goldin quickly established an ad hoc Mars Science Strategy Group under Dan McCleese of JPL to advise on what NASA should do in response. McCleese was a long-term Mars scientist who was now manager of exploration and space science at JPL. On August 14, 1996, one day before the McCleese group met, Huntress wrote Goldin a memo that the group would commence a reorientation in Mars strategy. "Our current strategy for the Mars Surveyor Program is driven by the goals of looking for water, resources, and evidence of climate change as well as life. This group will look at how the current strategy should be changed to focus on the search for evidence of life as the single most important priority." Huntress said he would also engage the National Academy of Sciences Space Studies Board (previously known as the Space Science Board) in the reorienting process.[30]

Goldin, on August 15, met with the McCleese group. McCleese remembered Goldin discussing the Mars meteorite and what its implications were for NASA in making "life" a focus for science strategy. "How do we follow-up" on the meteorite? Goldin asked McCleese and his team. "We have to have a sample return," they responded. Well, said Goldin, "can you do it in 2001?" The scientists stated that they did not think that would be possible. There were issues of money, technology, and "we don't know where to look." They also told Goldin that the science since Viking had centered on "habitability" as a theme. It was critical, they said, not to repeat the mistake of Viking, which had framed the life question in "yes or no" terms. The answer had seemed to come back "no," and that had made it difficult for NASA and the Mars science community to talk about Martian life ever since as a rationale for exploration.[31]

Goldin wanted to know if the missions currently on NASA's drawing board, specifically Mars Global Surveyor and Pathfinder, would get at the life question. No, said the scientists. It would take different technology, and NASA would have to look at Mars in a planet-wide context. Ok, said Goldin, "what's the next step?" McCleese and his team said they would try to answer him, but it would require some study.[32]

Goldin told the group to forget the politics, ignore the aerospace contractors, and concentrate on "what is the right thing to do" from the standpoint of good science. He asked for three options for accelerating progress toward MSR: "relaxed," "nominal," and "fast." He indicated that NASA might have to have international cooperation to do the job. In any event, he declared that while science could provide the direction, "the political process will say how fast we can go."[33]

Three days later, McCleese's panel made recommendations. McCleese told the media that NASA could reduce the amount of research on climate and other topics and develop more land rovers and subsurface drilling equipment. As early as 2001, robots could gather samples of Mars soil and rocks at especially promising places, with those samples brought back to Earth in 2003. Money was obviously critical to the "fast" option. The group suggested that international collaboration might well be needed. McCleese commented also that one problem his group had in making recommendations was the absence of exobiologists. Since Viking, he said, "there has been a turning away from biology as an active part of the NASA program." Now, "we are looking for a resurgence of a field." NASA would have to find specialists and "convince them NASA is serious."[34]

The 2001 option for a crash project for MSR did not last. McCleese and NASA wound up with 2005 as more realistic.[35] Not everyone agreed with this date, which seemed to fit the "nominal" option Goldin had requested. Some thought sooner, some later—but virtually everyone involved in NASA's decision process wanted to accelerate the quest for Martian life.

On September 19, Clinton issued his National Space Policy. The policy had been in the works prior to the Mars meteorite excitement. It was broader than civil space. However, it reflected the recent meteorite discovery and gave clear emphasis to Mars exploration as the top NASA and White House priority in space science. For the first time, an administration committed NASA to "support a robotic presence on the surface of Mars by the year 2000." Calling for sample return, the policy also endorsed the Origins initiative. It directed that NASA should look for "planetary bodies in orbit around other stars."[36]

What the Clinton policy did not do was call for a human Mars program, even as a long-term goal, as Goldin and advocates such as Zubrin might have hoped.

The Clinton administration wanted NASA to finish assembling the much-troubled space station before considering a human Mars decision.[37]

The Mars rock, still being debated scientifically, had already made a political and policy difference. The driver in Mars exploration was now established at the NASA Administrator and presidential level: the search for evidence of past or present extraterrestrial life. Gore decided that before the end-of-year policy summit took place he would bring together scientists, philosophers, and theologians to discuss the broader implications of the Mars meteorite. Gibbons, in turn, asked NASA and the NAS National Research Council to convene an interdisciplinary group of scientists to better delineate the Origins theme and Mars strategy. On October 28–30, NASA and the NAS-NRC convened three dozen leading scientists to consider the concept of Origins as a unifying strategy for shaping NASA's scientific future, including that of Mars exploration.

NASA intended Origins as a "big tent" under which many space scientists could gather and which could unite them and also attract public support. It was about studying origins of the universe and the beginnings of life wherever it could be found. Mars was central, but since Viking, space exploration had broadened and extended to the outer planets and solar systems of stars beyond. There was speculation about life under ice at Europa, a Jupiter moon, and at "other Earths" around distant stars.[38]

The aim of the meeting was to gain a scientific consensus on Origins as a theme for space science and—from NASA's perspective—"to convince the Clinton Administration that further cuts to NASA's science budget will endanger efforts to understand how life emerged."[39] Mars was integral to Origins. Ultimately, the question of priorities and money would come up, but the meetings, as well as the upcoming Gore conference, were primarily about "what" and not "how much."

Meanwhile, on November 7, NASA launched MGS. The cost was approximately $250 million, a fact that gave Goldin the opportunity to say that "Global Surveyor will give us 80% of Observer's science at one-quarter of the cost."[40] As MGS lifted off, Huntress beamed: "These are the kinds of days you . . . live for in space science and exploration." He commented that "it seems to have come together for Mars this year." But "to put a nail on that"—the question of life—he said, "NASA had to return samples."[41] "We will go to Mars," Goldin stated unequivocally at a George Washington University Space Policy Conference on the Mars rock.[42]

NASA Reprieved

OMB, leading presidential budget-balancing policy, had had NASA going down drastically in ensuing years, and that included draconian cuts for space science. Both Gore and House Speaker Newt Gingrich went on record after the Mars rock discovery, saying NASA would get more money for Mars exploration. There was political agreement on that. But from where would the money come? House Republicans argued for taking the money from Gore's favorite program—space-based global environmental observation—which they believed provided ammunition for Gore in the global warming debate.[43]

Goldin did not wish to rob the budget of other important missions to fund Mars. He pressed OMB to keep the overall NASA budget stable, at least, and not have it decrease in following years. On November 19, he wrote T. J. Glauthier, OMB associate director, pleading for budget stability in the out-years. He followed this up the same day with a letter to Glauthier's superior, the OMB director Franklin Raines, declaring that NASA was "at a crossroads . . . level funding is critical."[44] Huntress complained publicly that NASA's "productivity has been going up, but our budget has been going down." Gibbons acknowledged that NASA's financial situation was a "dilemma [that] is coming to a head."[45]

The Russian connection had helped "stabilize" NASA's space station budget in 1993, and the Mars Together initiative might have been an additional help with regard to the robotic Mars program if it proved viable. But Russia had monstrous financial problems and continued its string of robotic failures. On November 16, Russia launched "Mars '96," but it went awry soon after launch and crashed into the Pacific.

Russia's more recent failure reminded NASA and its Mars constituency how difficult implementing big plans for the Red Planet could be. "You always have to stay humble in this business," said Tony Spear, project manager for the Pathfinder lander/rover mission.[46] On December 4, Pathfinder soared into space successfully, on a trajectory that would take it to Mars ahead of MGS. Huntress exclaimed, "Pathfinder will establish the technological basis for missions of the future. Each mission will learn from its predecessors to pry loose the secrets of Mars."[47]

In early December, Sagan paid what would be his last of a number of visits to Goldin. His hair gone and appearance gaunt, he showed the ravages of a bone marrow disease that would soon take his life. The two men spoke for hours, during which Sagan "laid out a series of visions about the future of space exploration." "He was talking with intensity," Goldin recalled. "A man on his deathbed. This is the Carl Sagan I love, a man so full of hope and optimism that he never gave up."[48]

On December 12, Gore held his much-anticipated meeting to discuss the implications of the Mars meteorite. This was presumably preparatory to the policy summit Mikulski had demanded and Clinton had promised. The Gore meeting was about ideas and philosophy, not programs and budgets. Scientists, philosophers, theologians, NASA, and administration officials attended. Not present, and sorely missed, was Sagan. The scientist most identified with the quest for life was himself now so gravely ill that he could not come. He would die on December 20.[49]

The Gore meeting extended almost three hours, more like an academic seminar than a government hearing. Most present were impressed with Gore's knowledge of the subject. Although he "brushed off" one participant's suggestion of an Apollo-style approach, he clearly was eager to do more on Mars and Origins.

He said it was important to seize the moment, because the combination of Mars and extraterrestrial life fired the human imagination. He said he personally believed in the "ubiquity of life." At the end of the meeting, OMB's Glauthier—the same OMB official to whom Goldin had recently pled for funding—declared he would try to find ways to insulate space science from cuts. He would not have said that if it was not clear that Gore and Clinton were supportive of such a statement. Finally, after the meeting, Gore declared his desire for a "robust space science program."[50]

An unidentified "administration official" told *Science* magazine that prior to the Mars meteorite, space science had few advocates in the White House. "Clearly," this official said, "space scientists have more leverage now than they have ever had."[51]

In January 1997, Clinton began his second term as president. He retained Goldin, now regarded as a valuable member of his administration, the poster boy for federal reinvention. Clinton gave his State of the Union Address on February 4 and—in a rare presidential nod—specifically mentioned his support for space and the Mars program. He declared, "We must continue to explore the heavens—pressing on with the Mars probes and the International Space Station."[52] Then, on February 6, Clinton rolled out his budget proposal for the upcoming fiscal year.

As expected, the budget was a reprieve for NASA. Overall, the budget reflected Clinton's desire to make progress to achieve a balanced budget during his time in office. But it also showed a conscious effort to support research and development as key to the nation's future. NASA was helped by this government-wide decision. That did not mean a budget increase. It meant averting a huge cut. Within NASA, the budget also showed a particular desire to favor space science.[53]

The mark for NASA was $13.7 billion. This was $280 million down from its current budget. But this was far better than what NASA was going to get if OMB's earlier recommendation had had its way. At the White House's direction, the new Origins theme became the focus of NASA's budget submission, and OMB worked with NASA to add $1 billion over the agency's five-year projection for this initiative.[54] Goldin knew he had dodged a bullet, and he expressed delight to the media. "Holy Mackerel," he said, "this is a great program."[55] Space science got a 4% raise, bringing it to $2 billion. Ed Weiler, put in charge of the new Origins activity and destined to succeed as associate administrator for science when Huntress retired from NASA in 1998, declared, "I'm exceptionally happy. All the boats are going to rise."[56] This would especially be true of Mars exploration.

Senator Mikulski was particularly elated. When she saw the budget, she wrote Vice President Gore, declaring, "A space summit is no longer necessary." She had been the prime mover for such a summit, but she had achieved as much as she could have expected without one, thanks in large part to the Mars rock and the vice president's support. "How do you spell relief?" she asked. The answer she spelled out: "G-O-R-E."[57]

Exhilarated, Goldin pushed officials in NASA space science and human spaceflight directorates to work closely together, and—while there was friction over who would pay for what—there was serious effort expended by both sides.[58] Goldin believed that if all went well with the robotic and the space station programs, he or a successor could propose to the president the "next logical step," which Goldin regarded as human spaceflight to Mars. Goldin had come to Washington to set NASA on a trajectory to Mars. Like many advocates, especially Friedman, who had easy access to the NASA Administrator, Goldin believed that this was the place where the human and robotic programs converged. He shared Zubrin's view that humans to Mars required extracting resources while on Mars. But the "living off the land" philosophy required better knowledge of what resources existed that could be converted to human use.

Accordingly, NASA could equip robotic Mars probes with instruments to scout such resources, as well as detect radiation and other hazards to astronauts in the Mars environment. For Goldin, robotic Mars missions combined scientific and precursory rationales, and he made that clear to the agency.

Moving Ahead: Astrobiology, Pathfinder, and More

Although human exploration was the long-term goal, the search for life in the universe was the immediate driver for Mars activities. Goldin went on CNN after the December 1996 Gore workshop on the Mars meteorite and said he would provide money to nurture astrobiology.[59] Huntress had coined the term, a change from Viking-era "exobiology."[60] He wanted to convey a broader search for life in line with recent discoveries of extrasolar planets and possible water under Jupiter's ice-laden Europa moon. The new term also aimed to herald a new beginning in NASA's search for life.

Goldin designated Ames Research Center in California as home to a new Astrobiology Institute. Ames had been in jeopardy. It was an old aeronautics center whose role had diminished over the years. Senator Barbara Boxer (D-CA) had urged Goldin to secure Ames, and Goldin told her not to worry. NASA needed a lead center for astrobiology, and Ames was the logical place. It had always had an interest in life sciences and had played an important role in that respect in the Viking era through the pioneering work of biologist Chuck Klein. It had helped keep life-on-Mars research alive in the hiatus years after Viking. Goldin also asked Soffen, who had been lead scientist in Viking, but who had worked at NASA in the Earth observation field subsequently, to return to the life-search quest. He asked him to assist in planning for how NASA should rebuild the astrobiology field. The search for life on Mars *and* in the universe was now Goldin's vision and rhetoric for NASA.

What Soffen and others told him, and Goldin well knew, was that NASA and the planetary science community had few life scientists in their ranks. In July, Goldin spoke at the American Astronomical Society meeting and asked his large audience, "How many life scientists are in this room?" Practically no one raised a hand. If we are going to search for life, said Goldin, we are going to need life scientists. He announced that NASA was creating an Astrobiology Institute that would bring traditional planetary scientists and life scientists together.[61] Even though NASA's budget was constricted, Goldin proclaimed he would add astrobiology to his list of priorities. Although based at Ames, the institute would enlist an astrobiology community elsewhere, especially at universities. The intent was to rebuild a field of science which had become almost moribund after Viking.

Mars momentum was growing rapidly, the meteorite had been a catalyst, and then came the spectacular impact on the public of the Pathfinder mission. Launched in 1996, Pathfinder landed on Mars on July 4, 1997. For the first time in two decades, an object from Earth had made it successfully to the Red Planet. Pathfinder's task was not to search for life, but to demonstrate that a faster, better, cheaper mission could work at Mars. Its role was to establish credibility for the 10-year Mars Surveyor Program. Moreover, it carried a small rover, named Sojourner, and its goal was to show that such a vehicle could maneuver at Mars.

Everything about the Pathfinder/Sojourner mission was fascinating, including the way the landing was accomplished. Surrounded and protected by a cocoon of airbags, Pathfinder hit the ground and then bounced as high as a five-story building. Then it bounced again, 20 times, before coming to rest a mile from the initial landing point, on an ancient floodplain amidst rocks and boulders.[62] When Huntress, who was at JPL witnessing the landing, heard someone announce, "Full stop," he "jumped up and screamed." He ran to the mission team. Many were in tears, and one turned to him and said, "Thanks for giving us the responsibility to do this." Such a heartfelt statement of appreciation "broke me up," Huntress remembered.[63]

All the scientists and NASA officials at JPL rejoiced and then celebrated again the next day when Sojourner, a six-wheeled rover, rolled from its carrier and inched along the surface. It eventually met with rocks that got names such as Barnacle Bill, Yogi, Scooby Doo, and Boo-Boo.[64]

Pathfinder was an unalloyed triumph. Headlines everywhere proclaimed the success, as did appreciative editorials in leading newspapers. Both Clinton and Gore issued congratulatory statements to NASA, and Gore called JPL to praise all those associated with the mission. After so many years and a sequence of failures (Russian and U.S.), it was marvelous to have what was universally seen as a great success.[65] Striking pictures of Mars were shown on television, and Clinton admitted he couldn't get enough of watching them. Gore declared that the "validity" of faster, better, cheaper was being borne out by Pathfinder.[66]

Huntress was ecstatic: "This mission," he said, "has demonstrated quite clearly that we can in fact build and launch planetary missions for a low cost." And low costs "will allow us to continuously launch these missions and provide the American public with the excitement, the drama, and the knowledge that comes from our solar system exploration program."[67]

It was obvious that Pathfinder and Sojourner had hit a nerve with the public. NASA released images quickly not only to the media, but to the Internet. This decision to use the Internet brought about the largest virtual participation in exploration by people since the world watched the Apollo Moon landing in 1969. Indeed, no event up to this time had as many "hits" on the Internet—80 million a day in the first days, 450 million by the beginning of August. Various observers commented excitedly on the phenomenon: "It wasn't just the media that's picked up on this story," said Alex Roland, a Duke University history professor and former NASA historian. "People of their own volition are turning to it in incredible numbers."[68] What was especially impressive, said another NASA watcher, Jerry Grey of the American Institute of Aeronautics and Astronautics, was that this achievement came on "a shoestring" budget. Louis Friedman, executive director of the Planetary Society, said that the mission had "reawakened the image of NASA as 'the can do' agency." John Logsdon, space policy professor at George Washington University, called Pathfinder a "robotic folk hero" with the public.[69]

NASA made the most of the public's interest, emphasizing cost-benefit comparisons, pointing out that Pathfinder had cost taxpayers $250 million, whereas Viking would have cost, in 1997 money, $3.6 billion.[70] Viking employed thousands, whereas Pathfinder only a few hundred. Goldin personally gained enormous credit, and he said Pathfinder was just the beginning of NASA's assault on the Red Planet.

Pathfinder and Sojourner were destined to gradually cease operating in September, but while they were highest on the public consciousness, Goldin paid tribute to the late Carl Sagan. He held a special ceremony honoring the famous astronomer, writer, Mars advocate, and advisor to Goldin. With Sagan's widow, Ann Druyan, present, Goldin named Pathfinder a memorial station for Sagan.[71] Sagan thus joined former Viking scientist and NASA official Tim Mutch as having a memorial station on Mars.

On September 11, 1997, MGS, also launched in 1996, moved into Mars orbit. Its goal was to map Mars in unprecedented detail, almost as much as Mars Observer was to do. NASA again pointed to the difference in spending. NASA priced Mars Observer at $1 billion. This mission cost $250 million.

To get into proper lower orbit, MGS used aerobraking, a method by which it employed the friction of Mars's atmosphere to slow the descent. However, when it sought to do so, the air resistance caused one of the solar panels needed to power the craft to bend too far backward. NASA had to reposition the spacecraft to a higher orbit and replan the mission. The solar panel in question had apparently been damaged earlier in the flight; hence, there was serious concern of added harm. In November, NASA concluded that it could save the mission by very gradually lowering the orbit. This approach would minimize the atmospheric resistance, but it would take an extra year before MGS would be in its optimal orbit.[72] NASA decided to take the time; the process of a slow aerobraking began. The prognosis was positive.

The NASA budget Clinton proposed in February 1998 was $13.5 billion, a modest decline from the previous year. However, space science fared extremely well, getting another 4% increase.[73] With budget balancing continuing to be top priority for the president and Congress, this raise was impressive. Huntress used the good news for science as an occasion to announce he had decided to leave NASA after heading space science since 1993. "We seemed to be on a roll," he later commented. He felt it was the right time to retire.[74] He was also exhausted. Joseph Boyce, one-time NASA chief scientist, marveled that he had lasted this long. Huntress "had the highest threshold of pain I've seen," said Boyce. He saw Goldin "embarrass him in public. Rip him apart. But he knew how to get things out of Goldin. He kept his eye on the ball."[75]

Huntress gave way to Weiler, who had run the Origins initiative. Age 49 at the time, Weiler was an experienced science manager who had honed his internal and external political skills earlier as science leader of the Hubble Space Telescope. Pugnacious in style, he got along with Goldin. Because he was a single parent of a child with health problems, he had to leave his office at 4 p.m. Goldin gave Weiler the OK for this need but kept in touch with Weiler via a pager. He called Weiler at any time, day or night, seven days a week.[76] Weiler took over at a time of consensus in the White House and Congress that space science, especially Mars, should be protected from budgetary vagaries. The Mars rock was obviously the chief reason for this view.

In the time since the Mars rock announcement, scientific skepticism about the claims had grown, however. A University of Arizona–Scripps Institution of Oceanography study contended that 80% of the organic materials in the rock came from terrestrial contamination. JSC's McKay found the new report "interesting," but said the team stood by its original contention. Richard Zare of Stanford, the most prominent scientist on the team, said the research "cast doubt" but was not "a refutation" of the life hypothesis.[77] He did not believe minds had changed one way or the other since the claim was first announced. What was different, he said, was that prior to the Mars rock, "if you talked about searching for life on another planet, you were considered a nut. It has now become a huge topic that is attracting the best scientists." Weiler said there would be no settling the Mars debate "until we go there and get some samples."[78]

Whatever the scientific debate, the rock, combined with Pathfinder's public impact, gave NASA's Mars exploration program much greater momentum. Goldin saw search for life as the kind of exciting vision that could unify activities in the agency and build support outside.

Origins was a compelling theme for all NASA missions beyond Earth. Astrobiology was now an ongoing activity at NASA, with Ames the lead center. Scott Hubbard, the senior space scientist at Ames who had conceptualized Pathfinder in its formative stage, was working to relate astrobiology to flight missions.[79] In 1998, NASA formally established its Astrobiology Institute. This was seen as a "virtual" organization, with many institutions involved in government and the university world. Soffen assisted Goldin and worked with others to get the institute started. Soffen was at an age when he could have retired, but he wanted to help fulfill his own much-delayed dream.

In May, NASA announced the selection of 11 academic and research institutions as the first members of the Astrobiology Institute, calling it "a major component of NASA's Origins Program."[80] Goldin asked Hubbard to take over for Soffen, now that the institute was under way. Like Soffen, Hubbard was "interim." Goldin said he intended to recruit a "King Kong" biologist to head the new institute.[81] The next year Goldin hired the 73-year-old Baruch Blumberg, a biochemist who had won a Nobel Prize, to be its official director.

Also in May 1998, Goldin gave a commencement address at the University of Arizona. He urged the graduates to have a dream and follow it. "Mine," he said, "is an astronaut on Mars—in a nice, white spacesuit set against a red background, with a NASA logo on one shoulder and an American flag on the other." In August, he spoke at a memorial for Alan Shepard, the recently deceased first American to fly into space. "Alan," he promised, "America will go to Mars."[82]

Zubrin helped fuel this momentum from outside. The Mars Underground was gone, with Goldin acquiring one of the last remaining red identity buttons from its early days as a quasi-secret society. He "begged me for a button," Carol Stoker recalled.[83] In the Underground's place was Zubrin's newly organized Mars Society, which held its first meeting in Boulder in August. At least 750 people from 40 countries paid $180 to attend the four-day conference to make a case for sending humans to Mars. While emphasizing human exploration, Zubrin wanted the robotic program to scout the way. He urged that its budget be doubled.[84]

The sense of progress was surely felt at JPL. Charles Elachi, director of JPL's Space and Earth Science Program, headed a study for how to return samples of soil and rock from Mars, and Goldin approved plans he worked out. Norm Haynes, now Mars program director at JPL, spoke of returning four samples from four separate locations on Mars by 2011.[85] His boss, Ed Stone, JPL director, was caught up in the sense of optimism that permeated the agency, and Stone pressed Haynes hard for action.[86] Maybe it would be possible to go even sooner than 2005, some Mars advocates said.

The "yes, we can" mood was embodied in Elachi. Elachi called for sending two MSR landers, perhaps one as early as 2003 and another in 2005. An orbiter would collect samples in 2007 and return them to Earth in 2008. Asked to advise NASA, a panel of the SSB, while applauding the goal of MSR and endorsing the Elachi plan, nevertheless expressed some concerns that it was "aggressive" and entailed "risk." It stated "low confidence" that NASA had the money for such a multistage mission. It called for a more "comprehensive" approach to understand the context of Mars as an abode of life, past or present.[87] NASA's scientific advisors did not wish to deter the agency from speeding toward a goal the Mars community had long sought, but they clearly were worried that NASA might be going too hard, too fast, too narrowly with insufficient resources.

Goldin pushed, and there were some doubters, but most connected with Mars in NASA and at JPL shared Goldin's enthusiasm and longing. Doubters within NASA tended to keep quiet. No one wanted to be associated with what Huntress had called the "old guard." NASA was launching two missions to Mars every two years under its Mars Surveyor Program. These were faster, better, cheaper missions. They were now geared to the accelerated goal of MSR. So far they were successful. In December, NASA launched Mars Climate Orbiter (MCO) and followed it up a month later with Mars Polar Lander (MPL). These were half the size of their predecessors (Pathfinder and MGS). The polar lander mission was in part a fulfillment of Lederberg's desire to "go north" for landing in the Viking era. This was where Lederberg had thought life was most likely to be found. Unfortunately, Lederberg, who had helped pioneer the search for life on Mars, had died in February 1998.

NASA added two penetrators to MPL which would bore as deep as three feet below the planet's surface. MGS, meanwhile, was gradually wending its way into an optimal orbit, and already sending back striking images. Ironically, one of its first findings was to prove that a "face" on Mars some enthusiasts still believed to have been carved by intelligent beings, and which Viking had detected, was a mesa.[88]

America was going to Mars. And so were the Japanese. Japan, in July, had successfully launched its first Mars probe, Planet B. Like the U.S. spacecraft, it was scheduled to arrive in 1999. The excitement and ambition among Mars advocates were palpable. As people got to see Mars, even vicariously, they would start to comprehend that there was a fascinating world out there, Zubrin said. It was time for "political action," he proclaimed.[89]

In late 1998, NASA sent the elderly ex-astronaut, Senator John Glenn, back into space on a shuttle. It was a media extravaganza, as well as an occasion for national celebration and nostalgia for past glory. Walter Cronkite, who had covered the Apollo landing for television news, came out of retirement to interview Clinton at Cape Canaveral at the time of the launch. Clinton said that he was open to more financial support to NASA for the International Space Station. However, human spaceflight to Mars would have to wait. "Let's get the Space Station up and going and [then] evaluate what our long-term prospects are," he told Cronkite.[90] Where Mars was concerned, the robotic program held center stage, it seemed to be performing exceptionally well, and there was political support up to the president.

CHAPTER NINE

Overreaching, Rethinking

As 1999 began, NASA surged forward. NASA now planned an advanced rover, the kind analogous to one first discussed after Viking, which would traverse great distances and aid in identifying and collecting soil and rock samples. The sample would be reclaimed later and returned for analysis in Earth laboratories. The multistage mission was complicated and demanding. NASA knew it, but the collective attitude was exceedingly positive. The Jet Propulsion Laboratory appointed Bill O'Neill as Mars Sample Return project manager. An experienced leader, he enthusiastically began planning for the actions ahead. O'Neil called MSR "the most exciting, complex robotic space mission ever," a mission that was "historic."[1]

Clinton's policy toward NASA continued to be mixed. His budget, announced at the beginning of February, cut NASA by 1% from the preceding year. However, space science got another boost of $3.6%.[2] Mars exploration was obviously NASA's lead planetary program. For Goldin, it was much more. Mars was the destination about which he had been thinking since he was a boy.[3]

Goldin set up a "Decadal Planning Team" and enlisted the space science and human spaceflight directors in its support. He also established an activity called HEDS—for Human Exploration and Development of Space. This enterprise aimed at getting various parts of NASA to think about robotic and human Mars exploration. They would address, for example, the kinds of sensors that robotic spacecraft to Mars might carry to help future astronauts.[4] He instructed the Decadal Planning Team to think beyond the space station. "I want to get people to Mars for the right reasons," he said.[5] He truly believed that human spaceflight to Mars would be possible in the not-too-distant future, and it was time to plan for that eventuality. Toward that end, he expected the robotic and human programs to join forces. But Goldin's vision was cut short by unexpected and painful reality. NASA suffered a major setback in its Mars program and had to step back, rethink, and formulate a different strategy—in fact, a new program.

Designing Mars Sample Return

Wesley Huntress was now with the Carnegie Institution of Washington, a research organization. He remained a strong Mars advocate. *Space Times*, an aerospace trade journal, published in its May/June issue an article in which Huntress highlighted the importance of MSR. For him, as well as many space scientists, the search for life was indeed a prime motive for the space program. The Mars rock was provocative, but not persuasive enough evidence. Only a properly designed MSR would provide the evidence NASA needed. NASA believed that MSR would provide a "smoking gun" that would boost public support for space exploration.[6]

The Huntress article came at approximately the same time that Charles Elachi and Louis Friedman discussed their views about MSR in a magazine published by the Planetary Society. If Huntress explained "why" MSR was critical to NASA, Elachi and Friedman commented on the "how" question. They elaborated on the technical strategy Elachi had developed at JPL by which MSR could be accomplished.[7] There were various ideas about how to carry out MSR. There were obviously prodigious technical challenges, even if implemented in stages. As Kathy Sawyer wrote, for MSR to succeed, NASA would have to produce a "robotic package that was:

a. Lightweight enough to be practical;
b. Smart enough to do the job (make a sophisticated selection of desirable rocks and soils, for example);
c. Able to land safely on the rugged Martian landscapes designed as most promising for biological clues;
d. Able to take off again; and
e. Able to deposit the treasure safely and cleanly back on Earth."[8]

Then there was the issue of Earth contamination. The problem of contaminating the life prospect on Mars was always there, but the MSR mission raised the potentially emotional issue of contaminating Earth with Martian organisms. The science fiction book and movie *Andromeda Strain* depicted death on Earth from extraterrestrial microbes and would no doubt be used by opponents of MSR. Articles appeared stressing the dangers. NASA had its top planetary protection official, John Rummel, involved in MSR planning. It also asked its scientific advisory bodies to study the problems of contamination intensively.[9]

Finally, there was the question of costs. Goldin moved NASA ahead on MSR, with costs being calculated and recalculated as the agency learned more of the technical challenges. While faster, better, cheaper principles would be used as much as possible, the expense could be considerable, with $2 billion being one figure mentioned. However, NASA also used a $750 million and even $500 million cost estimate.[10] Exactly what would be the expense was left unclear as NASA charged ahead. The mood was akin to that preceding Viking, in that the "leapers" rather than the "gradualists" were in control. Goldin was in the Sagan camp, rather than that of Murray.

Goldin looked for ways to pay for MSR. He saw it in NASA-wide terms, an agency priority. MSR could be useful to human space planning, as well as the Science Directorate's search for microbiological life. Human spaceflight managers wanted to know about Martian soil in terms of possible hazards and resources available for astronauts.[11] Costs could therefore be shared within NASA. Also, Goldin sought international help. The long-standing herald of "Mars Together" was renewed and broadened. Italy, the European Space Agency, Russia, Japan, and France all expressed interest. In June, Goldin concluded an agreement with the French space agency under which the two nations would work together on MSR.[12] Other nations might follow. There was consensus about the goal which spread across spacefaring nations.

Goldin believed that NASA had to adapt organizationally and in personnel to the new vision of search for life. He went out of his way to hire life scientists. In addition to Blumberg, his Nobel Prize winner for the Astrobiology Institute, he hired another life scientist, Kathie Olson, to be NASA chief scientist.[13] And he spoke out frequently on the subject as opportunity arose: life—the search for microbiological life, as well as the eventual extension of human life beyond Earth. Goldin worked indefatigably to evangelize for space. Mars provided the chief focus of this effort. Robotic flights would come first, then humans, maybe in 20 years, he predicted. The media never quite knew when to take Goldin seriously. Goldin was an able salesman, and he was persuasive because he believed his own rhetoric. The media found him both perplexing and captivating. So did many in the Clinton administration and Congress.

In June, he spoke to a meeting of astrophysicists and accused them of representing the past, while biologists presaged the future in space science.[14] He spoke to a group of physicists at Fermilab in Illinois, a Department of Energy facility. The physicists were still smarting over the cancellation by Congress of their flagship project, the superconducting super collider, in 1993. How could that have happened? The reason, Goldin admonished them, was that they had failed to connect their machine to a vision the public could grasp. NASA had done that with its search for life, and that was why NASA was rapidly moving forward with its Mars program.[15]

Mars Climate Orbiter Fails

The Mars surge built on success and assumed more technical success. On September 23, the Mars Climate Orbiter was supposed to slip into a proper position to do its work. It did not do so. Instead, it flew off course and either burned up in the Martian atmosphere or missed Mars entirely and wound up circling the Sun in a useless orbit. What went wrong?

NASA convened a board of inquiry under the chairmanship of Arthur Stephenson, director of the Marshall Space Flight Center in Huntsville, Alabama. What the panel found seemed too embarrassing to believe. A young and inexperienced engineer at NASA's contractor, Lockheed Martin, had failed to convert navigation data from English to metric units. When the data went to JPL, the NASA center assumed that the conversion had taken place. No one checked the facts over the nine-month period during which the $125 million MCO sailed from Earth to Mars. It was an inadvertent mistake with dire consequences. The fact that no one checked the information was viewed by many in NASA as the most serious finding. The Stephenson panel pointed out that inattention, miscommunication, and overconfidence played roles in the mishap.[16]

What some critics, especially outside NASA, wondered was whether FBC management bore some responsibility. NASA rejected that view. Weiler pointed out that FBC "includes following rules and processes. Those rules and processes were ignored."[17] Hinners, who once had run space science and was now the responsible senior executive for the robotic probes at Lockheed Martin, thought there was an effect of FBC in this instance. "It's a matter of sufficient staffing—perhaps 10% or more—to make sure checks and balances work."[18]

John Pike, director of space policy for the Federation of American Scientists, put it this way: "They're basically trying to take 15-gallon trips on 10-gallons of gas." He urged NASA to take a hard look at FBC, a look the Stephenson panel had failed, in his opinion, to take.[19]

Everyone at NASA who spoke about the incident, especially Ed Stone, director of JPL, said that the mistake would not be repeated. Commenting on the next mission, the Mars Polar Lander, Stone said, "No one wants another mistake to go undetected. We are doubling and redoubling our efforts."[20]

A Second Failure: Mars Polar Lander

The next mission came quickly, in line with FBC principles and the requirements of an accelerated program. Mars Polar Lander was scheduled to land December 3. Although relatively small, MPL carried quite an assortment of equipment: "a pair of basketball-sized probes designed to shoot separately into the planet's surface, cameras, a weather station, a robotic digging arm, a minilab for analyzing soil samples, and the first microphone sent to another world." Also riding on MPL were NASA's reputation, schedule, and hopes for an early MSR.[21]

Goldin, Weiler, and other top officials joined senior JPL managers and the MPL technical team December 3 at JPL. They crowded around computers in a central control room and awaited the signal telling them MPL had landed safely. The atmosphere was tense. Because of the earlier setback, NASA needed a victory to assure itself, the media, the public, and politicians that the MCO accident was an aberration.

At 3:30 p.m. the signal was supposed to come, but there was silence. The NASA officials and technical team waited, "frozen as statues," as the minutes ticked on. Twenty minutes passed, and finally one NASA manager suggested a "leg stretch." There would be another window for communication a little later. There was hope yet.

The group convened again for an opportunity beginning at 5:24 p.m. Again, there was silence. The expressions on the faces of those straining to hear something from Mars were grim. After several more minutes, the window closed. There was now little doubt that something had gone terribly wrong. The third opportunity began at 11:08 p.m. that evening and closed at 12:30 a.m. The result was the same as the first two tries.[22]

The vigil continued episodically over the next several days. Senior NASA officials went back to Washington. After two weeks of repeated attempts to communicate, the agency and JPL admitted that MCO had failed.[23] The mission cost $165 million, not much by space standards, but the cost represented the new standard of FBC. There would have to be another board of inquiry. And this one would have to take a very hard look at the FBC approach to space missions.

Goldin had always said that FBC assumed that a small number of missions would fail. With risk and boldness came failure. But most missions would succeed, he predicted, and NASA would thereby push forward the space frontier. But here was a case of two failures in a row, and both had to do with Mars. Mars was special, of maximum visibility to the public. For many in the media and political community, it personified space exploration. For Goldin and many other space advocates, it was the heart of NASA's mission.

The year 1999 had begun so optimistically with a surge in the Mars program. It ended with the program immobilized.

The Investigation

The twin failures in 1999 resulted in much soul-searching at NASA, Goldin included. Immediately after the lander failure, Hinners called Goldin on behalf of Lockheed Martin. He "apologized," saying, "Sorry. We screwed up!" He fully expected to be "chewed out." But Goldin instead said, "Look, I don't want finger pointing. This is a management failure, not a technical failure." It was obvious to Hinners that Goldin was already seeing himself and FBC as part of the problem. But Hinners felt he had himself been blinded by a false sense of optimism.[24]

Goldin quickly launched a number of investigations. The most comprehensive inquiry was by A. Thomas Young, a retired aerospace business executive and former NASA official whose Mars experience went back to Viking. Known as the Young panel, the official name was Mars Program Independent Assessment Team (MPIAT). It was a blue-ribbon team with 16 members. After meeting with Goldin on the panel's mandate, Young on January 14, 2000, indicated that the word "independent" was serious: "Everything is on the table. There are no limits to what we can do."[25] Young said the panel's inquiry would take two months. The panel then got to work.

Meanwhile, Goldin worked inside NASA with Weiler. The associate administrator remembered one particular moment after the second disaster which helped define the recovery strategy. Weiler was at a basketball game with his son. He wore a pager. He received a message that Goldin was trying to reach him. "I went outside the building to return the call. It was a cold January evening, 9 p.m." When he reached Goldin, the administrator barked: "You've got one minute to tell me how to fix this." Weiler wasted little time in responding.

"I think we may have been too aggressive," he said. "We need a step-wise process. In addition to large missions, we may need a small mission or two. We could also have competitive missions proposed by the scientific community." Goldin listened and commented, "Good. Let's get together and think this through." Subsequently, Weiler noted, "Dan and I were arm-in-arm on Mars decision-making."[26]

As Weiler and Goldin assessed the damage and what might be done, various individuals were speaking out. In late January, Donna Shirley, a former JPL manager of NASA's Mars Exploration Program, now an assistant dean of engineering at the University of Oklahoma, laid the blame on Goldin and his push for MSR at the earliest possible moment. After the Mars meteorite excitement, she said, Goldin "urged a switch from gradual understanding of Mars to a rush to look for life. And the key to that was the sample return mission. Dan has always wanted a sample return mission because he believes it would attract public interest."[27]

David Page, a UCLA planetary scientist, also blamed Goldin, as well as his top NASA managers. He said the Mars program needed to get away from the process by which "leaders would propose lofty space exploration goals without consideration for the technical difficulties, hard deadlines, and funding. What was needed," he said, "was a grassroots process in which the goals, the schedule, the funding, and the risks are defined by the scientists, engineers, and managers who will be carrying out these projects. . . . In this model, ambitious goals such as a Mars Sample Return would eventually be accomplished, but only after the required technologies are in place and not at the expense of the much broader goal of studying Mars in all its diversity."[28]

The accusations continued in February, with some critics, including scientists, going beyond Goldin to Clinton and Gore for not giving NASA enough money. Others blamed JPL for poor management. Others targeted Lockheed Martin, the contractor, for underbidding and problems on the factory floor. Weiler retorted that "everyone was to blame," including the scientific community, for overconfidence. In 1998, the National Academy of Sciences National Research Council had called NASA's Mars exploration effort a "well-thought-out and rational approach to achieving NASA's programmatic goals."[29]

Taking Action

In mid-March, Young briefed Goldin privately on the panel's findings. Goldin subsequently told his top managers to prepare themselves for sharp criticism. The Young report, he warned, would be the Rogers Commission of space science, referring to the devastating critique delivered by a panel headed by William Rogers after the 1986 Space Shuttle Challenger disaster.[30]

One point Young made that hit home to Goldin was the need for precise lines of communication and authority. It was not clear who was in charge. The Mars program had expanded rapidly, and with many projects in different stages of development at once. There had to be a specific point of contact at the program level both at headquarters and at JPL. After conferring with Weiler, Goldin called Scott Hubbard, who was away from his Ames Research Laboratory base, attending a conference. Hubbard was with Ed Heffernan, Goldin's chief of staff. As Hubbard recalled, Heffernan's cell phone went off. "I was told it was the boss." Hubbard got on the line and Goldin stated, "I'll be in California, Manhattan Beach, on Saturday. I want to talk to you about something very important—can you be there?" Hubbard quickly responded, "Yes, of course." (It was not a good idea to say "No" to Goldin, Hubbard thought.) It was then Thursday.[31] Through Heffernan and "back channels," Hubbard soon ascertained what was going to transpire. That Saturday, Hubbard met with Goldin at a hotel in Manhattan Beach. The administrator explained, "I need someone who is not tainted, who knows a lot about the Mars program but is not part of it, and who knows about science and technology, to fix this mess. It will be a stain on my legacy. It is unacceptable."[32]

The two men talked for two hours, mostly Goldin telling Hubbard what he wanted him to do. Hubbard did manage to get the administrator to agree to give Hubbard a free hand in directing the recovery process. Hubbard said he had to talk to his wife and the Ames director first. "Fine," said Goldin. "Let me know Tuesday." The next Tuesday, Hubbard showed up in Goldin's office and asked, "When do I start?" "Right now," said Goldin. This was days from the rollout of the Young report. Goldin wanted Hubbard to be present when that happened so NASA could announce the appointment, showing that Goldin was already taking action in making repairs to a damaged program.[33]

President Bill Clinton was informed of the basic thrust of the report prior to its public release. On March 27, he wrote Goldin, in an obvious attempt to bolster the NASA Administrator's morale in view of the criticism he had already received and would receive when the report was announced. Clinton said he regarded robotic exploration of Mars and search for life "an important national priority." Further, he said, he "continued" to be committed "to a sustained and incrementally more aggressive program," as revealed in his proposed NASA budget for FY 2001. The use of the adjective "incrementally" may well have reflected Clinton's understanding of the report's basic criticism—that Goldin had pushed the Mars program too fast, too far, too cheaply.

Clinton seemed willing to admit that the White House had asked NASA to do too much for too little. His new budget, announced officially in early February, gave NASA its first increase in seven years. NASA's budget would go from $13.6 billion to $14.3 billion. Nearly half of the $700 million augmentation would go to science, including Mars research.[34] In his March 27 letter, Clinton directed Goldin to move forward and make changes necessary "to conduct a more robust and successful program of sustained robotic exploration of Mars."[35]

On March 28, the Young report was released. Weiler and Young announced the results to a packed auditorium at NASA Headquarters, and Weiler outlined the initial NASA response. Young said his team was reasonably certain that MPL failed because the engines were shut off prematurely as it came in for a landing and it crashed onto Mars at 50 miles per hour. The two microprobes that it carried never had an opportunity to demonstrate success, but had in any event not been tested sufficiently and were "not ready for launch." Weiler and Young both defended the FBC concept as basically sound and said it should not be abandoned. However, Weiler said that clearly NASA had "pushed the envelope too far." It was now time to reconsider the envelope, pull back, and regroup.

Beyond the technical causes and misuse of FBC, Young said his panel had found organizational issues. There was ineffective communication between NASA Headquarters and JPL and between JPL and Lockheed Martin. Moreover, the mission team was asked to do the impossible with what the Young panel felt was a budget 30% less than needed. The failed MPL was provided far fewer funds than Pathfinder. Weiler added that decentralization and headquarters downsizing had also been pushed too far. Also, part of the problem was that there were inadequate financial reserves, and that all the program funds had been sent to JPL and other centers. Weiler said he was now going to keep program reserve funds at headquarters. That would force center program managers to come to headquarters if they felt something was going wrong.

Young said his panel recommended it be clearer as to who was in charge both at headquarters and at JPL. Weiler announced that Scott Hubbard, present and visible for the occasion, was coming to headquarters from Ames Research Center to be a single point of contact for all NASA Mars exploration. He noted that JPL director Ed Stone would soon be naming Hubbard's counterpart at that center. Weiler said the ultimate goal of the program remained "the search for past and/or current life on Mars," but instead of MSR as the immediate driver, NASA would have the interim goal of "follow the water" as a focus. Water was pivotal to life as currently understood. He pointed out that Hubbard would report to him and lead in developing a new program strategy. He said the plan would cover a decade.[36] Almost immediately, the media referred to Hubbard as "the Mars Czar."

Young subsequently told interviewers that people were trying to do too much with too little and not adequately conveying their concerns to others, particularly upper management. He said, "No one had a sense of how much trouble they were actually in." He commented that "JPL managers did not come out as forcefully as appropriate" when discussing their problems with headquarters, and headquarters in turn may have misheard what JPL managers were trying to say.[37] Shirley told the media that JPL had tried to inform headquarters that problems were developing and that headquarters said "you have to do it." JPL then went back and tried, and in doing so took too many risks. She said that if you got to Goldin, you could get him to listen, but people at headquarters and centers feared approaching him. "When you laid out the facts for Dan, he would generally agree," she said. "But he would push you as hard as he could [and] when he pushed his people, they were afraid to say no."[38]

Hinners recalled a telephone exchange with Stone on one occasion which illustrated the issues as he saw them: "I said: 'We are out of money. We need $30 million more.' Stone replied: 'I committed to headquarters to do this project for the amount involved. I'm not going back to ask for more money.'" So Hinners accepted the refusal and the company did the best it could for the money it had. "We all fooled ourselves," he reflected.[39] John Casani, a senior manager at JPL, who led investigations of both Mars failures for JPL, put it this way: Goldin "said if you couldn't do it, he would find somebody who could. He was putting pressure on us, and we probably should have stood up to him, but we all got to believing we could do it."[40]

Goldin Shoulders Responsibility

On the day after the report came out, a chastened Goldin went to JPL. Scheduling an address to JPL employees and a press conference for the next day, he went to dinner that evening with Caltech president David Baltimore, Stone, and leaders of the Mars Mission Team. One of the younger men on that team pleaded, "Dan, don't let us go back to the old ways."[41] The next day he spoke at an assemblage of JPL employees. He took responsibility and stated, "The warning bells are sounding. The trend is very, very sobering, and one we can't ignore. It says we pushed too hard." Afterword, he spoke to the media. Again, he took responsibility. "In my effort to empower people," he declared, "I pushed too hard. And in doing so, stretched the system too thin. It wasn't malicious. I believed in the vision, but it may have made some failure inevitable." He told JPL and others he was not going to let the pendulum swing all the way back to the old days, before FBC, but that he was going to provide adequate resources to succeed.[42]

Stone addressed JPL subsequently. He said that the real motivator for FBC was to go more often to Mars, but that JPL had taken too many risks in trying to take maximum advantage of every two-year launch opportunity. He also told his personnel that downsizing was over at the lab.[43]

The congressional reaction to the Young report was critical of NASA but stopped short of calling FBC a failed strategy. Senator John McCain (R-AZ), chair of the Committee on Commerce, Science, and Transportation, commented that the report was an embarrassment to NASA and showed that the agency's leadership was "missing in action." He said Congress would have to exercise "more rigorous oversight" of the agency.[44] Congressman James Sensenbrenner, chair of the House Science Committee, expressed essentially the same sentiment, as did other lawmakers. The *Washington Post* issued a common media view, that NASA had to step back and review its actions. However, it advised the agency that it would be "dangerous" to abandon FBC and "swing back toward the old model of sky high budgeting." It noted that NASA and JPL simply had too many missions under way at once and too few experienced managers to handle all of them. It echoed congressional concerns that senior managers did not exercise proper oversight.[45]

Goldin, in congressional hearings on the Mars failures, faced "heat but no fire."[46] Again, Goldin took responsibility and admitted pushing too hard. Congress, while being critical of Goldin and NASA management, seemed anxious to also show support. The Mars program was popular, and lawmakers, like the president and media, wanted it to continue and succeed, while exercising fiscal constraint. Some Republican legislators blamed the White House for not providing NASA sufficient funding. Meanwhile, Louis Friedman, speaking for the Planetary Society, gave a view shared by many Mars enthusiasts. He cautioned that while NASA had to make program changes, "it is crucial that NASA not overreact and slow down the program too much."[47] Privately, Friedman also used the access he had to Goldin to argue for making the Mars director position a true czar authority, with power over both robotic and human flight. Friedman wanted a Mars focus for the agency, with the Mars program director under Goldin rather than Weiler. However, Goldin was not about to make that move.[48]

In the wake of the twin Mars failures and Young report, NASA began the arduous work of recovering its credibility and returning to robotic Mars flight. Hubbard was the man chosen to lead the activity, but he was part of a team that was assembled. The members were responsible for what NASA called the "architecture" of the revised program. Goldin wanted the group, which started in April, to complete its work by October, so that its recommendations could be incorporated into NASA's budget submission for the following year.

Designing a New Program

Age 52, Scott Hubbard was a longtime NASA official. With degrees in physics and astronomy, he had risen through the ranks of NASA-Ames and was associate director of that laboratory at the time he got his call from Goldin. A week after NASA announced Hubbard's appointment, Stone at JPL reluctantly named Firouz Naderi to be his counterpart as the single point of contact on Mars. Stone had wanted Naderi for a different assignment, but headquarters prevailed. It was made clear by Weiler's deputy, Earle Huckins, that "Dan Goldin wanted the very best talent applied to fixing the Mars program."[49] Like Hubbard, Naderi was a veteran scientist-manager but had not been directly connected with the two Mars failures. The two men—Hubbard and Naderi—did not know one another, but they "clicked." Hubbard, however, kept reminding Naderi that he was a "NASA-man," not a "JPL-man." Naderi went out of his way to come across in that way in his dealings with Hubbard—and JPL. There was much anxiety on Stone's part that JPL would be punished for the failures and lose its prized position as lead center for planetary missions.[50] He had reason to be worried, as there were severe critics of JPL in NASA Headquarters. One senior official urged Goldin to replace JPL top management.[51]

The third key member of the recovery team was James Garvin. Garvin was a scientist at NASA-Goddard who had been a graduate student at Brown under Tim Mutch, the Viking investigator and briefly NASA associate administrator for science. Garvin had worked on Mars Observer and had been leading the Decadal Planning Team for Goldin on NASA's long-term future. Goldin himself asked Garvin to join Hubbard and Naderi. Garvin was to play a major role in developing a strong connection between the team and the broader Mars technical community.[52] Hubbard wanted to reach beyond NASA and its various elite advisors to Mars investigators generally. Garvin was to help assemble a series of workshops with Mars scientists who thereby would provide input to the recovery team's decisions. These workshops would evolve into a novel mechanism, the Mars Exploration Program Analysis Group (MEPAG), which would become an ongoing connection between NASA and the broader Mars research community.

Finally, Hubbard asked two Viking veterans, Jim Martin and Gentry Lee, to serve as advisors. They would be the core of Hubbard's "kitchen cabinet." Weiler had already set some guidelines for the new program, but it would be up to Hubbard and his team to determine pace and specific missions. The quest for life—past or present—was still the ultimate goal of the robotic program. Getting to it, however, would be through a different strategy.

There were two Mars decisions that came up almost immediately in April for NASA which were important to recovery. One was forced on NASA by the schedule. The other was one Hubbard personally pushed to help free his team to develop a restructured program.[53]

Weiler took the lead on the first.[54] Under the now-suspended Mars Surveyor Program, NASA was to launch an orbiter and lander every two years. The next window was coming up very quickly—2001. JPL and contractor Lockheed Martin were building the spacecraft. What was to be done?

Weiler reasoned that the orbiter had failed in 1999 owing to the bizarre mistake in communication between contractor and JPL over metric/English navigation units. The orbiter had been technically sound as far as anyone knew. On the other hand, the lander, which had crashed, required significant modification.

After getting advice from Hubbard and others, Weiler gave a go-ahead to develop the orbiter and cancelled the lander. The termination decision did not please Lockheed Martin or JPL, but served to send a message Weiler wanted to transmit—that he was taking more authority over the program.[55] The days when decision making was largely delegated and headquarters stood back, downsizing as it did so, were over. Weiler was intent on building up a more robust Science Mission Directorate, and he did not intend to be a passive manager. The 2001 orbiter mission was called Odyssey, in honor of Arthur C. Clarke's book and screenplay for the film *2001: A Space Odyssey.*

The next decision was the indefinite postponement of the MSR mission. Weiler had indicated that it was coming, but Hubbard made sure it came right away. An early MSR had driven decision making on Mars since the program reorientation following the meteorite excitement. Hubbard did not want it to drive the recovery effort. When would MSR come? Hubbard believed that MSR was the right goal for the robotic program—but it had to come only when "ready," and that would probably be beyond the 10-year program he and his associates were designing. Naderi called the proposed accelerated MSR mission "science fiction."[56] Hubbard detected at least four significant technological barriers to MSR success. He went to JPL and confronted MSR's project manager, O'Neil. He pointed out the technical challenges. He asked, "What makes you so sure you can overcome these problems?" In addition, Hubbard challenged the MSR cost estimate. Then at $750 million, it was hopelessly low in his view.[57]

Hubbard subsequently went to a large workshop of Mars scientists, armed with a new sample return estimate. There may have been 65 to 80 people there, as he recalled. Most of them fervently embraced MSR as a goal as soon as possible. But Hubbard posed this question to the group: "The current estimate we have is that MSR will cost $2 billion plus. Where do I go to get a sample worth that?" No one in the audience had an answer.[58]

The next step for Hubbard was to go to Weiler, who agreed with his position to put MSR on indefinite hold. Finally, Hubbard and Weiler met with Goldin, who had been the champion of the accelerated MSR mission. Before we go for MSR, Hubbard told the administrator, "you've got to have scientific understanding." But, he emphasized, the Mars community, including the astrobiologists, did not have that understanding. Moreover, to get scientific answers would take the development of "four missing technologies." There simply wasn't the knowledge or time to make an early launch. MSR would have to be deferred for some years, he explained. Typically, Goldin responded to statements about difficulties with "You're not trying hard enough," But not this time. Goldin could not have liked what he heard, but he did not object. The message subsequently went to Stone: "You must go along!"[59] The MSR project was cancelled.

It fell to Hubbard to break the news about MSR to the French. Goldin had enlisted them in planning for an MSR mission. The French were not happy, nor were other potential international partners. Hubbard left the door open to possible later participation, but not in the near term.[60]

Goldin's Surprise Decision

The Hubbard team now had the full flexibility it sought to plan a new program, with no specific MSR goal with a hard or even soft deadline forcing an arbitrary pace. Weiler made it abundantly clear that the FBC approach was no longer gospel. Success was foremost, although costs were also important. He declared that the Mars Surveyor Program, as previously constructed, had ended.[61] Hubbard was to design a new program. As Hubbard saw it, he had a Mars funding line "without a workable strategic content."[62] It was up to him to give that content. In early conversations with Weiler and others, Hubbard decided that NASA would use the following expression to convey their overall new approach: "follow the water." Weiler had already used that expression publicly. Now it became "official rhetoric." Specific missions would connect through following the water and lead, hopefully, in the direction of life.[63]

The time for another big decision was looming. With MSR out as a short-term driver, what were Hubbard and his associates to do about 2003? That mission, Hubbard knew, had to fit into the new strategic approach. The basic strategy he and his team formulated was to alternate orbiters and landers over the decade—one mission for each launch opportunity. (Mars Surveyor Program had featured two missions every opportunity.) Hubbard spoke of them as a "ladder to Mars."[64] Since NASA was sending Odyssey, an orbiter, up in 2001, the strategy would seem to call for a lander in 2003. However, there were those at NASA who worried about advancing too far beyond what the agency had done, as the previous lander had been a failure.

Hubbard saw NASA's options for 2003 as (1) do not fly at all, (2) fly an orbiter, or (3) fly a lander. The lander mission was increasingly seen as carrying a rover, with the rover playing the dominant role in the mission. Such a mission would be more challenging than Pathfinder, with its tiny and short-lived rover, Sojourner. This rover would last longer and go farther. A meeting took place to discuss these options, attended by various Viking "graybeards," including Martin, Lee, and Mike Carr. There were also representatives from the "new generation," such as Steve Squyres. Sixteen people attended this meeting. Advocates of different approaches spoke.[65] At the end of the discussion, Hubbard called for a vote on the orbiter/lander options. The vote turned out to be split, virtually even. Hubbard announced the result. Someone shouted, "Well, then, you can do what you want."[66]

While Hubbard and his associates considered the 2003 mission in the context of a long-term strategic approach, rumors circulated that Mars Global Surveyor had made a tantalizing discovery. It may have spotted "evidence of liquid water" on Mars. NASA hastily called a press conference to report that MGS had not "seen" water but had detected images that appeared to look like springs or seepage from underground sources.[67] This information bolstered the strategy for the new program of "follow the water." It also strengthened the argument for a 2003 option that could follow up on this orbiter-based report with surface study.

By late July, Weiler had the benefit of Hubbard's counsel (which favored the lander/rover option), JPL and Lockheed Martin studies, and other sources of information about what to do in 2003. He also had the intriguing MGS findings. It was obvious that another orbiter following Odyssey would not carry as much public interest as a lander that released a plucky rover that could go a considerable distance and last longer than Sojourner. There was something about a rover that seemed to capture the public imagination. The question was, would JPL be up to the challenge of building one? Doubts were raised by JPL critics. Naderi told Stone that JPL would "have to prove our merit."[68]

Weiler, Hubbard, and Garvin now went to see Goldin. Weiler advised the NASA Administrator that the 2003 mission should be a repeat of Pathfinder, but with a larger rover that could have greater range and survive longer. Goldin was highly receptive. Then, Goldin asked, "What about two rovers?" He meant two lander/rovers. The suggestion came as a complete surprise to his three subordinates.[69] As Goldin saw it, this mission *had* to succeed, and adding another lander/rover lowered the risk of failure. That was the "old" NASA way, not the FBC way. Goldin asked Hubbard to check on the cost issue. "You study this," he ordered Hubbard. "Tell me the pros and cons of such a mission."[70]

Hubbard called Naderi. It was 9:30 a.m. in Washington and 6:30 a.m. at JPL in Pasadena, and Naderi was in his office. "Could you let me know in three hours how much an additional lander will cost?" Hubbard asked. Naderi immediately sought out Pete Theisinger, an engineer and highly regarded project manager, who also was in his office. Together, they estimated a one-third increase over the cost of one lander/rover combination. Their estimate was a total of $600 million.[71]

Hubbard had some additional studies undertaken, while Goldin traveled abroad. Garvin added more justification for sending two spacecraft: the rovers could go to two different places, making it more likely to find something scientifically important. Hubbard kept Goldin informed, sending him faxes of pro and con arguments. It was obvious Goldin was eager to move forward and do what was necessary to succeed. Weiler was also fully engaged in the decision-making process. He grilled Hubbard. Weiler worried about how he would come up with more money.[72]

On July 27, Weiler announced publicly that NASA had decided to go with the lander/rover option for the 2003 launch. He said the rover would be larger than Sojourner and far more capable of going great distances. He also revealed that NASA was considering a second lander/rover possibility, but that no final decision had been made on that front. He noted that NASA would decide in a few weeks.[73]

When Goldin returned from his trip, he met with Hubbard, who had gotten estimates for a second lander/rover from various sources.[74] Hubbard recalled that the number he used was $700 million, an inflated figure from the original $600 million estimate.[75] Hubbard noted, "Goldin got right up in my face, pointed his long finger at my nose, and questioned, 'are you absolutely sure that we can do this for the amount of money you quoted?'" Hubbard responded "yes" or "I am absolutely sure." Hubbard really wasn't certain, but later wrote that "sometimes you just have to play to win."[76]

Goldin subsequently decided he could not ask Weiler to pay for the entire increase out of a science budget already overextended. He checked with the White House, but got no financial help from that source.[77] He then gathered his senior managers together, including the various associate administrators responsible for all NASA programs. He said that it was critical for NASA to recover its reputation fully from the two Mars failures. He stated that Mars was a NASA priority, not just a Science Directorate priority. "Do you not agree?" he asked. The senior managers concurred. That being the case, he asked them if they also agreed with the idea of sending two rovers to narrow the risk of failure. They replied, "Yes." "If you agree," continued Goldin, "will you put up some money to help support the dual mission? You said it was a good idea!" Most of the managers went along.[78]

That was that. On August 10 NASA announced the decision. The agency would send twin rovers to Mars in 2003. The first mission would go up in May; the second, in June. The journey of each to Mars would take seven and a half months. They would land in different places. The $600 million figure was stated as the estimated cost of the duel mission. Goldin called the mission an "agency priority," a designation NASA made public.[79]

If the dual mission was an agency priority, it was even more critical for JPL. In August, Stone retired as director of JPL. Before doing so, he committed JPL to the two-rover mission, but defined success as having at least one rover make it to Mars.[80] Charles Elachi, 53, succeeded him. Elachi was closely connected to Mars research, especially MSR planning, but had not been blamed for the 1999 failures. He was a hard-driving space enthusiast who had spent his career at JPL. He initially questioned the wisdom of the two-rover approach. There were issues of time, risk, and money. He told Goldin of his reservations. Upon further thought, however, and his sense that headquarters would provide sufficient resources to enable JPL to succeed, he called Goldin and said he agreed.[81] Elachi's assent was critical in view of his new position as chief implementer of the decision.

When Elachi sat down with Naderi to discuss what needed to be done to make the two rovers effective, Naderi had a message for his JPL director. Naderi told Elachi that the lab would have to put its full force behind this mission. "What do you want?" Elachi asked. Naderi said JPL had to put its best technical personnel on the project, and Naderi told him who he thought they were. Elachi concurred.[82] The reputation and role of JPL in NASA were at stake.

Hubbard and his associates worked furiously to finish the new Mars architecture by October, the deadline Goldin had set, and the month when plans had to start getting into the next year's budget. As October approached, Hubbard found Goldin constantly intervening in his deliberations with his team. Goldin would call frequently and ask Hubbard about this or that fact or option. He even called him at 2 a.m. Weiler tried to buffer Hubbard, but to no avail. Hubbard developed different tactics to avoid Goldin so the administrator would not know he was around when he was at NASA. He did not want to meet him on the elevator. "I took to going up and down the fire escape steps and using the freight elevator in the back of the building," Hubbard admitted.[83] Finally, the architecture was completed and vetted by Weiler and Goldin.[84]

On October 6, Hubbard and Garvin discussed the new program with the Office of Management and Budget. Given positive signals from Clinton and Gore, OMB was helpful. Hubbard had kept the political side of the White House informed, via Leon Feurth, Gore's science advisor.[85] OMB's criticisms were constructive. It was clear that the White House wanted NASA to recover and get the new 10-year Mars strategy off to a good start. It shared Hubbard's view that it was best to start slow, move incrementally, and postpone MSR. There was no fixed date for MSR, but OMB wanted to make sure the missions in the new design moved systematically in the MSR direction. The OMB discussions were led by Steve Isakowitz, OMB's chief budget examiner for NASA. Isakowitz had an aerospace engineering degree from MIT and was personally interested in the Mars program.

Announcing the Mars Exploration Program

No budget decisions were made at this NASA-OMB meeting. Subsequent to it, OMB conveyed administration guidelines for the official announcement of the new Mars strategy. Specifically, NASA was to make it clear that MSR would take place later, possibly in the following decade; that the U.S. program was not dependent on international partners; and that there would be no mention of the program's connection with a possible human exploration mission.[86]

On October 26, after briefing staff of relevant congressional committees, NASA made known its revised Mars strategy in a press conference. Designated the Mars Exploration Program (MEP), there would be six major missions in the first decade of the twenty-first century. In addition to the previously announced Mars Odyssey (2001) and rover missions (2003), NASA planned an orbiter far more advanced than the existing ones called Mars Reconnaissance Orbiter (MRO) (2005). It would be able to see objects on Mars the size of beach balls. This project would be followed by a fourth mission, a "smart lander" carrying a long-range, long-duration mobile laboratory. Initially designated the Mars Smart Lander, this mission came later to be called the Mars Science Laboratory (MSL), keeping the same acronym and thereby causing some confusion. The Mars Smart Lander would have a rover, but this rover would be much more sophisticated than the two 2003 rovers. MSL would go up in 2007. The aim of Mars Smart Lander would be to reach difficult sites of compelling interest and conduct science at these sites.

In many ways, Mars Smart Lander would be the most spectacular mission of the new decadal plan, the de facto flagship. It would advance beyond the 2003 rovers to a very significant extent. It would be followed quickly by a relatively small "scout" mission proposed by the scientific community and selected competitively. One scout mission could be launched in 2007, the same year as MSL. This would be the fifth mission of the decade. The 2009 window was open for the sixth mission, and this mission could be a preparatory mission for MSR. Depending on how the previous missions went, an MSR might be attempted immediately beyond the 10-year plan, perhaps as early as 2011 or perhaps in 2014, with a second launched in 2016. The missions after the Mars Smart Lander were left deliberately vague, since they would build on what the earlier ones accomplished.[87]

Weiler called the new program "a watershed in the history of Mars exploration." Hubbard declared, "We will establish a sustained presence in orbit around Mars and on the surface with long duration exploration of some of the most scientifically promising and intriguing places on the planet." He said the effort was directed toward a fundamental question: "Did life arise there, and is life there now?"[88]

Weiler and Hubbard insisted that the program's organizing principle— "follow the water"—was the right one, given the state of knowledge. It was slower and more systematic than its predecessor program. Its theme was clearer. It was not driven by FBC strictures, budget caps, or an arbitrary deadline. There was flexibility built into the new program. To those who were disappointed about the delay in MSR, Weiler replied that, given the billion-dollar estimates for the mission, NASA had better know where to look.[89]

Others complained about the lack of connection with human spaceflight. The Planetary Society gave the new strategy tepid support. "The U.S. government [in 1996] made a national space policy for a 'permanent robotic presence on Mars,' that now seems lost," said Friedman in a written statement. "More disappointing . . . is the failure to connect the robotic program to the popular interest in the eventual human exploration of Mars." Weiler again countered that "before you send humans, you like to know where you're going." The robotic program "is doing the groundwork for the eventual human missions to Mars."[90]

Whatever the reaction, what everyone associated with the program knew was that there was additional uncertainty coming in the national political environment. The presidential election of early November had yielded a virtual standoff between Vice President Gore and Texas governor George W. Bush. Congress would also be closely divided on partisan lines. The Mars program had a new scientific strategy. But what would be its political context? Who would be the next NASA Administrator? Would this leader endorse the new strategy or even care about Mars?

CHAPTER TEN

Adopting "Follow the Water"

Formulated at the end of the Clinton administration, the new "follow-the-water" Mars Exploration Program was adopted under President George W. Bush. The question at the conclusion of the Clinton years for Mars advocates was whether the political consensus favoring Mars would hold. Aiding in the transition to a positive decision was the Office of Management and Budget, which provided continuity in policy as administrations changed. Goldin's successor as NASA Administrator, Sean O'Keefe, came from OMB and supported the new program. The Space Shuttle Columbia disaster of 2003 led to a decision by Bush to initiate a human spaceflight effort to the Moon and Mars. O'Keefe gave even more priority to the robotic Mars program as a result. As Mars became a higher priority, rival programs suffered in the budget wars, and there was significant conflict within the space policy domain. Opponents of what they regarded as too much Mars at the expense of other space science needs rose in rebellion. Implementation of the existing MEP progressed well, however. Mars advocates used the favorable NASA setting under O'Keefe to transform the flagship mission of the program, the Mars Science Laboratory, into an even bolder project.

Change and Continuity

On January 20, 2001, following the tumultuous aftermath of one of the closest presidential elections in history and a Supreme Court decision favoring him,

George W. Bush became president, with Dick Cheney as his vice president. Bush accepted Dan Goldin's offer to remain as NASA Administrator until he found his own appointee. Goldin's own power, so great under Clinton, diminished quickly under Bush.[1] Mars, however, would do well in the presidential transition.

Isakowitz pushed Hubbard and Garvin hard to refine their ideas. "The idea was to make our logic unassailable," Garvin recalled. They spoke about funding and the sequence of missions in the 10-year program. "What would it take to get to Mars Sample Return?" Isakowitz asked. The Mars Smart Lander emerged as especially important in the progression of missions toward MSR. If the Mars Smart Lander was so important, Isakowitz said, "Why not send two?" Hubbard and Garvin, conscious of costs, did not accept the invitation to go that direction.[2]

In the long meeting, NASA and OMB officials discussed and debated various issues. It helped that Hubbard and his associates had gotten support for the new program from authoritative scientific advisory groups. The science endorsement had to be clear. But so did possible benefits in terms of successful missions for the new Bush administration. As always, MSR was central to the meeting, but it would not take place under even two Bush terms. OMB, responsive to the president, wanted to know what Bush would get for the money.

Congress appropriated one year at a time, but OMB planned in five-year budget intervals. Hubbard was proposing a 10-year program. Budget options for the five-year projection included continuing the existing baseline, which was running approximately $350 million to $400 million a year; an increase over five years adding up to more than a half-billion dollars beyond the baseline; and expenditures even more ample that might accelerate MSR.

After the meeting, there were more interactions between NASA and OMB, with Weiler now deeply involved, as the new officials took power at the White House and OMB. Isakowitz and his political superiors negotiated the Bush budget for NASA. Far and away, the most serious issue with which they dealt was a huge overrun afflicting the International Space Station. The NASA science budget was consequently an area where Isakowitz and his civil service associates had more leeway in decision making. In early February, Weiler informed Hubbard that the signs were positive from OMB for the Mars budget.

On February 28, the Bush White House gave a preview of its budget plans, promising more details in April. The major consequence for NASA in general was the new administration's decision to tackle the space station overrun through major cuts in that program. However, the White House left space science relatively untouched. For Mars, the verdict was to give its advocates everything for which they had realistically hoped. As Hubbard later wrote, "The celebrations in the Mars Program office could be heard to the end of the [NASA building's] hallway at 300 E St. SW. We had received, over the period from 2002 to 2006, a total increase of $548 million, or more than one hundred [million dollars] a year total." He declared, "Now it was time for the champagne."[3]

In April, the details of the Bush FY 2002 budget were clarified. NASA would go up to $14.5 billion dollars, a modest raise from the Clinton budget. Space science would grow from $2.3 billion to $2.4 billion dollars. The Bush White House fully adopted the new MEP and the more than half-billion-dollar raise developed by NASA and OMB for its first five years. To add money for Mars and other priorities at NASA, the Bush administration cancelled proposed missions to Pluto and the Sun.[4] For Mars enthusiasts, the Bush transition was off to an ideal start—albeit at the expense of other space options. Moreover, the Mars advocacy coalition seemed now to include OMB.

Mars Smart Lander Becomes Mars Science Laboratory

On April 7, the first mission in the new program, Odyssey, was successfully launched. It began its long journey to Mars as the Bush budget details became known. In a speech he gave before the National Press Club a few days after the Odyssey launch, Hubbard alluded to the White House support for Mars exploration. He said the additional funds promised were assurance that future missions had resources they needed to succeed. He noted that the budget would make sure the 2005 mission, the Mars Reconnaissance Orbiter, "has the full set of science instruments that we've been talking about," and would allow NASA even to begin developing technologies for MSR.

He said NASA would in 2007 launch its first scout mission conceived by the scientific community and industry. In 2009, NASA planned to launch another sophisticated orbiter, possibly with international partners. This mission was present for the first time as a result of favorable budget trends, Hubbard declared. Indeed, he said that MSR might be possible in a 2011 launch.[5]

Garvin, meanwhile, wasted no time in taking advantage of what he saw as a window of opportunity for a Mars initiative. He was a man whose office answering machine extolled, "Have a great Mars day!" Garvin matched Goldin in his zeal for the Red Planet. As soon as he was sure that OMB and the White House supported the Mars program, he moved to clarify scientific and engineering requirements for the Mars Smart Lander. He did so by organizing a Science Definition Team. NASA was not going to have two MSLs. But maybe the one MSL NASA could send to Mars would be even more capable than what Hubbard and his team had originally conceived. What was implicit in the thinking of Garvin and other Mars enthusiasts was how to move forward more rapidly toward MSR. In discussions with OMB and Mars scientists, Garvin found agreement that Mars Smart Lander was especially important in developing technologies for MSR. If MSR was an end, then each mission leading to it had to be justified as means. Each mission had to build on a predecessor and show real progress. That was the quid pro quo between NASA and OMB in garnering White House backing for its long-term program.

While Garvin maneuvered in Washington, as well as with the Jet Propulsion Laboratory and the Mars science community, Hubbard held discussions with possible international partners, especially Russia. These did not lead very far.[6] Tired but elated, Hubbard decided it was time for him to leave the nation's capital. Satisfied that the new Mars program was off to the best start possible with a good funding prognosis, Hubbard on April 19 announced he was stepping down as headquarters Mars director. He and his wife, he said, wanted to return to California and Ames. He said he had promised Goldin and Weiler a year to fix the Mars program, and that year was up. Weiler praised Hubbard effusively and said he had taken on "mission impossible" and converted it into "mission accomplished." Orlando Figueroa, a veteran NASA manager, would succeed Hubbard.[7]

Figueroa inherited a Mars program moving quickly, with significant change under way where Mars Smart Lander was concerned. There was growing consensus among Mars advocates that Mars Smart Lander could and should be augmented. The original emphasis of Mars Smart Lander, as the name implied, was precision landing. It was seen initially as a technology pathfinder.[8] In the first half of 2001, and later into the summer and fall, it became increasingly clear that the Bush White House was supportive of robotic Mars exploration. In this political context, the Mars Smart Lander gradually morphed into the Mars Science Laboratory. Garvin provided leadership within headquarters for the change of emphasis, but the shift reflected widespread support at JPL and in the Mars science community for augmenting the science laboratory—that is, rover—component of the mission. As the orientation of the device altered, so did its name.

What evolved was the sense that MSL should be substantially more capable in performing science than Pathfinder and the two rovers that would be launched in 2003. One way MSL could be significantly better would be if it were nuclear powered. The 2003 Mars Exploration Rovers (MER)—known as Spirit and Opportunity—would be solar powered. That limited their range and capacity to work at night. A nuclear-powered rover would not have that limit and could do so much more, so much longer, planners reasoned.

Garvin and his associates took their case to Weiler. They argued the merits of a nuclear-powered rover. Weiler had been a scientific leader of the Hubble project. He was a telescope man. The Mars advocates called MSL an "observatory on wheels." Weiler saw the merits of nuclear technology. He agreed.[9] And so did OMB and the Bush White House. Sean O'Keefe, the politically appointed deputy director of OMB, Isakowitz's boss, had been secretary of the navy under the first Bush. He had seen the value of nuclear propulsion for ships. A nuclear proponent, he now also saw the value of nuclear propulsion for spacecraft.

O'Keefe Succeeds Goldin

In mid-October, Goldin announced he would be leaving NASA in November after his record-setting nine-and-a-half-year tenure. While departing with controversy over the space station's $4.8 billion overrun swirling above his head, Goldin won plaudits from friends of space science. He had seemed personally to favor space science over human spaceflight (Space Shuttle, ISS), and that fact showed up in the increased percentage of NASA's budget that went to space science under his leadership. Wesley Huntress, Goldin's former science chief, commented that working under Goldin could be "brutal," but praised him for raising the position of space science in the agency.[10]

No administrator in NASA's history had been more passionate about Mars than Goldin. He had prioritized the robotic program and given it close attention. He could leave believing that the 1999 problems of the Mars effort were being alleviated in 2000 and 2001. As Odyssey sped into the orbit NASA had planned for it, Goldin left NASA. Meanwhile, Congress completed work on the NASA budget, raising the overall level to $14.8 billion. Mars did well, and some of the missions Bush had tried to kill, particularly the Pluto project, were back in the NASA program.[11] The process of adoption for the new MEP was complete.

Bush announced that Sean O'Keefe would become NASA Administrator. O'Keefe got the job principally because the White House believed that NASA's major issue was to get the ISS budget under control. As deputy director of OMB, O'Keefe was already deeply enmeshed in ISS issues. O'Keefe had also been supportive of Mars interests, as indicated by OMB's actions.

O'Keefe, age 45, came aboard at the beginning of January 2002 and brought a consolidating and incremental-innovating style to NASA. He was seen as a competent, nontechnical manager, not a space enthusiast. While not visionary like Goldin, he brought political connections Goldin lacked to the Bush White House and was especially close to the powerful vice president Dick Cheney. As expected, he concentrated on the space station in his first year, bringing NASA's major project under better financial control. Weiler found he had much more autonomy under O'Keefe than Goldin to set space science policy.[12]

The Bush budget proposal announced in February emphasized priorities associated with changes in national policy following the terrorist attack of "9/11," 2001. NASA was a modest winner among domestic agencies, and its budget reflected O'Keefe's desire for continuity, with a few, targeted changes. NASA got a 1.4% raise, elevating its budget to $15 billion. The most striking change was $125 million for a nuclear systems initiative—nuclear electric propulsion systems and nuclear electric power generation systems. Seizing the opportunity O'Keefe's personal interest provided, Weiler contended that nuclear systems were the best choice for long-term missions to Mars and beyond, allowing for greater use of a more complex array of instruments. Weiler was thinking well beyond Mars—that is, the outer planets. However, the major immediate impact of this new stance was on MSL. What had been discussed and planned among scientists and engineers now became NASA and national policy. MSL would be powered by nuclear batteries, thereby allowing it to operate continuously. But the launch of MSL would necessarily be postponed from 2007 to 2009 in part to take advantage of new nuclear power systems and other technical improvements now foreseen for it.[13]

Aside from augmenting MSL, O'Keefe generally maintained the Mars program he inherited. The White House's interest in the search for life triggered by the Mars meteorite in 1996 had led to a relatively stable and politically sustainable MEP costing more than $500 million annually. "Biocentric arguments had tended to do well," commented Steve Isakowitz. As a part of the Mars-friendly continuity in the Clinton-Bush transition, Isakowitz stood out. His influence soon became even more obvious when O'Keefe brought him to NASA to work with him as comptroller.[14] For O'Keefe, policymaking and budgeting went hand in hand, and Isakowitz and O'Keefe were on the same wavelength when it came to decision making.

The continuity between the Clinton administration and the Bush administration, seen generally in Mars policy, extended to support of astrobiology.

Ironically, the new astrobiology program NASA had launched under Goldin, and which O'Keefe inherited, ran into resistance from various scientists. *Science* magazine reported that astrobiology got "little respect from many traditional planetary scientists." They saw it "more as a creation of Washington politicians than as a legitimate research area." Bruce Jakosky, an atmospheric physicist at the University of Colorado Boulder, gave an astrobiology briefing to a National Research Council panel in November 2002. He likened the experience to "teaching freshman geology." He complained that as he spoke, panel members leafed through newspapers or chatted quietly with other participants. Weiler's response to such academic snobbery was, "It's really scary when OMB may have more vision than scientists."[15]

On April 12, O'Keefe traveled to his alma mater, the Maxwell School of Syracuse University. There he gave his own "Vision" address. He declared that NASA's mission was "to improve life here, to extend life to there, and to find life beyond." In doing so, he said NASA would not be destination driven. Instead, science would drive NASA, and NASA would invest in technologies (like nuclear propulsion) to better enable science to advance.[16] The O'Keefe strategy facilitated what the Science Mission Directorate and Mars program wanted to do.

Keeping Spirit and Opportunity on Track

O'Keefe gave general support to the MEP and left it to Weiler to make major decisions. It was up to Figueroa, the new Mars director, to take those decisions and implement them effectively on a day-to-day basis. One issue that had to be resolved in connection with the 2003 MER was the tension between maximizing chances for successful landing for Spirit and Opportunity and also pursuing the most exciting science. In May, a major meeting took place among scientists and engineers at a hotel near JPL, where NASA and the research community narrowed down the number of possible sites for the 2003 rover missions.[17] A final decision on where to land would be made at a later date.

A second large issue affecting the rover mission was budgetary. Inexorably, the budget for the rovers rose significantly to deal with a host of technical problems that came up as development proceeded. Reluctantly, Weiler provided more resources. He did so more than once. By December 2002, the budget was pushing $800 million and Weiler was not happy. In fact, he was close to a decision to kill one of the rovers. Steve Squyres, the principal scientist, was deeply worried about how Weiler would decide. He pled to Naderi and others at JPL.

They pled to Figueroa. Figueroa beseeched Weiler for yet one more chance to keep the project as conceived alive. "Convince me why we should not cancel one of these rovers," Weiler demanded. Weiler trusted Figueroa and weighed his opinions of others involved. "In the end, administration comes down to people. We had to succeed. The whole world had its eyes on us," said Weiler.[18] Weiler and Figueroa worked out an arrangement whereby Figueroa could borrow money to cover the extra costs against reserves intended for 2004 and 2005.[19] MER survived as a two-rover project, but the decision had been a close one.

While concentrating on Spirit and Opportunity, Figueroa dealt with other aspects of the overall MEP. He initiated competition for a Scout mission scheduled for 2007. The Scout concept, modeled on Discovery missions, was geared to smaller projects proposed by the academic community. They would be capped at $325 million, the approximate cost of Odyssey. These were the new program's version of faster, better, cheaper—a term utterly out of favor in the era of O'Keefe.[20]

Exactly what would be the cost of MSL, still very much in a planning stage, was unknown. At NASA's suggestion, the NRC was conducting a decadal survey of future NASA planetary missions. As it did its work in 2002, it endorsed MSR as a long-term goal, with MSL as a critical enabler. It put cost estimates on missions, with MSL listed as moderate at $650 million. The scientists involved in the survey did not have the benefit of independent cost expertise. Moreover, NASA was still determining what instruments should go on MSL and grappling with engineering questions. NASA could not understand how the NRC arrived at a $650 million figure. That might have been suitable for MSL when it stood for Mars Smart Lander. But MSL now stood for Mars Science Laboratory, and this was a much more ambitious mission. Garvin was at this point estimating costs for MSL at $1 billion.[21]

The cost was not an issue at this point. The design of MSL was still unfinished. The NASA budget, overall, seemed to be doing well, allowing for expansive thinking about MSL. At the beginning of 2003, NASA was preparing to announce an FY 2004 budget that was especially good for space science. The O'Keefe vision called for a science-driven NASA. The agency was scheduled for a 2% raise, putting it at $15.4 billion, but space science was set for a 19% raise, going from $2.9 billion to $4 billion. The five-year projection showed steady increases moving NASA to $17.3 billion overall in FY 2008. The biggest percentage winner would continue to be space science, which would jump to $5.6 billion by FY 2008. Mars exploration, as the dominant planetary program, would gain accordingly. Weiler gave credit to O'Keefe and his clout with OMB and the White House. "Without [O'Keefe's] support," he said, "these increases never would have gotten through."

The Columbia Disaster

But the fanfare that ordinarily would have greeted the formal announcement of the president's proposed budget February 3 was missing. Just two days before, disaster struck the agency and nation: Space Shuttle Columbia disintegrated as it prepared to land, killing seven astronauts aboard and scattering debris across a number of states. The shock numbed all of NASA.

What would be the impact of Columbia on NASA's robotic MEP? O'Keefe had been appointed Administrator chiefly to deal with the space station financial problem. All of a sudden, he was cast in the role of a disaster manager. This role dominated O'Keefe for the remainder of the year and influenced most of his decisions the following year. The shuttle and its future took center stage. Moreover, the media gave saturation coverage to the Columbia investigation, carried out by the Columbia Accident Investigation Board (CAIB), a body appointed by NASA but given maximum independence by O'Keefe.

In this time of crisis, Weiler and his SMD managed the Mars program with an extra burden. They knew how crucial to the agency it was that the twin rovers, Spirit and Opportunity, succeed. From the outset, the rovers had been more than a science priority—they had been a NASA priority. In the wake of Columbia they took on even more significance. They would symbolize NASA's technical credibility. As the CAIB investigation extended over a seven-month period (February to August), it uncovered evidence of not only technical but also organizational flaws. These added to the blemishes on NASA's record stemming from the 1999 robotic Mars failures. The media highlighted these multiple indicators of management weakness. All those connected with the upcoming Mars flights worked harder than ever to make them show that NASA was still a "can do" agency. JPL in particular became increasingly focused on the MER. Never before had Naderi seen JPL come together behind a project so intensively as it did on the Spirit and Opportunity rovers.[22]

Those involved at JPL and in the Mars science community worked incredibly hard. Squyres, the principal scientist, recalled his own experience: "I taught [at Cornell] on Monday and Wednesday. Then, Wednesday night I flew to the West Coast. I worked at JPL Thursday and Friday. Friday night, I stayed at a Los Angeles International Airport motel. I flew back East in time for dinner with my family on Saturday evening....I lived on West Coast time when in the East. I went to bed at 2am, got up at 10am....I usually arrived in Pasadena at 11pm." Squyres kept to this routine over and over again as necessary, for six years spanning before and after the rovers launched.[23]

Success for the Mars rovers depended greatly on where they landed. As Weiler put it, NASA had to balance "science value with engineering safety."[24] NASA looked at 155 potential places to land, involving 100 Mars scientists in the decision process. On April 11, NASA announced its choices. It selected two sites. The first rover, scheduled for launch in late May, would be sent to Gusev Crater, 15 degrees south of Mars's equator. The second, to go up in late June, would go to the Meridiani Planum. Gusev was a giant crater that appeared to have once held a lake. Meridiani was a broad outcropping with deposits of an iron oxide mineral, usually associated with water, 2 degrees south of the equator, halfway around Mars from Gusev.

NASA and the Mars community grew increasingly tense as the date for the first rover launch approached. Their preparations were accompanied by the din of media attention not only to their work but, even more loudly, to the CAIB investigation. Ironically, Scott Hubbard, architect of the new Mars program, was on the CAIB panel and playing a leading role in determining what had gone wrong. He knew that everything he had done to put the Mars program in shape could go down the drain if both Spirit and Opportunity failed. One of these missions had to work. He declared, "I will be holding my breath with everybody else."[25]

On June 11, Spirit, the first of the two rovers, rocketed into the sky. It became clear shortly afterward that all had gone well. The next month, Opportunity went aloft, again successfully. Each would take seven months to reach its destination. Figueroa was elated and told the media so. Weiler tried to explain where this particular set of launches fit into the program. "We're not searching for water this time," he pointed out. "We know there's water on Mars; we know there was water on Mars" in the distant past. "What this mission does is try to understand how long water preserved at any one point. That's the key question for life." Where water has been around for thousands, even millions of years, "life seems to spring up," he said. Then, he contained his enthusiasm with a note of caution. He pointed out that there was no guarantee of success. "Mars is a death planet," Weiler lamented. "It's a graveyard for many, many spacecraft. Despite all these efforts [to eliminate risks], the rovers remain high-risk missions."[26]

Using Columbia to Advance

On August 28, CAIB released its report on the Columbia disaster. It found that the immediate, technical cause of the shuttle accident was a chunk of foam that had been jarred loose during takeoff and hit a vulnerable part of the shuttle with sufficient force to cause a rupture. On entering Earth's atmosphere, the enormous heat that built up penetrated the shuttle and caused it to disintegrate. CAIB went beyond the technical explanation to score NASA on numerous organizational fronts, all of which revealed the agency to be less vigilant than it should have been. Finally, it went beyond even NASA to criticize the "failure of national leadership" in space policy. National leaders had not had the will to replace the aging shuttle or provide the vision and money a robust human space program required. CAIB wanted a national policy response—a new vision for the space program. CAIB urged the president and Congress to give NASA a higher purpose for risking human lives, one that was greater than sending people around and around in near-Earth orbit.

Following the publication of the CAIB report, Congress held hearings, making its own inquiry about what had gone wrong and what specifically NASA was doing to improve the safety situation. The congressional hearing showed that many lawmakers wanted NASA to have a bolder goal and grander "vision" than it had. Exactly what that might be was undecided, however.[27]

In his first year, O'Keefe had not wanted to talk about destinations. After Columbia, and particularly the new pressures for a bold and clear vision, he was open to possibilities. He understood that that vision would ultimately have to come from the president.

Prior to Columbia, Bush had shown little interest in space. After Columbia, he said "our journey into space will go on." But what did that mean? O'Keefe, using the leverage he had owing to his connections with Vice President Cheney, organized a small but high-level interagency group of White House and cabinet officials to recommend an answer to that question. The chair of the group was Steve Hadley, deputy director of the National Security Council.[28] It was deliberately a "trans-NASA" body, an attribute that would potentially help it make a recommendation with a more "national policy" base.

The group met periodically behind closed doors in the summer and well into the fall. It considered a range of possibilities. O'Keefe wanted a big decision, but also one that was affordable. Over time, the group decided that a return to the Moon made sense technically and financially. Bush, informed of the committee's preliminary thinking, indicated that the Moon was not exciting enough. He wanted to add Mars, much as his father had, in his aborted Moon-Mars initiative. The culmination of the planning effort came on December 19. O'Keefe, Cheney, Hadley, presidential science advisor John Marburger, top political advisor Karl Rove, and others gathered in the Oval Office with Bush. After looking at decision papers and budget numbers, Bush noted that the decision stressed return to the Moon. "This is more than just about the Moon, isn't it?" he asked. With some prompting from Cheney, the group responded with "yes." "Well," said the president, "let's do it!" He told Hadley to work out the time and place for the official announcement.[29]

Spirit Sets the Stage

The timing and substance of the president's announcement could not be disconnected from what happened with the MEP. If Spirit succeeded, it would be much easier to herald a new human program to the Moon and Mars. But there was reason to be wary. Other countries recently had joined the United States in the Mars quest. They were finding the Red Planet as daunting as the pioneering nations, the United States and Russia, had. The Japanese on December 9 had to declare a Mars mission they had sent a failure. They were unable to put their probe into its intended orbit. On Christmas Eve, the European Space Agency did achieve Mars orbit with its Mars Express, but the Beagle 2 lander/rover it carried failed the next day.[30]

Weiler's comment about Mars being "a death planet" had justification. O'Keefe, Weiler, and Elachi were all present on Saturday night, January 5, 2004, at JPL's mission control room as Spirit made its long-awaited attempt to land. Because of the distance between Mars and Earth, there was a gap of several minutes between what happened on Mars and signals of what happened were received on Earth. "I'm scared," admitted Weiler. "An awful lot of things have to go right . . . it's up to the Gods now." Carrying its 384-pound rover NASA described as a PhD field geologist in capability, the spacecraft began its harrowing descent to Mars. It entered Mars's atmosphere at 12,000 miles per hour and had six minutes to carry out a series of automated maneuvers that would lead to either a safe landing or a disaster. Weiler called this period of time "six minutes from Hell."[31]

Spirit made it. When the signal arrived that the spacecraft had safely concluded its bounce-after-bounce landing, joy erupted at JPL mission control. Scientists, engineers, and NASA officials cheered. Naderi cried. He "ran down the corridor to see Theisinger [the project manager]." He was emotional too. Everyone "hugged one another."[32] Sean O'Keefe opened a bottle of champagne. The celebration was one of immense relief. "There are probably several hundred people here for whom it's the best day of their lives," one scientist told a *Washington Post* reporter. At a news conference a little later, O'Keefe stated, "This is a big night for NASA." "We're back!" he exclaimed, "and we're on Mars."[33] O'Keefe later told Elachi, "You saved the agency."[34]

The significance of Spirit's achievement for the White House was indicated by Bush's science advisor, John Marburger, who was also among the notables at JPL. "This is going to give everybody a big boost," he commented. "It gives a big boost to the American people. Obviously, this helps a lot to instill confidence in any policy step that you make."[35]

The Bush Decision

On January 14, 2004, President George W. Bush came to the NASA auditorium to announce his "Vision for Space Exploration." NASA, he said, was going back to the Moon by 2020. It would eventually go on to Mars and beyond. It would go first with robots, then humans. NASA would retire the Space Shuttle in 2010 and bring on a successor in 2014. "The vision I outline today," Bush said, "is a journey, not a race."[36] He indicated that NASA would get a significant increase in funding to jump-start the initiative. The Bush "Vision" appeared to augur well for the robotic Mars program.

The immediate reaction to Bush's Vision for Space Exploration was generally positive, mixed with concerns about funding. The political mood favored getting out of near-Earth orbit and back to exploration as NASA's central role in human spaceflight. If America was going to risk lives, it had to be about a goal worth the risk. That was the message of CAIB and subsequent media and congressional importuning. It was also the message of the National Academy of Sciences' Space Studies Board. Len Fisk, one-time NASA science associate administrator, was now chair of this body. He strongly believed that it was imperative for NASA to get human spaceflight out of Earth orbit, while also fully supporting space science and applications. Under his leadership, the NAS SSB produced a report that was released on the same day as Bush's address. There was "synergy" between the report and Bush's vision, said Fisk.[37]

O'Keefe had carefully crafted a budget strategy that avoided the sticker shock that had killed the Moon-Mars program of George H. W. Bush in 1989. The strategy was to emphasize the Moon first, as a stepping stone, and the lunar goal seemed manageable the way O'Keefe explained it. Mars—and its prodigious expense—was downplayed and pushed for future discussion in regard to human exploration. Conversely, robotic Mars exploration was emphasized for precursory missions. The term "exploration" was inclusive insofar as rhetoric was concerned.

But not everybody believed the NASA rhetoric or budget numbers for the first five years, certainly not everybody in Congress (or at NASA). And even if the numbers were "right," they required NASA to reprogram money to make resources available for a new mission. There was worry on the part of scientists who did not do lunar or Mars research that they would be losers as NASA refocused. Weiler tried to calm such fears. "This is not a flags-and-footprint program," he said. "NASA intended the Moon as a stepping stone, and scientists are excited about going to Mars," Weiler declared. So was the president, he added.

What Weiler understood, as did Fisk, was that the vision could be good for NASA as a whole if the necessary resources were forthcoming. Also, it was critical that the scientific community support the vision for it to have a chance to succeed.[38] What sent a shock wave through the scientific community, and seemed to belie the reassuring words of the vision rhetoric, was the news on January 15, the day after the Bush announcement, that O'Keefe was killing a planned shuttle servicing mission to Hubble. The information came via an inadvertent leak and appeared in a *Washington Post* story about the Bush vision.

The decision was about safety, not budget, as far as O'Keefe was concerned, but was interpreted by many astronomers as a trade-off with Moon-Mars. "This is a kick in the teeth," said one. Hubble became a major distraction as O'Keefe tried to promote the new mission to NASA's various constituencies in 2004.[39] While the selling of the new human space program got off to a rocky start, the robotic Mars program continued to shine amidst the uncertainty and scientific controversy. Moreover, O'Keefe gave it a priority as he dealt with the NASA budget, too much so in the view of O'Keefe critics—Mars's rivals for space funding.

On January 24, at 9 p.m. (PST), the second Mars rover, Opportunity, entered the Red Planet's atmosphere, beginning the complicated set of procedures that would enable it to land safely. Opportunity aimed at a Martian plain called Meridiani. Once again, O'Keefe and other top NASA officials were present at JPL. So also were former vice president Al Gore and California governor Arnold Schwarzenegger and his wife, Maria Shriver. Mission controllers ate traditional "lucky peanuts" as they intensely watched the telemetry. Six minutes after starting its descent at 12,000 miles per hour, Opportunity reached the surface. When the signal was returned to Earth that Opportunity had come to rest at a final place, safely, the landing chief at JPL announced, "We're on Mars, everybody!" Everyone in the room erupted in applause, backslaps, and hugs. As before, O'Keefe got out the champagne. He lauded the Mars science team as "the best in the world." NASA and its partners had known that the stakes were high. They had put in an "extraordinary effort," and it had paid off.[40]

Prioritizing Mars

In early February, the president's budget was announced. The new mission proved extremely helpful for the robotic Mars program, which would get a 16% increase over the previous year, with the prospect of its doubling over a five-year period. Although the possible 2011 date for MSR had succumbed to reality, NASA still wanted to launch this mission as soon as possible. NASA's comptroller, Steve Isakowitz, said that NASA might be able to go for an MSR mission in 2013.

The total budget for NASA would rise from $15.3 billion to $16.2 billion—the big initial boost Bush had promised. But the raise in space science—from $3.9 billion to $4.1 billion—mostly went to lunar and Mars priorities and required holding spending down in other areas of space science. The outer planets took a hit.[41] Fisk and many other leading space scientists were aghast. The budget was not what they had expected. It emphasized specific priorities, not balance. It was a "bifurcated policy," stated Fisk. The SSB was very much pro-Mars and had in 2003 endorsed the new MEP. But the pendulum was swinging now too far in the Mars direction. Said Fisk, it was "Moon-Mars vs. everything else." He recalled hearing O'Keefe say that Mars would get an extra $1 billion over ensuing years. But, argued Fisk, an astrophysicist at the University of Michigan, "The money came from the rest of us."[42]

The most striking impact of the Moon-Mars decision for the robotic program was seen in the creation of a new line of "exploration"-relevant missions in parallel with the existing set of science-driven missions. Beginning in 2011, there would be a series of "Mars test bed" flights. These would be intended as mainly precursors for human flight. They would demonstrate specific technologies, such as improved aerodynamics entry, orbital rendezvous and docking, high-precision landing, and resource extraction and use. In addition, Mars test bed missions would characterize the local radiation environment at potential landing sites to ensure safety of future Mars astronauts. To prepare, NASA initiated a precursory mission "Safe on Mars" budget line for research in 2004.

Further, O'Keefe announced a reorganization of NASA in June to accommodate the Moon-Mars mission and streamline other activities. An Exploration Systems Mission Directorate for human spaceflight was created along with an expanded SMD for robotic missions. SMD now embraced both the Office of Space Sciences and Earth Science Office. SMD would implement the "Safe on Mars" activity, but that work would be funded by the Exploration Systems Mission Directorate. MSR was seen as benefiting the human program as well as the science quest to find life, and could presumably also involve cooperation between SMD and Exploration Systems. Beginning in fiscal year 2005, NASA intended to spend $50–$70 million annually to prepare for a 2013 MSR mission.

The robotic Mars program thus was being augmented not only for science but also to meet needs of the new human exploration mission. The science and human precursory goals were projected as two parts of the same robotic program. Never before in history had these two rationales for Mars exploration been made so explicit in planned missions and with funding provided to accommodate them. Weiler declared, "I have a feeling that all [Mars] missions are going to be a mix of those two things." Indeed, Figueroa indicated that NASA in 2011 would launch three probes to Mars—two for science and one to initiate a test bed line for human exploration.[43] The quests to find life on Mars and to launch life to Mars were joined.

There was an air of great expectation for the robotic Mars program, not only because of the Bush decision and promised infusion of funds, but also because of Spirit and Opportunity. Opportunity, in particular, provided exciting findings almost as soon as it commenced roving. The initial reconnaissance Opportunity performed yielded discoveries that were more striking than Mars researchers might have imagined. Geological indicators suggested that a "salty sea" had once been in the place Opportunity had landed.[44]

On March 2, NASA held a press conference at which Weiler announced what Opportunity had found, and he called the discovery profound. He pointed to "indicators for astrobiology," commenting that "if you have an interest in searching for fossils on Mars, this is the first place you want to go." Squyres, the principal science investigator, declared, "We believe that this place on Mars, for some period of time, was a habitable environment . . . it would have been suitable for life." Weiler noted that Opportunity had landed where liquid water once drenched the area. Opportunity, in short, had found the "right" place where life was possible, at least in the past. Those who wanted to renew the push for an early MSR mission now pointed to at least one locale where they would want to go.[45]

Criticism from Mars Rivals

The euphoria of the Mars science community was not shared by other scientists who saw both their status and resources diminished. In spite of Weiler's denials, they saw a zero-sum game of winners and losers. Astronomers and advocates of planetary missions other than Mars were vocally unhappy. They spoke of "collateral damage" and pointed to O'Keefe's Hubble cancellation decision as evidence that Bush and O'Keefe had no real understanding for science. Referring to funding choices, Fisk remarked, "Some of us feel like lesser species."

Fisk warned O'Keefe that NASA was creating first- and second-class citizens, by splitting the agency's scientific constituency into haves and have-nots. This was "an unnecessary distinction which I think will work against the program." What non-Mars scientists saw as a problem, others viewed as good management. Dave Radzanowski, the White House OMB official overseeing the NASA budget, applauded O'Keefe's decisions and spoke of "setting priorities and showing leadership." NASA was getting the largest increase in budget among agencies, other than Defense and Homeland Security, and making decisions in accord with presidential preferences, he pointed out.

Weiler, who was a telescope astronomer, not a planetary scientist, tried to assuage the self-identified losers by arguing they were seeing their wishes deferred, not cancelled. Over the long haul, if NASA gained, everyone would benefit. "I love all my children," he avowed.[46] However, in August, O'Keefe shocked many scientists by moving Weiler from his SMD leadership and making him director of the Goddard Space Flight Center, a shift many observers regarded as a demotion for the dynamic science chief and possibly related to differences with O'Keefe over the Hubble service termination issue. O'Keefe replaced him with Al Diaz, who had been director of Goddard. Diaz was an engineer by training and had been Fisk's deputy when Fisk ran the Science Office. Although an able manager and interested in Mars, Diaz was not regarded by the science community as "one of us," as Weiler had been so regarded. Nor did Diaz seem comfortable with the science advocacy role that came with the job.[47]

Criticism of the budgetary choices also came from Congress. Politicians were generally supportive of the Moon-Mars destinations in principle. But several worried about costs, and Sherwood Boehlert (R-NY), chair of the House Science Committee, expressed concern that NASA could become a one-mission agency.[48] He especially worried about the future of the Earth Science activity.

Throughout most of 2004, lawmakers sparred over the new NASA mission in committee hearings, awaiting the November election results, and postponing most federal spending actions. John Kerry opposed Bush for the White House, and neither presidential candidate had much to say about space. Kerry agreed in general with the human exploration goal but claimed he could manage the space program better. Neither said anything about paying for the mission. When the votes were counted in November, Bush prevailed, as did Republican control of Congress. Stability in political leadership seemed to presage relatively smooth sailing for Moon-Mars and NASA, at least for a while.

But there were technical issues complicating funding which had little to do with the election. The repairs to the shuttle fleet, grounded since February 2003, were proving both quite difficult and very expensive. It looked like the shuttle return-to-flight bill could be $2.2 billion. At the same time, O'Keefe had bent with the avalanche of criticism he had received for his Hubble decision. He still refused to use a shuttle for the Hubble servicing mission, but he did agree to seriously consider a robotic-repair mission. He had announced that fact in June at an American Astronomical Society meeting, to loud applause. But work since then was showing that a robotic-repair mission was exceedingly complex and could also cost $2 billion or more.

Even with the raises contemplated at the time of Bush's decision, these unanticipated shuttle and Hubble costs would prove a serious burden. O'Keefe pled for help for these unforeseen costs from the White House and was refused. Comptroller Isakowitz admonished Congress to give the agency what Bush had requested. He said that anything less in money would not only affect human spaceflight but also have a "negative" impact on science.[49]

Those scientists and their allies who did not identify with the Bush vision were more wary than ever of the trends they saw. It did not help their cause that many leading U.S. scientists had vehemently and visibly opposed Bush during the election. James Hansen, NASA's best-known climate change researcher, had been especially vocal in opposition to Bush. Marburger himself admitted that Bush and the scientific community had differences.[50] Mars was doing extremely well under the vision—but other fields were perceived as suffering. However, Mars scientists were not entirely pleased because of Diaz. Diaz asked Figueroa to be his deputy. Figueroa in turn had to relinquish his Mars program director role and asked Doug McCuistion to take on that task. McCuistion was a 48-year-old systems engineer with a background as a manager in Earth sciences. In effect, Diaz appeared to have "demoted" Mars organizationally since McCuistion reported to a deputy rather than directly to the associate administrator.[51]

Congressional Action

Just before Thanksgiving, Congress put virtually all spending bills for FY 2005 into a massive omnibus appropriations measure. The Bush administration lobbied Congress hard to ensure that its priorities would prevail. Where NASA was concerned, the White House and O'Keefe had needed help from two extremely influential legislators. One was Tom DeLay (R-TX), majority leader and representative from the Houston-area district where the Johnson Space Center was located. The other was Senator Ted Stevens (R-AK), chair of the Senate Appropriations Committee and a former O'Keefe mentor.

The result was to hold all discretionary spending not related to Defense and Homeland Security to a collective 1% increase over what the agencies had received in 2004. Other agencies were literally "taxed" to provide the larger increase Bush sought for NASA. NASA rose from $15.3 billion to $16.1 billion in spending. Moreover, the bill was written to provide O'Keefe maximum flexibility to reprogram money, to make sure the new mission got off to a strong start. Diaz stated that the new budget was good for his Science Directorate. It would provide for a "very robust science program," he said.[52]

O'Keefe was elated with the financial victory and directed his troops to "deliver." To his regret, the Hubble controversy still festered, mightily. In December he received an interim report from an NAS panel that strongly urged a shuttle repair mission for Hubble, saying the robotic mission O'Keefe favored was so technically demanding that it was unlikely to be possible before Hubble's crucial equipment expired. O'Keefe, however, would not be dealing with Hubble—or the Moon-Mars program and alleged "collateral damage." On December 13, he announced he was resigning, effective in February 2005.[53] He was headed for Louisiana State University as its chancellor.

O'Keefe Departs

One of O'Keefe's last acts as NASA Administrator in early February was to announce Bush's proposed budget for NASA for FY 2006. It was $16.5 billon. This was a raise of $400 million from the congressional appropriation. Ominously, it was only half the amount Bush had promised when he made the Moon-Mars decision. The president's desire to trim the budget deficit and put more money into the war on terrorism (especially in Iraq) and defense generally trumped virtually all other federal programs. In addition, Clay Johnson, the deputy director of OMB, was personally close to Bush and a harsh critic of the Moon-Mars vision. He persuaded Bush not to put his political capital behind space in his second term.[54] It was more O'Keefe's influence than Bush's support which allowed NASA to fare better than most domestic agencies in the budget process. Speaking of the raise NASA got, O'Keefe commented, "It's rather remarkable under the circumstance."[55]

O'Keefe had continued to prioritize sharply. He put the money he had behind the new mission, and this policy worked to the advantage of the robotic Mars program. Overall, the NASA science budget was slightly down from the previous presidential budget, from $5.5 billion to $5.4 billion. The lunar science program, which had been suffering benign neglect for a long time, tripled in size. Mars projects also gained, jumping from $681 million to $723 million. O'Keefe was creating a budget wedge intended to raise the robotic Mars program to the $1 billion level in 2010.[56]

But what was good for Moon-Mars science was bad for every other field. The outer planets and Earth observation satellites were suffering, and Hubble on the way out altogether. If NAS said that robots could not fix Hubble, then Hubble would not be fixed, at least under the O'Keefe policy.[57]

Prioritizing meant winners and losers. There was little question that non-Mars scientists envied the money going to Mars, and not just money—glory! The successes of Spirit and Opportunity on Mars were high profile, giving the scientists associated with the rovers, especially Steve Squyres, what a *Science* magazine editorial called "the astronomical equivalent of rock star status."[58]

CHAPTER ELEVEN

Implementing amidst Conflict

Implementation of the Mars Exploration Program, elevated to a more favored basis, and projected to grow substantially by O'Keefe, ran into an unfavorable environment soon after he left. On March 11, 2005, the White House announced that Michael Griffin would be replacing O'Keefe. Age 56, Griffin was a lifelong space enthusiast who had started his career at the Jet Propulsion Laboratory and later headed the ill-fated Moon-Mars initiative of George H. W. Bush. He had an engineering PhD and several other degrees and was viewed as arguably the most qualified man in the country to implement Bush's Vision for Space Exploration, from a technical standpoint. Griffin had coauthored a technical book, *Space Vehicle Design*, and had also written about the policy need for NASA to go beyond the shuttle and space station and get back to its true mission: exploration. He came to NASA from Johns Hopkins University's Applied Physics Laboratory, where he headed its Space Division. While his orientation was human spaceflight, he was also an advocate of robotic exploration and had, as a young engineer, worked on robotic missions to Mars at JPL.[1] One of his reasons for leaving JPL and NASA was the erosion of robotic Mars activity after Viking. Although very different in style—Griffin was shy and taciturn whereas Goldin was outgoing and manic—he shared Goldin's passion for space. Like Goldin, he returned to NASA to fulfill a life's dream.

A big difference, however, was that Goldin favored the robotic Mars program and looked for savings in human spaceflight, particularly the shuttle. For Griffin, human spaceflight took precedence in his mind, even if science had to suffer as a result. Griffin wanted to focus on getting the shuttle back to flight, completing the International Space Station, and especially implementing Bush's Moon, Mars, and beyond human spaceflight vision. This required an emphasis on building an expensive new rocket and other equipment relevant to the Moon. He looked to his Science Mission director to run the robotic science program, including Mars. The science directors who served under Griffin during his tenure had their own problems with implementing the "follow-the-water" strategy and particularly technical and cost issues with Mars Science Laboratory development. As much as he might have wanted to concentrate on the human program—and felt he had to do so because of national policy decisions—Griffin found that contentious decisions regarding science and Mars kept coming to his desk. A man who savored rationality and disliked politics, he found himself embroiled in the politics of space science.

Pressuring Griffin

As he prepared for his Senate confirmation hearings, Griffin was lobbied by scientists who disagreed intensely with the priorities he inherited from O'Keefe. A total of 17 scientists signed a "manifesto" they delivered to Griffin, and some of them personally spoke to him. "The balance between the two modes of exploration, human and robotic, is now threatened," the manifesto declared. It was not just the balance between human and robotic exploration that worried these scientists. They were concerned also about the balance between Mars and non-Mars robotic science. They called the concept of "exploration" O'Keefe used too narrow. "Should other forms of space exploration be cancelled or curtailed to make this new, but limited, exploration vision possible? We think and hope not," said the paper.

Among the 17 signers of the petition was Fisk, who was converting the chairmanship of the National Academy of Sciences Space Studies Board into a position for leadership in opposition to the O'Keefe priorities. Al Diaz, who had the NASA job Fisk once possessed, called the Moon-Mars science orientation a result of "strategic" decision making. Fisk called that kind of thinking bad strategy and warned, "There is a firestorm coming."[2] Griffin tried to put the concerned scientific community at ease at his confirmation hearings on April 12: "We as a nation can clearly afford well-executed, vigorous programs in both robotic and human space exploration as well as aeronautics," he stated. However, Griffin was also clear about his priorities, which were to return the still-grounded shuttle to flight, finish the space station, and begin building the technical systems (called "Constellation") that would take America back to the Moon by 2020. He especially declared his intent to accelerate the transition from the space shuttle to its successor. He regarded the four-year gap he inherited as too long and a challenge he wanted to remedy.[3]

Griffin was easily confirmed and in office by April 14. In taking command, he was aware of an understanding reached between Bush and the Office of Management and Budget during the hiatus between February (O'Keefe's leaving) and April (his arrival) at NASA. This would cut NASA's five-year projected budget by $2.9 billion as its contribution to deficit reduction during Bush's second term. Griffin knew that budget games were played constantly in Washington, and he thought he could reverse this move. He believed that Bush was serious about supporting NASA and his vision, and that he could counter OMB. With the original budget projected in 2004, he felt he could deliver a program that could get the Moon-Mars vision off to a sound start while also supporting science and other NASA programs. He wanted "balance" in NASA and said, "There is no inherent conflict between manned and unmanned space programs, save that deliberately promulgated by those seeking to play a difficult and ugly zero-sum game." But one former NASA Administrator, who was quoted without attribution by *Science* magazine, predicted, "He's going to have to choose sides; he can't make everyone happy."[4]

A Rebalancing Act

Griffin got off to a rapid start, making a series of decisions, technical and organizational, relating primarily to human spaceflight. He was determined to narrow the four-year gap between the shuttle and its successor to one or two years and took close control of decision making regarding that front. He ordered relevant officials to plan in accord with the Moon-Mars priority, including the head of the Science Directorate. Although he said he would review O'Keefe's Hubble servicing termination and ultimately reversed O'Keefe on this matter, he gave the science enterprise far less attention than he did human spaceflight. However, he did listen to what he had heard from critics of the O'Keefe-Diaz "strategic" approach.

As the Griffin-led review process of NASA programs proceeded in Washington, NASA's Spirit and Opportunity continued to perform remarkably well on Mars. They were already well beyond their 90-day prescribed lifetimes and still going strongly. They were traveling further and reaching higher than expected. They were viewing scenes never before observed and sending these images back to Earth for scientists to analyze and the general public to witness.

Spirit's and Opportunity's achievements were based on decisions made years before. There is a long latency between decisions and their execution in space policy. Thus, decisions made at NASA years before would influence what Griffin could or could not do in his tour. In May, in his characteristically direct and laconic way, he said that NASA "can't afford to do everything on its plate." Looking for efficiencies, he sent a document to Congress on May 11, describing sweeping changes affecting most NASA programs. He told Congress that human space exploration would trump the science budget only "under the most extreme budget pressure." At a Senate hearing the next day, he stated, "We have tried to be sensitive to the priorities of the affected research communities and have listened carefully to their input." He said NASA had responded accordingly.[5] The overall spending level for space and Earth science was unchanged. However, he said there would be shifts within the science envelope.

What the "shifts" to which Griffin alluded meant for Mars became clear in July. Figueroa had "filled out" the 10-year MEP with a 2009 telecommunications satellite. NASA now killed plans for that satellite. McCuistion explained the decision as driven by a diminished need for a dedicated relay at Mars and enlarged funding requirements of astronomy, Earth science, and planets other than Mars. The "core" program that Hubbard and his team had designed in the era of Goldin remained. But the "augmented" program that came with Bush's human exploration initiative under O'Keefe was deferred. The "safe on Mars" element that linked robotic with human spaceflight died. The funding wedge O'Keefe had instituted for the Mars program to reach $1 billion in 2010 was cut 50%. Diaz made the cuts under Griffin's direction.[6]

Joseph Alexander, staff director of the NAS SSB, applauded the rebalancing decision. He pointed out that Mars exploration had long commanded a large share of NASA's space science spending and said he was pleased to see Griffin's restoration of balance to the science program. "In the grand scheme of things there were people who felt that the emphasis on Mars was starting to come at the expense of other areas," Alexander noted.

The Planetary Society's Friedman was not happy with the downplaying of Mars. He protested that "taking Mars out of the exploration program, as was done in the budget cuts, and pulling back on the infrastructure for the eventual Mars outpost, could create another dead-ended program with no destination and no public support."[7] What Griffin was doing was sacrificing the longer-term Mars requirements for more immediate needs of other science programs. Mars Sample Return, for example, was pushed farther into the indefinite future.

Even as decisions went against more distant Mars interests, Mars spaceflights already scheduled moved ahead. Due up in August was the Mars Reconnaissance Orbiter. "It's the most powerful suite of instruments ever sent to another planet," said McCuistion. "The MRO spacecraft is many things," commented Richard Zurek, the JPL primary scientist for the mission. "It's a weather satellite, it's a geological surveyor, and it's a scout for future missions." James Garvin, now NASA's chief scientist at headquarters, called the August 12 launch "utterly stupendous." However, while excited about MRO, Mars advocates could not help but express regret that MSR was "still only a dream."[8]

Mary Cleave as Associate Administrator

At the time NASA launched MRO, Griffin made a series of across-the-board changes in his management team. They affected most aspects of NASA, mainly human spaceflight. However, they also had impacts for Mars. Griffin replaced Diaz, who retired, with Mary Cleave as associate administrator for the Science Mission Directorate. Cleave came out of the Earth science division of the SMD, a fact that some observers considered indicative of Griffin's rebalancing effort. She was also a former astronaut, as well as an engineer. Finally, she was someone with whom Griffin had long-term connections. Whatever her merits, she was not an advocate of robotic Mars as her directorate's top priority. That brought her into conflict with Mars proponents.

Figueroa, not happy with Griffin's policies, moved to a new position at the Goddard Space Flight Center in nearby Greenbelt, Maryland. McCuistion stayed, frustrated with the trends for Mars. Garvin, however, soon joined Figueroa at Goddard. Griffin told Garvin he had no need for a chief scientist. Garvin believed that his outspoken advocacy for Mars as a priority made him especially expendable.[9] Another individual who had been central to the Mars buildup was Isakowitz. He was also a casualty of Griffin's actions to replace O'Keefe appointees with individuals of his choosing. Isakowitz wound up at the Department of Energy. The upshot of the various personnel moves was the distinct weakening of the Mars constituency at headquarters.

Cleave, meanwhile, inherited a program that had far more budget challenges than easy solutions. One huge issue was the James Webb Space Telescope, Hubble's eventual successor, suffering an overrun of $1 billion.[10] Her problems were also those of Griffin, but he had many additional and more pressing ones, most connected to human spaceflight. He was anxious to delegate responsibility for science decisions to her. In July, the shuttle returned to flight, and while it was successful in some respects, it still experienced foam-shedding problems that had caused the Columbia accident. It would need further repair work. Delays meant that more money had to be diverted from other activities to the shuttle, NASA's most troubled (and most expensive) human spaceflight program. What Diaz had begun (cutting Mars), Cleave would have to continue.

Space News, a leading trade journal, had a suggestion for Griffin—and Cleave—which it claimed might help with their money troubles. Why not defer the MSL, scheduled for a 2009 launch? In an editorial entitled "Mars Science Lab Can Wait," it made this argument as one way to get funds for more urgent needs.[11] Mars proponents were appalled. MSL was the flagship of the MEP. Friedman wrote an impassioned response in a letter to the editor. He reminded readers of an earlier debate he had had with a senator who had suggested Mars could "wait."

"Without Mars as a target," Friedman said, "there will be no sustained program of space exploration. Mars is the compelling goal that drives the 'Vision for Space Exploration'—not just for robots, but for human space flight as well." He added up the recent decisions against Mars and said they came to $2 billion in long-term diversions and cancellations. The money extracted included funds from the SMD, as well as what the Exploration Systems Office had for precursor missions. "You take away the vision [of Mars]," he declared, "you lose public support."[12]

The public was surely engaged with what was then happening robotically on Mars. The public seemed to identify innately with the intrepid rovers. Spirit made it to the top of a summit in August, a feat that NASA hailed as an unexpected milestone. William Farrand, a researcher connected with the project, noted, "When we started the mission, if anyone had told us that we would not only drive all the way over to the Columbia Hills, but also drive to the highest point there, I think we would not have really believed it."[13]

NASA seemed to be doing better on Mars than on Earth.

Katrina and "Apollo on Steroids"

As September came, so also did disaster to the United States. Hurricane Katrina slammed into the Gulf Coast, wreaking havoc and causing floods, horrific damage, and loss of life. The impact on New Orleans, in particular, was devastating, and television pictures of that city's forlorn victims turned the disaster into a public relations calamity for the Bush administration. NASA was affected in many ways, most notably in the damage to two Gulf Coast facilities, Michoud and Stennis. Estimates of damage to these facilities hit $1 billion.[14]

Katrina was obviously going to cost the United States a fortune in recovery money. The collective attention of the nation was trained on New Orleans and adjacent territory through most of September. Nevertheless, on September 19 Griffin chose to unveil NASA's plans for how it would return to the Moon. Based on extensive study, Griffin pointed out, NASA's intent was to use a shuttle-derived system he called "Apollo on Steroids." There was to be a capsule atop a rocket, as was true of Apollo. This would be the basic approach of Project Constellation, the overall Moon-Mars effort. Constellation would develop first a rocket (Ares I) and capsule (Orion) to replace the shuttle. This system would pave the way for developing a larger, heavy-lift rocket (Ares V). The heavy-lift rocket could carry astronauts in the Orion capsule, along with newly designed landing equipment, to the Moon. Griffin estimated that the cost of going back to the Moon would be $104 billion, and that NASA could get there before the president's deadline of 2020. He set a goal of 2018.[15]

The media, congressional, and public reaction was quiet, given the concentration of the nation on Katrina's aftermath. Some critics questioned Griffin's political sensitivity in terms of timing. Griffin noted that Moon-Mars was a long-term program and would have to take place in an environment of many national setbacks of one kind or another.[16] At the press conference during which Griffin made his announcement, a reporter asked whether the money for a return to the Moon would require taking funds from science. Griffin responded that "not one thin dime" would come from science.[17] That point seemed important not only to scientists but also to some of their congressional allies.[18]

While thinking about future missions, Griffin had to deal with day-to-day funding issues. He was getting frustrated as he negotiated the NASA FY 2007 budget with OMB. He was having problems fixing the space shuttle and therefore could not finish ISS. He was finding getting resources from the Bush administration to match its Moon-Mars goal impossible. Then, there was the scientific community, which seemed to have no understanding that he had limited finances.

In October, Griffin met with NASA's Astronomy and Astrophysics Advisory Committee and said he was "fed up" with the conflicting advice and pressures from NASA's various science constituencies. He had a finite budget, he explained, and the scientists should come up with priorities. He noted the over-runs on science projects and also scorned scientists for lobbying for specific projects—like Hubble repair—without considering the financial implications. He did reiterate that he would continue to back science and would not divert science funding to human spaceflight. "The good news," Fisk commented, "is that Griffin was going to give us our fair share. The bad news is that we can't execute the programs we have with the money we have available."[19]

The Crunch Comes

In November and December, there were ups and downs for Griffin. The positive event was that Congress passed legislation that gave the agency the funds Bush had requested for FY 2006—$16.5 billion. It also, for the first time, formally endorsed the Vision for Space Exploration (i.e., Constellation) in the NASA Authorization Act of 2005.[20] Congress also directed NASA to retain the general balance between human spaceflight and science, a directive not so welcome for Griffin in view of the setbacks he experienced in this period.

Griffin desperately wanted to speed up the development of Orion–Ares I, as the shuttle successor and initial capsule-rocket component of the Constellation Program was called. But what he discovered was that the shuttle cost projections were now running $3 billion to $5 billion higher than estimated in 2004. One reason lay with the lengthy and expensive repairs. Also, instead of launch expenses declining as the shuttle neared its rescheduled 2010 retirement date, they would require virtually what they had cost in earlier years.[21]

Griffin went to OMB in November and asked for a realistic budget that returned the money taken away from NASA's projected five-year budget just before he became administrator. He also wanted a substantial raise for NASA. The request was for almost a 9% increase. OMB responded with less than half that figure, with most of that going to Katrina-related facility repairs. Griffin warned that unless he got substantial additional program funds, he would be forced "to hold science's budget fixed at FY 2006 levels for the next five years." That would hurt all science programs, including the robotic Mars effort.[22] *Science* editorialized that NASA was "back to eating seed corn." It commented that the issue was not just human spaceflight versus science. When push came to shove, Griffin would have to go with lunar research over Mars research because the Moon came first in the Moon-Mars initiative.[23]

Many of the individuals who were involved in the post-Columbia interagency committee meetings in late 2003 and who assisted O'Keefe in getting the Vision for Space Exploration adopted were no longer in government or had moved to positions in government different from that decision period. Bush and Cheney were still around, but both preoccupied with issues other than space. Neither had given any indication, at least publicly, of wanting to use political capital to help NASA. Nevertheless, Griffin appealed beyond OMB to the president to get the money he believed the agency had to have to do its job.

Bush—like most presidents—did not customarily intervene in agency-OMB disputes, but Griffin was forcing the issue. In December, the day of decision arrived. The key protagonists at the meeting in the Oval Office were OMB director Josh Bolton and Griffin. Bush, Cheney, presidential science advisor John Marburger, a state department representative, and various other high-ranking aides were present.[24]

It was quickly clear that the big raise Griffin wanted would not fly when the administration was still scrambling to pay for higher priorities, including the Iraq War and Katrina recovery. So how was NASA to pay the huge and unanticipated shuttle costs? Bolton gave the OMB position, which was to end the shuttle early, arguing that it was a dead-end program. But terminating the shuttle meant also abandoning the space station and breaking a host of international commitments, as Griffin pointed out. Bush's legislative aide told the president that Congress would not let him kill the shuttle even if he had wanted to do so.

The next option discussed was to take money from science—the five-year flatlining strategy Griffin had warned might happen without a significant NASA raise. Griffin felt forced by his desire to protect Constellation and narrow the shuttle succession gap to defend the science-flatline option. Bolton opposed him, and Marburger spoke up for science. In the end, Bush sought a measure of compromise. He gave NASA a modest raise. The shuttle would continue, as would the space station. Science would get a small increase, perhaps an average of 1% a year for five years. Exploration systems would also go up, but they would have to help fund the shuttle costs.

Griffin had said "not one thin dime" would be taken from science to support human spaceflight, specifically the Constellation Program. Now, thanks to the White House funding decision, money would come from *both* science and human exploration systems to pay for the shuttle shortfall. Griffin's goal to narrow the four-year shuttle successor gap could now not be achieved.[25] For the robotic MEP, these presidential choices created an environment that virtually guaranteed austerity for at least the remainder of Bush's term.

In February 2006, Bush's budget request was announced. NASA got additional funds for Katrina repairs. Science at NASA received a 1.5% raise, but the five-year projection showed that mark declining to an average of 1% a year. Mars funding was cut in half. The *Science* magazine news report on the budget stated, "The prospects for NASA-funded scientists are among the bleakest in the federal government."[26]

The reaction from scientists and their congressional supporters was immediate and negative. Huntress, a former director of NASA's Science Directorate, criticized NASA for "using money intended for science programs to fund continued operation of the shuttle." Like OMB, he could not agree with that trade-off, since the shuttle was "a program scheduled for termination." Congressman Sherwood Boehlert said he was "greatly concerned" with the science cutbacks.[27]

Among the scientists, the planetary researchers were most alarmed. Rita Beebe, a New Mexico State University researcher and member of the NAS SSB, complained that the planetary scientists had been especially hit by the cuts. "The proposed budget transforms an existing, vibrant program into a stagnant holding pattern," she declared. Beebe suggested that NASA was "reenacting the events of the 1970s," a time when the planetary program went from an active series of missions to the doldrums. She called the damage "immediate and increasingly irreversible."[28] An astrobiology researcher, Rocco Mancinelli, called the budget "a disaster," coming just when "instruments aimed at understanding the fingerprints of life . . . are being built for the Mars Science Laboratory."[29] The Planetary Society (of which Huntress was the current president) launched a Save Our Science (SOS) campaign among its members.[30]

Griffin called the scientific response "a hysterical reaction, a reaction out of all proportion to the damage done."[31] There was still a great deal of money for science, he pointed out, and the 1% five-year raise was better than the 0.5% cut borne by federal nondefense discretionary programs generally. Griffin's words seemed to make many scientists all the angrier, and they reminded him of his "not one thin dime" pledge the year before. Fisk pointed out that 1% was no raise; it translated into "a major retrenchment" given inflation.[32]

Fisk and his SSB decided that they should take the lead in building a united front among space scientists. Unless they did so, the various disciplines and specialties would fight among themselves and weaken their position. At best, a united community might be able to appeal to Congress to enlarge the science budget. The lawmakers made their decisions regarding the president's budget proposal over the course of 2006. At minimum, the scientists would try to agree among themselves as to what were top priorities that had to be protected, rather than having those priorities determined by NASA managers or Congress.

Cleave, Griffin's embattled science chief, bore the brunt of the scientists' ire. She was pilloried for cancelling a small mission in March—an asteroid project called Dawn—shortly after she had promised a congressional committee that she would be attentive to such missions. These kinds of smaller projects benefitted academic scientists and their graduate students. Critics charged that NASA seemed intent on protecting the larger "flagship" missions and the institutions behind them, such as JPL. They were flagships, Griffin insisted, because the scientific community, via the NAS, had helped make them so through NAS decadal surveys of science priorities.[33] The "big" versus "little" science balance clearly was an issue, as was Mars versus non-Mars planetary science.

Debating Priorities

Fisk and the SSB were able to elicit from Griffin and Congress approval to try to reach a consensus among space scientists which would provide guidance for policymakers in funding decisions.[34] Thus, in March, an ad hoc group of two dozen senior scientists from various disciplines convened in a Washington, D.C., conference room to see if they could work together rather than as rivals in setting priorities. After hours of debate, the group found basic agreement on certain principles, such as the importance of smaller missions. What that meant for big missions, such as the MSL, was not clear.

The group did not identify specific trade-offs. The vice chairman of the SSB, George Paulikas, did not promise that the scientists could do that. Asked about how the scientists would spread pain among themselves, he told a reporter, "Stay tuned."[35] The ad hoc group would meet again to try to be more specific.

While NASA and the Mars scientific community fretted and argued about cuts to future opportunities, implementation of the existing Mars program continued to go smoothly. On March 10, MRO successfully sailed into the correct orbit to take the next step in Mars exploration. Launched in April 2005, the $720 million spacecraft carried six state-of-the-art instruments, including ground-penetrating radar.

MRO's goals were to provide new knowledge about surface features, reconnaissance for future landings, communications for future rovers and landers, and clues to life, past or present. In line with the scientific strategy, it would "follow the water." It had flown 300 million miles by the time it swung into Mars orbit. Then it disappeared for a half hour, dropping out of radio contact. When it emerged, it signaled that all was well.

"Look at that!" yelled an engineer at JPL mission control. "Right on the money!" shouted another.[36] McCuistion announced his delight with the way the mission was going. MRO would take over for Mars Global Surveyor, whose mission had extended far longer than expected. MRO, along with the European orbiter and two U.S. rovers, solidified the fact that spacefaring nations were creating what JPL director Elachi called a "permanent [robotic] presence" around another planet.[37]

On March 29, McCuistion came to a meeting of the SSB to present a possible Mars program that had been scaled back to fit into a $600-million-a-year expenditure stretching into the future. This was still the major program in the planetary sciences, he explained, but NASA simply could not do all that it had hoped to do. The agency was developing plans for the period extending from 2011 to 2016, and he wanted the board's views as soon as possible. In 2011, a relatively small Scout mission would go up under the reoriented effort. In 2013, NASA proposed to send the Mars Telecommunications Orbiter, which had been killed earlier. The next launch window would come 26 months later, falling in 2016. NASA was considering a successor to MSL called the Astrobiology Field Laboratory. This mission was aimed at finding evidence of life. If that mission did not work out, perhaps a repeat of Spirit and Opportunity, but with more sophisticated rovers, might be an option, he said.

Actually returning samples to look for life was not on the agenda at all, unless NASA could get other spacefaring nations to help finance such a mission. McCuistion tried to look on the positive side—$600 million a year was still a lot of money for one planet. But he could not contain his frustration: "Are we a little fragile?" he asked. "Yeah, we are a little fragile. But we still have a program that's viable for the next decade."[38]

In early May, the SSB again convened a group of leading scientists to help it determine priorities to recommend to NASA. Seventy attended, meeting at the University of Maryland at College Park. Speaking to the group, Griffin admitted, "I made a mistake. I made commitments in advance that I wasn't able to keep."[39] The scientists divided into four groups. One was devoted to the planetary sciences. The cut in science funding had fallen heavily on the planetary scientists. This was in part due to the nature of the field. Missions were separable, whereas astronomy consisted chiefly of a few large telescope projects, one of which—the James Webb Space Telescope—was suffering a huge overrun destined to grow. Also, astronomers were relatively cohesive on priorities, whereas planetary scientists seldom united.

As the largest subgroup among the planetary scientists, the Mars researchers especially needed to cohere around their preferences. The Mars scientists, however, were unable to agree on much. One mission with which they grappled was a small Scout mission scheduled for 2011. Should it go as planned or be modified, deferred, or cancelled? When the Mars scientists began to discuss the 2011 mission, six members of the group had to recuse themselves because they had proposals pending on this mission. "We can't very well make a decision to cancel the Scout mission after all the qualified people have left the room," said the chair of the group, Sean Solomon, of the Carnegie Institution of Washington. "We're going to punt. Our hands are tied by legal restrictions."[40]

Legal restrictions were not the only issue. Many of the planetary scientists were fierce competitors. When all was said and done, they could not get beyond general guidelines. The SSB called the existing NASA program "fundamentally unstable [and] seriously unbalanced."[41] The balance to which the SSB referred primarily involved the ratio of big and small science. But critics of this view at the College Park meeting noted that many "big science" missions provided a substantial number of subcontracts to individual investigators and their graduate students. Also, in the Earth observation field, many satellites were characterized as "moderate" in size. Where did such a concept fit in the overall scheme of expenditures?

The effort to involve a larger body of scientists in setting science policy at NASA was of limited help to agency officials as they struggled to define what NASA would do in the future. The next scheduled mission, to go up in 2007, was a Scout mission called Phoenix. A stationary lander, Phoenix made use of concepts that were intended for the ill-fated Mars Polar Lander. Sensors that had been devised for that failed mission could be put to use with Phoenix. Its very name came from the mythical bird that died in a fire, only to be reborn from the ashes. Phoenix would arise from the ashes of MPL.

Leaping Ahead with MSL

Then, there was MSL, the biggest and most important project of the MEP, set for 2009. In June, after years of planning, exploratory design and research, and continuing debate, MSL came up for Preliminary Design Review (PDR). This is a critical milestone in any NASA mission's evolution. It is the point at which NASA decides either to go ahead with full-scale development to meet a launch deadline or not to do so. It is the point when sufficient technical agreement is reached on design, schedule, and cost for top NASA officials to say, in effect, "go forward and implement."

Richard Cook, the project manager, and his team made their case to a review board headed by Figueroa during lengthy meetings at JPL.[42] Various technical experts and managers were present, and there was considerable participation. The PDR process took the whole week. Getting to this decision point had been tortuous. As with all landers, matters of safety and science had to be equated. In the years leading up to the PDR, NASA had reconsidered the question raised in 2001 by Isakowitz, when at OMB, about whether it would be better to send two MSLs to lower risk of failure. As Spirit and Opportunity were performing so remarkably well, some scientists also suggested sending a number of Spirit-Opportunity–scale missions to various sites.

JPL engineers recommended that NASA go beyond Spirit-Opportunity technology in a major way with one flagship mission.[43] There were also scientists who felt that breakthroughs were more likely to come with a single bold mission than with a number of smaller missions having incremental improvements over Spirit and Opportunity. Both scientists and engineers wanted to go beyond following water to detecting chemical building blocks of life. Out of the myriad technical discussions and debates prior to the PDR meetings, a consensus emerged—that NASA go for a truly significant MSL mission. The performance of Spirit and Opportunity was so outstanding that it gave confidence to those who favored the more ambitious approach to MSL. Moreover, MSR and the quest for life seemed to many to require a leap forward with MSL, rather than a more incremental strategy.

JPL and the Mars science community had devised a host of cutting-edge scientific instruments. They had conceived a new mechanism (called sky crane) to deliver MSL to the Mars surface. MSL had by now grown to be the size of an automobile and was thus heavy—and also very vulnerable to damage. It had to land with precision and delicacy, and JPL believed that retro-rockets or the kind of cushions Spirit and Opportunity used would be inadequate. But would this sky crane device work?

Many of the scientists and engineers employed on MSL had migrated from Spirit-Opportunity once its development was done. They brimmed with confidence, perhaps bordering on hubris. In 2005, Griffin had visited JPL and met with Elachi and the MSL team. "Can you do this?" he asked the scientists and engineers on the team. They said "Yes."[44] "Do you believe you can launch MSL in 2009?" Griffin asked Elachi. Elachi responded, "No problem."[45] Now, in 2006, the time for formal decisions had arrived, and grilling at the PDR sessions revealed at least two key problems that remained to be solved for mission success. One had to do with the motors that moved various parts of MSL. They were called actuators. NASA had opted for an advanced technology that would improve the capacity of the actuators to perform in extremely harsh places scientists wanted MSL to go. The other pertained to sample processing technology, critical for scientific understanding of the ingredients enabling life on Mars.[46]

There was considerable debate at the meetings, and Gentry Lee, who worked at JPL and had advised Hubbard when the MEP was planned, stood up and expressed skepticism with what the project team was saying. He believed that MSL could be built, but not by 2009. "Impossible!" he exclaimed.[47] But the consensus view was more optimistic that the problems could be solved quickly enough. "Yes, we can" was the mood at the end of the PDR process, and it was widely shared. MSL got the go-ahead from Figueroa and his review team. The message from Figueroa to the project management team was, "We trust you. Don't get cocky!"[48] Reviews up the chain of command led to the necessary approvals from NASA leaders, and MSL advanced to the next stage: full implementation.

Lee did not relent. He was convinced that the MSL team, as outstanding as it was deemed to be, was asking for too many technical miracles to launch in 2009. He went to Elachi to express his reservations. He was not able to persuade Elachi to hold back. Elachi believed that JPL would come through and had conveyed his confidence in MSL success to Griffin.[49] The PDR process established the best estimates technical people could make about design, schedule, and cost, as a basis for NASA decision making. The MSL cost estimate that followed PDR, and which became the number NASA used in dealing with OMB and Congress at the time, was $1.6 billion.

At the same time that the PDR process was under way in 2006, NASA was also initiating a process to determine where MSL would land. In late May and June, NASA gathered 120 Mars researchers together to initiate siting discussions. It was noted that Viking had initially been wrong in its siting decisions and was forced to make last-minute changes. Indeed, luck had played a major role in its successful landing. Spirit's initial site was predicted to be scientifically interesting, but it turned out not to be so. Only its roving capability saved it from disappointment. Opportunity's site was one that quickly proved fortuitous, however. With only one expensive rover and an exceptionally ambitious goal, NASA had to pick the right place. The group discussed a long list of possibilities and narrowed the list. However, this assemblage was just the first of several meetings.[50] Existing NASA missions at Mars would provide further information as time went on.

The question of NASA's post-MSL future was also still open. A panel of the SSB in July released a report responding to the plan for 2011–2016 which McCuistion had provided earlier. While generally supportive of McCuistion's proposal, the NAS panel suggested that NASA consider postponing the 2016 Astrobiology Field Laboratory mission until 2018 and instead use the 2016 opportunity to build a seismic network. It was concerned that NASA research not neglect Mars's structures and evolution in the agency's quest to find past or present life. The panel backed the notion of resurrecting the telecommunications orbiter that had been killed. Further, it called for beginning technology development leading to MSR. The panel said that given the financial constraints, future plans as presented by NASA were "not optimized."[51]

Griffin Splits with Scientists

What the SSB and scientists generally wanted to do cost a great deal of money, and Griffin was increasingly annoyed with scientific criticism and calls for NASA to do more when it had a flat budget. He had reconstituted the top-level NASA Advisory Council to reflect his and Bush's priority, the Moon-Mars mission. He had appointed Harrison Schmitt, former Apollo astronaut and New Mexico senator, as NAC chairman. Schmitt was not happy with the role some leading scientists played on the council. He wanted them to advise Griffin how to carry out existing priorities. They disagreed with the priorities. He complained they were not being useful. Griffin backed Schmitt, and on August 21 he sent a message to the NAC that revealed his frustration not only with the scientists on NAC but with his scientist-critics generally.

"The scientific community . . . expects to have far too large a role in prescribing what work NASA should do," Griffin charged. He noted that the community spoke of effectiveness in NASA policy. "By 'effectiveness,'" said Griffin, "what the scientific community really means is 'the extent to which we are able to get NASA to do what we want to do.'" He said that if NAC members wanted to have NASA take a different course than it was taking, "the most appropriate recourse" was "to resign."

The chair of NAC's scientific subcommittee, Charles Kennel, a former NASA Earth Science Division leader, currently director of the Scripps Institute of Oceanography, did resign. Griffin then personally requested that two other members, Wesley Huntress and Eugene Levy, the provost of Rice University, step down. Huntress countered, "This is a different NAC. Our advice was simply not required nor desired." The current council, he added, "has no understanding or patience for the science community process."[52]

While Griffin battled over policy with scientists in Washington, including Fisk and Huntress, former associate administrators of NASA's Science Directorate, NASA's operations on Mars continued to go extremely well and provide remarkable discoveries. In early October, Opportunity began complicated maneuvers 242 million miles from Earth at the massive Victoria Crater. This was potentially the most spectacular and significant target of the entire $800 million twin-rover mission. "We are frankly feeling a little overwhelmed by what we see so far," said Squyres.[53]

Later in the month, the first results from MRO yielded new evidence of diverse watery habitats capable of supporting life eons ago. MRO also found evidence of recent climate changes only hundreds of years apart.[54] As November began, MRO's predecessor, MGS, reported technical problems. Launched in 1996, it had been the longest-lived Mars mission in history, and one of the most productive. On November 21, Michael Meyer of NASA's SMD said at a press briefing at JPL, "We may have lost a dear old friend and teacher." He declared that MGS had "surpassed all expectations."[55] Its most important findings had come in its waning days of operation and were not announced publicly until NASA had carefully confirmed them in December.

A comparison of photos taken several years apart by MGS found that two gullies, at least, had apparently experienced flash floods between the photo shoots. "Water seems to have flowed on the surface of today's Mars, said Meyer at a December 6 news conference. "The big question is how does it happen, and does it point to a habitat for life?"[56] "This is the sort of thing you dream about, what everybody's been waiting for," said planetary scientist Jennifer Hellmann of NASA's Ames Research Center. The discovery lent support to the existence of liquid water so near the surface, at least in places, that it could spurt out on rare occasions.[57] MGS also found evidence of recent high-velocity impacts from meteorites. This finding was critical in that it pointed up hazards for human exploration. Either way, MGS could not have ended its life on a more significant note.

The Political Environment Grows Toxic

The macropolitical context of Mars exploration changed significantly at this time, making it even more difficult than before for Griffin and the scientific community to reach an accommodation. The Democrats recaptured control of Congress in the November elections. With Bush in the White House, relations in policymaking grew toxic. Fights between Congress and the president in November and December made it almost impossible to get budget bills through. This was bad news for NASA, space science, and robotic Mars exploration. The lobbying campaign by the Planetary Society and others had paid off in getting congressional bills that added money to science at NASA. Moreover, friendly senators had gotten the Senate to pass a $1 billion supplement to NASA, above its regular program appropriation, for shuttle recovery and Katrina-related repairs. Such legislation—if it had become law—would have been a great help to the agency in restoring some of the cuts to science, including Mars exploration.

But legislation of this kind was not to be. The best Congress could do was to pass a continuing resolution to keep most agencies, including NASA, funded at their current year's rate. This meant no raise at all, at least until after the Democrats had taken charge of the new Congress in January 2007. If the continuing resolution held the entire next fiscal year, NASA would have to cut deeper into its programs. Mars research could be further damaged.[58]

The year 2007 opened with NASA getting decidedly mixed signals from its political masters. In February, the new Congress, led by the Democrats, extended the continuing resolution that funded NASA and various other federal agencies at the 2006 level through the end of the fiscal year, September 30. In doing so, Congress gave some agencies small increases at the expense of other agencies that received modest cuts in order to maintain the overall figure. NASA was one of the agencies cut. However, the president's FY 2008 budget, also announced in February, gave NASA a raise, to $17.3 billion. That was a 3.1% increase over the president's 2007 request, which Congress did not grant thanks to the continuing resolution. So NASA had to do the best it could, Mars science included. The president's science advisor, John Marburger, suggested that space scientists curb their appetites and turn off missions before launching new ones.[59]

Griffin focused all the more acutely on his own priorities under the circumstances. Everything narrowed down to his view of the core mission, and programs were weighed in terms of their value to that mission. For Griffin this meant concentrating on the first phase of Constellation, Orion–Ares I, which promised a shuttle successor and technology development relevant to Orion–Ares V, the heavy-lift Moon rocket that would come later. Mars research gave way to lunar research. The Mars budget was far from what it had been projected to be when O'Keefe left the NASA Administrator's post. But Mars was treated better than many other planets or science projects. Astrobiology was especially hard-hit, not just by budget reduction, but also by Griffin's words. "If they [astrobiologists] want to work for government money," he declared, "they must look at what the government wants—not what they think it should want."[60]

Implementation of the existing MEP continued. The next mission in line for launch in the Mars program—Phoenix (2007)—was experiencing an overrun. NASA considered killing Phoenix, the first Scout mission, but wanted to stay with its launch-at-every-26-months strategy. NASA decided to meet the additional costs. McCuistion indicated that the Mars program had very little flexibility and that "the overrun on Phoenix was going to have some effect on us," which meant that NASA would have to take money from elsewhere in the Mars budget to pay for Phoenix.[61]

In 2007 MSL also revealed overrun issues. Solving the overrun problem for MSL was going to be much more difficult than for Phoenix given MSL's size, the criticality of the mission, and the scale of the potential overrun. Dealing with MSL would not be Cleave's problem, however, as she had retired in December 2006. Colleen Hartman served in her place on an acting basis. Griffin in February announced that S. Alan Stern would succeed Hartman in April, as associate administrator for science. Stern was executive director of the Southwest Research Institute in Boulder, Colorado. He was a well-known, respected planetary researcher and the principal investigator of the Pluto mission.[62]

A Stern Approach to Mars

On April 2, 2007, Alan Stern, age 50, joined NASA. A one-time astronaut candidate, Stern had a $5.4 billion budget to manage and a constituency up in arms. He pledged to wring more good science out of his budget and to stop "management by checkbook," that is, constantly adding money to projects beyond their original cost estimates. Either principal investigators would manage projects within costs, or they would risk project cancellations, he said.[63] "There are going to be things I do that cause pain."[64]

Griffin had appointed Stern in part to help him deal with the scientific community. Cleave had never been truly accepted by the community. Stern bolstered his office's status by appointing John Mather to be his chief scientist. Mather was a cowinner of the Nobel Prize for Physics in 2006 for his discoveries connected with the big bang. With Stern (planets) and Mather (telescopes) in charge, space scientists had to take the NASA science leadership seriously.[65]

Griffin, meanwhile, continued to criticize the scientific community. In May, he accused NAS of failing to take account of realistic costs in its decadal surveys of space science needs. The NAS SSB, he charged, routinely—and dramatically—underestimated costs and then complained when NASA scaled back or cancelled projects the science body favored. An NAS spokesman acknowledged that Griffin had a point.[66]

Stern reinforced Griffin, explaining that scientists were involved in a zero-sum game. Echoing Marburger, he noted that to make room for new projects, NASA would have to turn off long-running projects. One of the longest-running and most celebrated projects under Stern's aegis was that of Spirit and Opportunity. In May, Spirit made a major discovery. It analyzed a patch of Mars soil that was extremely rich in silica. This provided some of the most convincing evidence yet that ancient Mars was quite wet. The processes that generally produced such a concentrated deposit of silica required the presence of water.

"You could hear people gasp in astonishment," said Squyres, the principal investigator. "This is a remarkable discovery. And the fact that we found something this new and different after nearly 1200 days on Mars makes it even more remarkable. It makes you wonder what else is still out there."[67] Obviously, long-running, still-productive projects like the Mars Exploration Rovers would not go quietly, particularly when led by a scientist with a public relations sense like Squyres.

Stern was a change agent in temperament but, unlike Huntress, was uncomfortable and unskilled in bureaucratic politics. Stern was impatient with the routines and constraints of operating in a complex organization. Huntress was a quiet entrepreneur. Stern was overt and public. Stern called his administrative philosophy "pragmatism," and what he meant by that was "exchanging" more perfect solutions for more practical ones by using existing systems, modified to the least extent practical, to accelerate the pace of exploration.[68] He applied that approach to Mars. The goal was MSR—this is what counted. To move the date for MSR closer would mean modifying the schedule of missions he inherited and converting MSL into even more a means toward MSR than it already was. In the long-standing tug-of-intellectual-war between leapers and gradualists, Stern was emphatically a leaper in the sense of wanting to get to MSR quickly.

The problem he had was resistance to the changes he wanted to impose and suspicions on the part of many Mars advocates about his motives. This was especially the case because he was seeking to innovate in a major way in a time of budget stasis. To do the new, he had to cut back on the old, including Mars projects that did not fit into his "pragmatic" philosophy. And he was in a hurry.

Griffin focused on human spaceflight and was reluctant to cope with the scientific community—a group he found vexing. He told Stern he would have a relatively free hand to run SMD. Stern took him seriously and told his staff he "had the keys to the program."[69] He wanted to reshape it in accord with his priorities. Many worried Mars advocates saw him as an "outer planets" man, but he vowed he also had a strong interest in Mars. However, he intended to move Mars research in a better direction. For example, one month after he arrived, JPL came to him asking for more money for MSL. He wanted that practice to stop. What he really wanted to do was speed up MSR. As he recalled, "What I wanted was to give Mars Sample Return a higher priority. Since I was a boy, people have been talking about Mars Sample Return. I wanted to move it forward. . . . I wanted to get the first sample back as soon as possible—unlock the door. An imperfect sample return would be better than none at all."

He spoke to Griffin and told him about his MSR priority. Griffin responded, "Let's go." And Stern was off and running.[70] On July 10, Stern used a telephone hookup to speak to some 500 Mars scientists attending the 7th International Conference on Mars. He said that it was time for NASA to target MSR in a serious way. A new Mars astrobiology strategy recommended by the SSB set "analysis of a diverse suite of appropriate samples" as the highest-priority Mars science objective.[71] In keeping with this recommendation, Stern said NASA needed to reorient the existing program as soon as possible in spite of the constrained budget. He proposed to begin by attaching equipment to MSL which would allow it to capture a sample of soil and rock as it moved across the Mars surface. "I think there's something concrete about putting your stake in the ground," he declared. Retrieving the sample would come later. Such a return mission would be costly, he stated, perhaps $3 billion to $4 billion. To get that kind of money would require skipping a mission between MSL and MSR. However, he said the lost mission would be worth it, given the significance of MSR. MSR would build support for the planetary program, he argued, in the scientific community, public, Congress, and OMB.

He pointed out that even at the present scaled-back level that the Mars program had undergone, it still absorbed almost half of all the money the planetary program had—46%. The Mars community, he urged, should thread the needle. He warned that if the community did not opt for concentrating resources in the manner he described, the Mars budget would shrink. "That's my analysis," he said, "not my wish . . . that's my analysis of the way the politics will go." He called for an MSR mission in 2018. "Let's get this done . . . make some history," he exhorted.[72]

The Mars community reacted with considerable wariness. MSR was indeed the holy grail of the robotic program. It was the goal toward which the sequence of missions designed by Hubbard and implemented by Figueroa and now Mc-Cuistion moved, the culmination of the "follow-the-water" strategy. What sent a shiver through the community was the trade-off. Stern's comment about dropping a mission to get the money sent a signal of alarm. Which mission? There were many Mars scientists who were not astrobiologists and had other technical interests. And would omitting one mission be enough?

Philip Christensen, a leading Mars scientist and professor of geological sciences at Arizona State University, spoke for many of his colleagues when he declared, "I am concerned that the sample return mission would take over the Mars program. If you put that mission too far in the future with not much in between, then you lose a lot of momentum . . . a lot of young talented scientists and engineers." He saw "a real serious challenge" in carving out enough money in the near term to pay for MSR and still maintain a dynamic program.[73]

Zubrin wrote an op-ed in *Space News* entitled "Don't Wreck the Mars Program." He indicated that Stern was possibly thinking beyond killing one mission for MSR, to the point of considering sacrificing all missions, including the 2011 Scout project, to get money for sample return. Zubrin defended the robotic program as an essential precursor to human flight. "Since the origin a decade ago, the existing fly-every-opportunity robotic Mars program has proven to be a brilliant success." He claimed to be no fan of Dan Goldin, but he gave the former NASA Administrator credit for launching a "sustained exploration program involving frequent launches" which created not only an infrastructure on Mars but a "proficient team competent to carry out ever more complex Mars missions."[74]

Stern was undeterred by the criticism. He directed Ames to design a caching box for the MSL. Chris McKay, one-time Mars Underground leader and now an astrobiologist at Ames, was one who supported the push for MSR by starting with changes in the MSL rover. Indeed, he wanted to go further. He called on NASA and the Mars community to think not of one sample return mission, but a program of missions. The first sample return, he said, should be a "simple, pathfinder-like sample return . . . a technology demonstration."

By using MSL to cache samples, NASA would get people to begin focusing on sample return as a goal, said McKay. "It ties sample return to the ongoing program. There's a tendency to think of sample return as something 'out there.' . . . It doesn't need to be. It can be something in the Mars program." McKay argued that sample return had to "connect, ultimately, with human exploration of Mars."[75]

The European Space Agency, meanwhile, expressed interest in cooperating with NASA on an MSR mission. NASA indicated openness to the possibility, and discussions began in a very general, long-range way.[76] Stern had wasted no time in putting his stamp on the implementation of MEP.

As Stern planned what the next development steps in the Mars program would be, the operating program he inherited continued to move forward. In August, NASA launched the first Scout mission. In contrast to the other projects, this mission was generated through a competition in the scientific community and largely run by non-NASA scientists. A Scout mission was intended to be smaller in cost and personnel than a typical NASA/JPL venture. However, it had to be relevant to the NASA strategy, i.e., follow the water.

The goal of Phoenix, as designed by the University of Arizona's Peter Smith, was to go near the Martian North Pole and "touch" water ice. From Lederberg's entreaties at the time of Viking, there had been the view that near the poles there would be water (in the form of ice) and evidence of possible life. Phoenix would not be equipped to actually determine life issues. Moreover, it was stationary, not a rover. But at $400 million, it did carry a digging capability. It would try to penetrate the ice and see what characteristics it had that might be favorable or unfavorable to life. MPL had had this objective, but it had crashed. Phoenix "rose from the ashes" to take MPL's place.[77]

Reshaping MSL

In June, MSL went through a Critical Design Review (CDR), the most significant decision point since the 2006 PDR. It revealed that some of the problems that had surfaced at the PDR, especially those of the actuators, had not been solved. NASA would have to go back to more conventional actuators, and that would add money and time. Figueroa, who chaired the CDR, warned Stern he would need to put more money into MSL and should not make any unnecessary changes in design. But Stern was determined to speed MSR. By September,

Stern faced the reality that if he wanted to add a sample-collecting capability to MSL, he would have to subtract certain other capabilities. The problem was that costs kept going up. He had taken office promising to end what he called management by checkbook.

The issue came to a head over an amount that was relatively modest—$75 million in a project now costing $1.7 billion. The issue was that this was the most recent of a sequence of cost increases. More importantly, Stern saw a need to hold the line, or admit defeat in his get-tough management approach. Thus, he ordered the MSL project manager to omit two instruments, cap others, and alter certain design elements. Doing so, in his view, would avoid the overrun, while also providing scope for his sample return addition. Stern called the changes "low-impact mission scope reductions." In discussing his decision September 19, he stated, "I've spent all the reserves for the Mars Exploration Program for next year. The next check I write results in cancelling a mission or mission extension." He warned that he had even considered terminating MSL.[78]

The Planetary Society, led by Huntress, decried the reductions in capability in MSL. The Society sent letters to U.S. lawmakers urging them to block implementation of the cutbacks until Congress could "evaluate them in the context of the overall NASA budget." It charged, "The loss to science on MSL seems out of proportion. The goal of MSL is to conduct science, and to throw out so much of the mission science objectives for less than 4% of the mission cost, and for assurance costs that have not yet been realized, seems penny-wise and pound foolish."[79]

The Stern decision caused particular dismay for those contractors directly affected. They sought to find ways to deal with the situation. Stern struck a hard bargain. In November, he announced the outcome of negotiations. The two primary devices to be deleted would be restored, he said. These were the Mars Descent Imager and the Laser-Induced Remote Sensing for Chemistry and Micro-Imaging Instrument.

Stern explained in a letter to the Mars science community that he agreed to restore the descent imager because its maker, the Malin Space Science Systems Company, "has agreed that there will be no additional costs to NASA." As for the other laser instrument, he said the principal contractors had found ways to significantly reduce its costs to NASA. Those costs were down to $400,000, a figure that caused Stern to declare "victory" in his negotiations. "The outcome," he said, "is even better than we had imagined possible in September."[80]

Decision making for the Mars program under Stern increasingly revolved about MSL. It was emerging as a flagship not only for the Mars program but for NASA generally. As its costs rose and debates swirled over what kind of equipment it should carry, the question of where it should land on Mars also simmered in the background.

NASA had a steering committee for the MSL site selection. This committee in late October brought together a large assembly of leading Mars scientists to narrow the number of places MSL might land. The group met for two days in a Pasadena hotel. The group represented various disciplines, including astrobiology.

There were 51 possible sites discussed. As in the past, safety and scientific potential were critical values to balance. Each site was discussed at length, and NASA said additional sites could be nominated. The meeting became "boisterous" as strong-willed individuals advocated their choices. Majority votes were taken. When the meeting concluded, the list stood at six. "A lot of people subverted their interests [in a particular site] to the science. This degree of community participation is one reason the Mars program has been so successful," stated David Des Marais, a geochemist at NASA-Ames.[81]

NASA said that with the help of the steering committee and other scientists, it would decide on a single site in nine months. It was still uncertain what would come after MSL, but preparations for this project moved ahead.

As 2008 began, Congress finally passed an omnibus budget bill to keep the government running. Its most important impact on NASA was that while continuing to keep NASA spending relatively flat, the Democratic majority made modest changes in the science budget reflecting a desire to raise the priority of Earth science.[82] For years, this NASA program had been constrained by the Bush administration's relative disinterest in the climate change issue. The Democrats increased spending on the field. But without major enhancements of NASA funding overall, the stress on Mars spending worsened.

The Mars program suffered another significant blow in early 2008 when NASA had to postpone its next Scout mission, an orbital project to study Mars's atmosphere, from 2011 to 2013. This decision, due to a conflict of interest discovered in the proposal competition, marked the first time in a decade that NASA would miss a Mars launch window.[83]

Fisk, as chair of the NAS SSB, declared that the way NASA was going generally, and in science particularly, was not "sustainable." As 2008 was an election year, he, Hubbard, Huntress, and many others outside the agency discussed ways they could get a message to the next president that NASA was in trouble, as well as suggesting what might be done as remedy.[84]

In early February, Bush released his FY 2009 budget. Once again, *Science* magazine reported, the president put NASA "between a rock and a hard place."[85] With virtually everything squeezed within NASA's $17.3 billion budget, the science program was held to a 1% increase, or $4.6 billion. Griffin knew that the budget was "painful" to scientists, but he had his gargantuan problems with the human spaceflight program and Bush's failure to support it. "You're only going to get so much," he told the community. "Suck it up and live with it."[86]

The Stern Strategy

Stern wanted bold projects in spite of the budget constraints. As the principal investigator of the Pluto mission, he was an outer planets advocate. He wanted an outer planets flagship mission to go forward under his watch, to either Jupiter or Saturn. He also, as he had made clear, wanted to depart from Mars incrementalism to a truly great leap forward—MSR. He thought big, envisioning a sequence of flagship missions: MSL, outer planets, MSR. How to pay for these larger missions was the question.

The NASA budget and its longer-term projection that appeared in February 2008 revealed his solution to the problem. Rather than continue the inherited strategy of missions every 26-month opportunity leading gradually to MSR at some indefinite future date, he decided that money for Mars would be drastically curtailed after MSL in 2009, thereby enabling other non-Mars initiatives to take place. The budget would then rise significantly later in the decade to finance an MSR mission in 2020. This was indeed a daring move, and Stern, confident and fast-moving in style, did not necessarily consult widely either in NASA or externally. He also riled Mars advocates by being quoted within NASA as saying "Mars is just another planet."[87]

The White House budget office had been sufficiently surprised by the radical change in approach that it asked a group of scientists independent of NASA to take a look at Stern's new program strategy. It was a marked departure from the steady "follow-the-water" sequential approach up to now. OMB acquiesced to the Stern strategy, as did Stern's superiors at NASA. When the projected Mars spending in the NASA budget became public, however, most Mars scientists were shocked. Many charged that Stern was hollowing out the Mars budget to get resources for his outer planets flagship. Robert Braun, professor of space technology at Georgia Tech, wrote that Stern was "gutting" the Mars program. With no missions slated for 2011 and 2018, and only tentative missions in 2016 and 2020, the budget "puts the future Mars program on a path toward irrelevance," he charged. He called on the Mars community to grill Stern vigorously at the next Mars Exploration Program Analysis Group meeting, on February 20.[88]

MEPAG was an outgrowth of the ad hoc meetings of Mars scientists which Hubbard and his associates sponsored to receive broad input from Mars scientists generally when they were designing the new program in the year 2000. It had been subsequently institutionalized as a two-way communications mechanism between NASA and the scientific community. MEPAG now met two times a year.

Philip Christensen of Arizona State University, chair of the White House–established panel, came to the MEPAG meeting, held at Monrovia, California, to give his panel's assessment. As Stern sat in the front row, he heard Christensen blast the new strategy. "The phasing [of missions] is just wrong," said Christensen. "Our assessment is that it just won't work," he said, pointing out that it would take substantial funding early in the next decade to develop the technology for MSR in 2020. MSR entailed "launch vehicles, a Mars landing system, an Earth return vehicle, a Mars lander, a Mars ascent vehicle, a rover, an Earth-reentry system, and a sample-receiving and curation facility on Earth."[89]

Indeed, Christensen said, to prepare and fly in 2020, NASA would have to cancel everything after MSL, including 2013 and 2016 missions, to do what Stern wanted to do. "You have to come clean," Christensen pointedly told the NASA associate administrator. "Either you fund the [Mars] program or you accept the fact that it will be significantly reduced for the next decade."

Stern defended himself and his strategy, stating, "No missions have been cancelled—none, zero, zip, nada." "The Mars program is really healthy," he told *Science* magazine, which reported on the meeting. But various Mars scientists countered him. "I don't think many people accept his budget," which went precipitously down and then abruptly up, said Bruce Jakosky of the University of Colorado Boulder. "We just don't see how you connect the dots," declared John Mustard of Brown University, who chaired NASA's Mars Advisory Committee as well as the MEPAG meeting. Moreover, Mustard criticized Stern's decision to add a $2 million cache to MSL. He said he did not believe it would be able at this early date to pick up enough scientifically interesting material to be worth the expense.[90]

Stern had few vocal supporters in the Mars community. McKay did stand up at the MEPAG meeting and defend what Stern was trying to do. He believed that the Mars program was "in a rut" and needed to "turn heads."[91] But Stern also turned heads with his money decisions. Just the year before, McCuistion had said there would be stability in Mars spending at the annual $600 million level. That number itself had reflected a substantial cut from a previous projection. Now it was falling to $300 million under the Stern strategy.

Where Stern had support was in the outer planets community, which was projected to receive $3 billion for a Jupiter or Saturn mission. As Frances Bagenal, of the University of Colorado Boulder, chair of the Outer Planets Advisory Group and co–principal investigator with Stern on the Pluto project, explained, "Alan is trying to do the right thing by offering something to keep everyone happy. But it's impossible." In any event, said Bagenal, "it's time to take a break from Mars and work on other things." Stern said he was trying to have a "balanced" program. Stephen Mackwell, of the Lunar and Planetary Institute in Houston, commented that there would be "winners and losers" given the budget constraints. Most Mars scientists definitely thought themselves "losers" under Stern's new approach.[92]

Charging Ahead

If Stern was deterred by the negative reaction from the Mars constituency, he did not show it by his actions. Moreover, MSL continued to go up in costs, complicating his ardent managerial effort to keep expenses down. More and more, MSL became a test case for Stern and his campaign for reform and desire to reshape the Mars program.

He went to JPL and came back convinced it could not make the 2009 launch deadline without raising costs even more. He looked at the NAS National Research Council Decadal Survey published years before, which had called MSL a "moderate" mission of $650 million. How had it gotten so expensive? The answer for Stern was "duplicity" on the part of JPL and the Mars science community. They had gotten away with an elaborate increase in instruments and sophisticated technology, and NASA's review committees either had not seen the problems or had ignored them. "It was Christmas in the middle of the year," Stern complained. He called MSL "a ticking time bomb" in terms of cost. He wanted to descope "profligate instruments." He wanted to stop spending money on a crash basis to meet a deadline that he did not believe JPL could meet. By extending the deadline to the next launch window, 2011, he believed he could make changes in MSL that might control costs while also making it more relevant to MSR.[93]

Figueroa—who was implicitly a target of Stern's charges about review committee failure—opposed Stern in the debates about MSL within NASA. So did Garvin, who was outraged by Stern's actions. These officials made the case that it was far too late in the development process of MSL to make radical design changes. Doing so would slow down the project even more and further raise costs. Moreover, they defended MSL against the charge that it had grown enormously from the $650 million Decadal Survey figure Stern used. MSL was not the Mars Smart Lander that was in the original Hubbard plan. Stern, they said, was comparing apples and oranges. Others entered the fray, including John Grotzinger, Caltech professor and chief scientist for MSL, who defended the project.[94]

The debate over MSL escalated to associate administrator (and general manager) Chris Scolese. Scolese conducted his own analysis and sided with Stern's opponents. He agreed it was too late in the development stage for significant changes other than those absolutely necessary. Beyond technical issues were institutional commitments with JPL, industry, and the Mars science community which could not be disrupted.

The issue of whether to delay MSL from a launch in 2009 to 2011 went to Griffin for decision. Stern argued that JPL would not make the schedule and hence NASA should hold up expenditures to the laboratory sooner, rather than later. That saved money could then be spent more wisely. Delay, however, would cost money in the long run, too, and affect other projects, so the decision to postpone launch raised extremely tough issues either way. Griffin decided to keep going with the existing schedule. He told Stern, "Run for the cliff." When Stern protested that JPL would not make it, Griffin responded, "Try!"[95]

The Spirit-Opportunity Decision

That decision by Griffin to "run for the cliff" meant that additional money had to be spent on MSL to add personnel and their time to make a 2009 launch possible. Exasperated, looking for savings in other projects wherever he could find them, Stern and his planetary director, James Green, sat down and went through a list of SMD expenditures. They started with Mars activities, since the MSL problem was within the Mars program. Green characterized the effort as a way to "sweep" through the budgets of various missions to collect unspent money in the current fiscal year which could be redeployed to speed MSL. Stern saw the exercise as making the innocent pay for the sins of the guilty, a practice Stern believed was unsavory. At the same time, he felt there were projects whose science return was less than others. In Stern's view, that group of diminishing science return included extended missions for Spirit and Opportunity.[96]

At Stern's direction, on March 19, Green sent a letter to JPL ordering it to hold $4 million in unspent money from the MER account. This was money NASA expected not to be used in the current fiscal year and that would carry over to the next fiscal year if unspent. The amount was trivial, but symbolic for Stern in the sense that "Mars should pay for Mars."[97] The letter explicitly stated that the purpose was "to provide additional funds for Mars Science Laboratory (MSL) in fiscal year 2008, while determining if MSL will be able to achieve the established 2009 launch date."[98] Immediately, the letter leaked to the media, causing a firestorm of protest. Squyres declared that the cut, small though it was, would put one rover into hibernation and the other into part-time operation. The congressman whose district included JPL came to Spirit-Opportunity's defense. This was Adam Schiff (D-CA), who happened to serve on the House Appropriations Committee that funded NASA.[99]

Griffin heard about the cut from the media, and he reacted quickly and publicly: "Closing down either of the rovers is not on the table."[100] Stern believed Griffin had given him "the keys to the program," and Griffin had indeed granted him great leeway to make decisions. However, Griffin was not happy with the way Stern had handled decision making on MSL, and he had certainly not helped smooth Griffin's relation with the scientific community. Most importantly, in this Spirit-Opportunity case, he had failed to inform Griffin of a decision with extreme public visibility and political sensitivity—even if out of all proportion to the money involved.

On March 25, Griffin and Stern met in the NASA Administrator's office, and Griffin, who was known to be "direct," let Stern know sharply of his displeasure. Stern tried unsuccessfully to defend himself. He also stated, "If you have lost confidence in me, I should go!" Stern then went back to his office. An hour later, Griffin summoned him to return. "I accept your resignation," he said.[101]

The same day, NASA released an official statement: "This letter [to JPL] was not coordinated in the Administrator's office, and is in the process of being rescinded. The Administrator has unequivocally stated that no rover will be turned off."[102] While Spirit-Opportunity was the immediate reason for his leaving, Stern attributed the underlying cause to pressure from JPL and the Mars science community. "The knives came out over MSL," he said.[103]

Weiler Takes Control

Griffin quickly asked Ed Weiler to leave the Goddard Space Flight Center, where he was director, to return to his former position of associate administrator for science. Weiler took the job initially on an acting basis, but soon was appointed permanent associate administrator. Weiler knew he faced a controversial situation and a program in disarray. His professional background had been with space telescopes, and hence he was viewed as having neither an "outer" nor "inner" planets bias. Moreover, JPL and the Mars scientists knew that Weiler had played a strong role in building up the Mars program under Goldin and O'Keefe.

Weiler appreciated the public's fascination with Mars. He shared the view that MSR was the long-term goal of the robotic program. He did not agree with Stern about the need to go there sooner at the cost of interim missions. He was conscious that MSL was a project in trouble, and that another decision to delay or not delay was likely to arise. Unlike Stern, he knew he had to have influential allies on a possible delay decision, if that in fact was required. A delay would indeed cost additional money, and MSL was already consuming much of SMD's planetary budget. A delay decision could also have cascading effects on other missions. A delay decision would not come easily. JPL was still pushing for a 2009 launch, and so was the NASA Administrator.

A seasoned bureaucrat who was politically astute, Weiler announced he would revisit the 2009 Mars budget plan, given "all the criticism." "When I left the SMD," Weiler stated, "we had a program in place of taking advantage of every [Mars launch] opportunity and someday, when we have the data we needed and all the science we needed, we'd spend the billions of dollars for a Mars Sample Return."[104]

What we need to know, Weiler said, is this: "What is the real cost of Mars Sample Return? Does the community want to pay that cost in terms of missions that could be done in the interim or not done?" Aware that Stern had taken an especially strong stand on the issue of cost control, Weiler promised he would watch costs very carefully and would cancel projects if necessary. "On the other hand, I'm also going to make sure that programs aren't nickel and dimed to save a few cents, because I have direct personal experience where cost was the only concern. And that was Mars '98 [the two Mars missions launched in 1998 that failed in 1999]. Do you remember that little baby? And what I got for good cost control on that program was two craters on Mars."[105]

He also commented on some other lessons learned from experience: "There are three things you don't do at NASA. That is cancel Spirit, Opportunity, or Hubble."[106] Finally, Weiler made it clear organizationally that he understood the status of Mars as "first among equals" in the planetary program. He inherited the organizational structure that placed the Mars director under another individual, James Green, who was in charge of planetary programs. Weiler dealt with this individual for planets other than Mars. Weiler wanted an integrated planetary program. But when it came to Mars, McCuistion could work with Weiler directly.[107] The perception—as important as reality in government—was that Weiler intended to restore Mars's status to where it had been during his first tour as associate administrator for science, organizationally and in funding.

While controversy abounded in Washington and the planetary science community over the future of the Mars program, Weiler found that the existing operations on Mars were going extremely well. Indeed, the Phoenix mission illustrated how a sequence of missions, coming every 26 months, each building on the one before, could succeed.

In 2002, the Odyssey orbiter had detected substantial amounts of water ice lying just beneath the Martian surface. It was Phoenix's task to follow up and dig for ice in this northern polar region. The principal scientist of Phoenix, Peter Smith of the University of Arizona, had written his Scout proposal as a clear follow-up to Odyssey.

On May 25, Phoenix descended. When it landed successfully, after several nerve-wracking minutes, the celebration began. "It was hugs, cheers, and high-fives all around." Griffin was there in the control room at JPL. He pointed out that Phoenix marked the first successful landing without airbags since Viking in 1976.[108]

In June, Phoenix was put to work, its robotic arm extending to shovel soil and ice to its lander instruments. "This is an incredibly science-rich location," exclaimed Smith. In short order, Phoenix "found proof" of water-ice on Mars which was away from the polar caps. Then, later in the month, Phoenix discovered evidence of mineral nutrients in Martian surface material which would be essential to life. Rather than being hostile, the soil results were "friendly" to life. In July, Phoenix detected water vapor coming off a scoop of Martian soil. A tiny oven on the lander heated the dirt until ice mixed with it and evaporated. For the first time, instruments touched water and did more. "We have tasted the water and it tastes good," said William Boynton of the University of Arizona. "It's something we've been waiting quite a while for."[109]

August began, and rumors flew that Phoenix had made discoveries that precluded life, and then that it had found something so profound about life that NASA had contacted the White House. The reality, as NASA and Phoenix leaders explained at a hastily called teleconference with reporters on August 5, was that Phoenix had indeed made an important discovery, although not one that was necessarily worthy of presidential notification. It was the presence of perchlorate, a chemical compound commonly found in the Atacama Desert of Chile, one of the driest places on Earth. Atacama was often used as an analogue for conditions found on Mars. Researchers had found "extremophiles" at Atacama. Some of them survived on perchlorate. The Phoenix team held that the finding was highly favorable to the possibility of microbial life on Mars.[110]

As Phoenix fascinated scientists and received considerable media attention, Weiler took increasing control of the science program. He confirmed Stern's plan to launch a flagship outer planets mission to either Saturn or Jupiter as soon as possible, but perhaps not as quickly as would Stern. He dropped the approach that would spend relatively little on Mars in the early and middle years of the second decade and maximize spending as 2020 approached. Weiler wanted missions at regular opportunities throughout the decade, although it remained to be seen what they would be. Like Stern, Weiler saw MSR as the goal of the robotic science program, but he was willing to let it recede beyond 2020 to allow for adequate funding of missions at periodic, nearer-term launch opportunities.

Meanwhile, he firmed up plans for the first flight beyond MSL. Because of the conflict of interest that obviated the 2011 mission, this Scout project would go up in 2013. It would be called MAVEN, for Mars Atmosphere and Volatile Evolution. An orbiter, MAVEN's task was to study Mars's atmosphere and inquire how it had dissipated. With so little atmosphere, there would be no water on Mars's surface. The cost would be $485 million.[111]

MAVEN might be followed in 2016 with a rover potentially costing $1.4 billion. This rover could test equipment useful for an eventual MSR mission.[112] An international planning group now estimated the actual MSR mission to cost up to $8 billion or more—an expense that would require international sponsorship.[113] Stern had said he would try to get partners to supplement U.S. expenditures for his 2020 MSR mission. Weiler stated that he would have to do the same for a post-2020 mission.

While the goal of MSR was clear, Weiler's date for this mission was not, and the projects leading up to it still had to be clarified. Weiler understood that the follow-the-water decade established when he was formerly associate administrator was ending and a new strategy had to take its place. If Stern's approach was not acceptable—and the Mars scientist, Mustard, called it "smoke and mirrors"[114]—Weiler had to develop a substitute long-term strategy, even as he helped make exceedingly difficult decisions about the immediate challenges facing the program.

On October 10, Weiler, Griffin, and others connected with the Mars program met to determine MSL's fate. JPL reported that it was still having technical problems that would cost perhaps another $100 million to resolve. Moreover, making the 2009 launch window would be quite difficult, but still possible, JPL said. The cost of MSL was now pushing close to $2 billion. If NASA had to delay the mission to the next launch window, two years off, NASA would have to spend even more. Cancellation was a possible option, but Griffin did not consider it seriously. Once again, he decided to retain the October 2009 launch date.

However, Griffin said he would meet again with Weiler and others in January 2009 for a final decision, based on program progress between October and then. Afterward, Weiler explained the rationale for the approach to the decision: "It's easy to say, 'Let's cancel it and move on,' but we've poured over a billion and a half dollars into this. This science is critical. It's a flagship mission in the Mars program and as long as we think we have a good chance to make it, we are going to do what we have to do."[115]

Stern, outside NASA but assertive in his views, told the media he would raise two questions about the decision: First, "How much new damage will this do to other parts of the Mars planetary and wider SMD program to continue on this path; and secondly, will the time for testing before the launch in '09 really be sufficient to guarantee that this [more than] $2 billion investment will work?"[116]

On October 31, Stern wrote the editor of *Science* that a 2020 MSR launch could have been possible had NASA supported his cost-control policies, especially early in 2008, in regard to MSL. But "higher levels" at NASA rejected his proposed options for addressing MSL and other problems.[117]

While JPL worked overtime, and NASA Headquarters pondered what to do about MSL, the national elections took place. On November 4, Barack Obama was elected president and given a Congress with a strong Democratic majority. During the campaign, he had said he was a strong supporter of space and would increase the NASA budget. However, his specific space policy priorities were unknown, and advocates of various positions were eager to influence him to favor their claims.

Stern looked to the new president to address the ills of NASA. He published an op-ed in the *New York Times* on November 24, lamenting NASA's "Black Hole Budgets." He asked Obama to extirpate the "cancer [that] is overtaking our space agency." The cardinal example of the overrun issue was the "poorly-managed, now over-$2 billion Mars Science Laboratory." He again pointed out that when he fought the cost increases, "I found myself eventually admonished and then neutered by still higher ups, precipitating my resignation earlier this year."[118]

Garvin, who had fought Stern inside NASA, countered Stern's outside campaign against MSL in a letter to the editor of *Science*, November 28. He pointed out that MSL was less costly in current dollars than Viking and would have far more capability. The cost had grown, he said, because of technical problems not foreseen. Nevertheless, thanks to MSL, he argued, NASA was "ready to assault the Martian frontier."[119]

That assault on Mars would not now include the $2 million sample cache Stern had added to MSL. In a decision widely viewed as a rebuke to Stern, NASA removed this device. A NASA spokesman said it was done after "extensive interaction with the science community" had indicated its low science value. A science advisory group told NASA that the cache, coming late in MSL's development, "was likely to complicate MSL operations, leading to interruptions in the activities related to the prime mission."[120] Deleting the cache, NASA stated, would also save money on MSL.[121]

Delaying MSL's Launch

Weiler and Griffin deeply wanted to launch MSL in 2009 and follow that up with MAVEN in 2013 and a possible rover capable of testing technology for collecting samples in 2016. Toward that end, they had wanted to hold to the 2009 schedule until the last moment possible for decision in January. JPL told headquarters, "Trust us." But Weiler said, "Trust and verify." In November, Weiler established an interim milestone that would have to be met in December. He did not want to take decision making to January—the point Griffin had set as the time to choose to go forward or not to do so.[122]

At JPL, in November, there was turmoil, as the laboratory struggled mightily to make the deadline. Elachi was like Griffin in deeply wanting to launch MSL in 2009, and he was holding out hope that JPL would come through. Gentry Lee and others had persisted in arguing that JPL could not make the deadline. The laboratory was working double shifts, pursuing technological solutions to various problems in parallel efforts, and spending money accordingly.[123]

Rob Manning was the chief engineer on MSL and emphasized that the attitude at JPL is such that when critics say "can't do," you show them "can do!" But Manning saw several problems with MSL in November which remained to be mitigated and probably could not be in the brief time left. Weiler sent Figueroa with a review team to JPL, and Manning was the bearer of bad news. Weiler and his advisors had established certain requirements that had to be met in December. JPL would be unable to meet those requirements, it was increasingly clear. "We had hoped we would come up with some tricks to get us through. We ran out of tricks," Manning recalled.[124]

The "bad news" went to Elachi and Weiler. Weiler knew what was happening through McCuistion. McCuistion advised Weiler to pull the plug. Telephone calls went back and forth. The principal parties agreed—Elachi, Weiler, and then the associate administrator, Scolese. A consensus view went to Griffin, who concurred.[125] In pushing for a decision sooner than later, Weiler made what he regarded in hindsight as his "wisest decision" since returning to head SMD.[126] He could have spent a very great deal of money pushing MSL to "the cliff," and then would almost certainly have retreated to a delay decision, or even worse—pushed ahead. It was better to admit reality sooner than later and use the money for other needs. Ironically, Stern would have made the same decision—to delay—even sooner. But what Weiler did was build a coalition of support for the decision to take to Griffin.

On December 4, Griffin, Weiler, McCuistion, and Elachi held a joint press conference in NASA's auditorium to announce the final decision. It was to delay MSL to 2011. That delay would add $400 million to MSL's costs, pushing it to at least as much as $2.3 billion, and causing financial distress to other Mars and Science Directorate projects. The media reported that the four managers looked "grim-faced" as each spoke to reporters and gave views concerning the decision.[127]

Griffin said the major issue was with actuators, gears that would move MSL's wheels on Mars. NASA could have continued to rush to solve the problems, and might even have succeeded on time, "but we've determined that trying for '09 would require us to assume too much risk—more than I think is appropriate for a flagship mission like Mars Science Laboratory." Griffin made it clear he regarded MSL as NASA's top priority in the science program and one of NASA's top priorities overall, right after the human spaceflight activities. Asked about the increase in costs, Griffin responded that the rise was a function of the technical challenges of MSL. "We know how to control costs—just build more of what you built the last time."[128]

McCuistion commented on the technical issues, saying that NASA could not take a chance with the actuators. If they did not function properly, the $2-billion-plus vehicle would turn into "a metric ton of junk." Elachi emphasized how hard JPL had worked to make the deadline and how disappointed he personally was, but he agreed with the decision to delay. Weiler noted there would be "impacts" of the decision that would be negative, such as postponing other missions. However, he also cited a possible silver lining. It would help force NASA to move in a direction that made long-term sense for the next stage of the Mars program. He had met the day before, he said, with his ESA counterpart, David Southwood. Southwood had money and schedule problems with ESA's top Mars priority, ExoMars, scheduled for 2016. Southwood needed to collaborate to help pay for this expensive rover mission. Thanks in part to the MSL overruns and delay, NASA also needed help with its Mars missions, beginning in 2016. Finding a way to work together on future missions, said Weiler, was a "no brainer."[129]

CHAPTER TWELVE

Attempting Alliance

On January 20, 2009, Barack Obama became president, and Congress tilted more dramatically in the Democrats' favor than was the case under Bush. With economic crisis at home and wars abroad, Obama had many priorities. Space policy did not appear to rank high on his agenda, although he extolled Apollo as an inspirational event. Obama's style was consensual, and he conferred with various senators, especially Bill Nelson (D-FL), in choosing a NASA Administrator. It took him a while to find someone.

As the Obama transition got under way in the early months of 2009, and as he waited for a new NASA Administrator to take command, Weiler plunged ahead with his own agenda.[1] He had two immediate Mars tasks: one was to reconstitute the Mars Science Laboratory for the next two years, and the other was to develop a robust program for the years after MSL which would get to the long-sought goal of Mars Sample Return. As far as he was concerned, he had inherited no real program beyond MSL. He had problems financially, in part because of MSL and its overruns. But he also saw opportunity if he and the European Space Agency's Southwood could mount an effort to Mars together. The challenge was to get NASA and the new administration to commit to a long-term program. He and NASA needed a multimission successor as the existing Mars Exploration Program ran its course. The bilateral effort seemed a "win-win" for the United States and Europe. But NASA and ESA were trying to sell this big science program at a time when resources on both sides of the Atlantic were hard to obtain. What worked internationally depended—at least for NASA—on what transpired domestically. Domestically, the space policy subsystem would face upheaval due to the impact of presidential and congressional political struggle. The larger conflicts in Washington were mainly about human spaceflight and overall budgets. But robotic Mars policy could not be protected from these macro events.

Reconstituting MSL

Weiler's first task was to consider any changes in managing MSL, delayed two years. Naderi, the Jet Propulsion Laboratory's Mars director, had been succeeded a few years earlier by Fuk Li. The consensus in NASA was that Li was hamstrung in overseeing MSL by Stern.[2] Indeed, JPL and the Mars science community pointed fingers at Stern for many of MSL's problems. The consequence was that the only significant personnel change Weiler made was to move the JPL project manager, Cook, to a deputy slot. Pete Theisinger, the project manager for Spirit and Opportunity, and one of JPL's most respected managers, was put in charge. The fact that Cook remained attested to the fact that NASA and JPL held him in high regard and believed that his expertise was essential. But NASA leaders also wanted to show Congress, the Office of Management and Budget, and others that it would not tolerate cost overruns.[3]

Cook soon developed a plan for Theisinger, JPL, and NASA showing how the various technical issues that had led to delay could be mitigated. The plan was approved up the line, and additional organizational and personnel adjustments made. Then, JPL got to work in making the technical improvements. Actuators were the prime culprit, but so also were avionics, sampling instruments, and other technologies. As Cook reflected in a paper he subsequently wrote, "As the project got bigger and more complex, the problems grew not linearly, but geometrically."[4]

It was not long before NASA and JPL began feeling fortunate that the decision to delay was made. But the cost increases in MSL went beyond the $400 million calculated at the time the decision was announced in December 2008. Soon MSL was up to $2.5 billion in total costs. The good news was that some of MSL's harshest critics, such as Lee at JPL, were seeing MSL as viable for a 2011 launch.[5]

As Weiler saw the prospects of MSL improve, he put more money into the Mars program. He tried to replenish it after years of cutbacks. But he needed to plan for the post-MSL future and how to pay for it. He saw a Mars Together strategy as imperative.

Mars Together with ESA

Weiler and Southwood were old friends who had been talking about collaboration for years. They saw eye to eye on the importance of Mars in planetary science and public support for space generally. Both had Mars programs with large ambitions and money shortages. From the time that Weiler had returned to head the Science Mission Directorate, the two men had speculated, with increased intensity, about linking their programs to a degree never before attempted. Both shared MSR as the goal of the two programs. In December 2008, when announcing the MSL delay, Weiler felt his discussions with Southwood had reached a point such that he could speak of them publicly. Indeed, Southwood had already informed his superiors in Europe of the Weiler interactions and received a go-ahead to keep talking.

Now, in early 2009, Weiler and Southwood agreed to go forward with a long-term program that would be jointly planned from the start, with one space agency taking the lead at one Mars opportunity and the other at the following opportunity. Every 26-month window would be used. Some characterized the approach as a "tag-team" strategy. Southwood termed the NASA-ESA conversations as a "courtship."[6]

Whatever the case, Weiler did not at first see much support for this collaboration strategy within NASA. The new Administrator, Bolden, would not take office until July 17. In the meantime, Weiler did what he thought made sense. "I may be the only person in NASA who believes that this is the right thing to do. My toughest job is to get my view understood at all levels below me and especially at certain NASA centers." He meant JPL in particular. He said that JPL would eventually conclude that it was better to lead one mission every four years than having control of a bankrupt program every two. Weiler felt he was fighting "psychology and nationalism." But if scientists really wanted a strong Mars program, they would have to realize that the "flag" on the mission did not matter.[7] It was more than psychology and nationalism for JPL; the Pasadena center needed major projects to maintain its workforce.

To help identify what those missions would be, while also gaining scientific support, Weiler looked to the National Academy of Sciences National Research Council, which was mounting the next Decadal Survey for the planetary program. The NRC panel was headed by Steve Squyres, the chief scientist behind

Spirit and Opportunity.[8] The study got under way in March. In April, Weiler asked Hubbard to chair an MEP Analysis and Review Team. The Hubbard group, as the name implied, would focus on Mars.[9] Weiler wanted both advisory bodies to think about priorities within a constrained budget. As he noted, "There is no greater thing than starting a sexy new mission. We all love it. The thing that prevents me is I've got new, sexy missions started five years ago that are costing more than they were supposed to."[10] MSL was one of his particular cases in point, the biggest project at JPL, and one undergoing technical and managerial change.[11]

Hubbard echoed many of Weiler's concerns. "The [Mars] program is now at a crossroads," he said, "with an indeterminate future for the next decade."[12] Friedman of the Planetary Society emphasized in his society's publication that the crossroads for Mars extended to human spaceflight as well as the robotic program. "Are we to take the road to Mars all the way to an MSR mission and then on to a human destination?" he asked. He complained that Mars planning had been eliminated from the Moon-Mars human mission and that the Mars robotic program had had cuts of more than half a billion dollars in the past several years.[13] For Friedman and other Mars enthusiasts, the hope was that the Obama administration would forge a national policy favoring Mars as a top priority. Under Bush, the Moon had become the overriding focus, as Griffin, with inadequate funding, had increasingly cut back to his immediate goal of narrowing the gap between a shuttle retirement and successor relevant to lunar exploration.

In May, Obama announced details of his budget plans, and these included modestly more funds for NASA—$18.7 billion for FY 2010—and continuity in all inherited space programs. However, he also created a blue-ribbon panel under the chair of Norman Augustine, a highly respected retired aerospace industrialist, who had led a similar panel concerned with space policy under President George H. W. Bush. The Augustine panel's charge was to assess NASA's human spaceflight program with particular reference to the Constellation Program. Should NASA stay on course or change direction? The Augustine committee's task was to provide options to the White House and the NASA Administrator.[14] The Administrator's name was also announced in May, Charles Bolden.

Transatlantic Alliance—and Concerns

On June 29–30, Weiler and Southwood met in Plymouth, England. The choice of Plymouth was at Southwood's insistence. He had met Weiler on his turf. In a joint venture, Southwood believed, there had to be reciprocity in every way possible.[15] Weiler recalled that the two men "argued" for hours. They were trying to forge a "genuine partnership," and that meant dividing costs fairly evenly over the course of a multimission program. Typically, joint programs were dominated by one agency, but because of both "mutual interest and mutual dependence," they realized they had to construct a different model. The goal and costs of MSR required it.[16]

They agreed to create a joint initiative. The basic road map they worked out called for the two agencies to design missions for 2016, 2018, and 2020. Missions would include "landers and orbiters, conducting astrobiological, geological, geophysical, and other high-priority investigations, and leading to the return of samples from Mars in the 2020s." A NASA spokesman said that NASA and ESA would begin to develop the initiative. Southwood said he would not comment, pending conversations with ESA member states.[17]

Southwood did indeed have a problem. The key nations in ESA whose assent he had to have were the primary funders: Germany, France, and Italy. Italy was not happy. The head of the Italian Space Agency, Enrico Saggese, declared that while he supported NASA-ESA collaboration in principle, he saw negative implications for ExoMars, the next ESA Mars mission, scheduled for 2016. The ExoMars project he had heard being discussed by Weiler and Southwood did not look like the ExoMars project "we subscribed to, and frankly, I'm not sure my national industry has much to gain from it."[18]

What bothered Italy was the report that Weiler and Southwood were discussing a shift in a significant part of the mission to NASA. NASA would assume responsibility for launching ExoMars on an Atlas rocket and supply an orbiter to relay data to Earth. ESA would continue to provide the descent, reentry, and landing module; rover; and drilling system. Some planned items for ExoMars would be removed. This arrangement would let ESA keep the mission under the $1.7 billion limit member states had set in November 2008.[19]

Clearly, Southwood needed assent from Italy as well as other nations. Weiler did not have a body of 17 nations to please, but he did have his own agency, JPL, the White House, Congress, and the scientific community to bring aboard. Toward getting scientists' support, Weiler met with the NRC decadal panel, which convened July 6 and 7. Weiler bluntly told the committee that because of the cuts Mars had absorbed in recent years, "we no longer have a viable Mars program." He stated he was trying to rebuild the program financially by allying with the Europeans. He also warned the panel that it should be careful about what it recommended. Money was going to be extremely tight. If the scientists wanted to add new ventures, they should suggest existing items to delete.[20]

Weiler's stark warning was reinforced by OMB's Amy Kaminski, who also advised the group, "Don't anticipate a lot of growth in the budget for science, particularly planetary science." Where policymakers wanted help from those looking ahead in the Decadal Survey was on scientific priorities.[21]

Bolden and Augustine

On July 17, NASA's new Administrator, Charles Bolden, was sworn in. A retired Marine Corps general and former astronaut, the 62-year-old Bolden was very much a Mars advocate. His orientation was human spaceflight, but his Mars interest and internationalist bent augured that there would be support at the top of NASA for what Weiler was trying to do. In his first address to NASA employees, Bolden declared that most NASA people wanted to go to Mars, but "there are a lot of ways to get there. The challenge for us in the next few months is to figure out the single most cost-efficient path to get there."[22]

Early in September, the Augustine panel reported to Bolden and the White House on its preliminary findings.[23] Its most striking comment was that NASA was on an "unsustainable trajectory." There was too great a gap between what it was supposed to do and the money it had to do its job. The Augustine panel had been asked to concentrate on human spaceflight. But what it said about sustainability in this area had applicability to everything NASA did.

The panel was charged to come up with options for NASA and the president, and it did so. Two that fit within NASA's existing budget profile it did not believe worth pursuing, since they kept the United States where it was, in low-Earth orbit. The three it believed were potentially viable for an exploration agency were "Mars First," "Moon First," and "Flexible Path." All would require more money, $3 billion more than NASA currently had, to be added over the next few years, with annual cost-of-living increases each year subsequently. It emphasized that Mars was the "ultimate destination," but there were too many uncertainties about risks to make it the first place to go for humans. "Moon First" represented the current pathway, which Obama had inherited from Bush. An alternative, "Flexible Path," meant going to "free space" destinations, such as asteroids, Lagrange points ("special places in space of particular significance for scientific observatories and future space transportation infrastructure"[24]), and moons of Mars. In this option, the Moon base that was part of the Bush plan would be postponed, but not necessarily abandoned. The Flexible Path option was akin to the strategy many Mars advocates (such as Hubbard and the Planetary Society) proposed. They wanted to go to Mars as soon as possible and feared that the United States would get mired building a base on the Moon.

The Augustine panel did not explicitly recommend a choice between the Moon First and Flexible Path options, stating they were not mutually exclusive and both were relevant to Mars. But in subsequent congressional testimony and interviews, Augustine came across as favoring the Flexible Path. The Augustine panel noted that human and robotic spaceflight could be integrated well under the Flexible Path option. Thus, humans "could rendezvous with a Moon of Mars, then coordinate with or control robots on the Martian surface, taking advantage of the relatively short communications times" this proximity provided. There were "a lot of exciting things one could do [via the Flexible Path] along the way to Mars," Augustine told a House committee.[25]

In October, the Augustine panel produced its final report.[26] There were no surprises. It provided much greater detail on all aspects of human spaceflight. However, its basic message was consistent with the interim document when it came to future exploration: Mars was the "ultimate destination," and the most viable and desirable routes "worthy of a great nation" were Moon First and Flexible Path. The problem was that the current spending was far inadequate for either approach, again calling for an additional $3 billion by FY 2014, with cost-of-living raises each year afterward.

The report went to Bolden and the White House science advisor, John Holdren. What they would recommend to President Obama remained to be seen. What he would decide remained to be seen. What stood out for many observers was the panel's emphasis on Mars as a destination for scientific and exploration purposes, as well as the call for using a "first among equals" strategy for NASA to achieve its goals with international partners.

The Augustine panel's report seemed to strengthen Weiler's hand as he worked to secure the future of the robotic Mars program. Weiler commented, "I think the trend [in Mars space policy] is more and more collaboration . . . we're talking about getting extra value . . . out of a mission."[27] While the United States remained the dominant space power, there were many spacefaring nations active in the twenty-first century. The United States was talking with ESA about Mars, but there were additional alliance possibilities. Significantly, Russia was collaborating with China on a mission to Phobos, the Mars moon that had long been a goal of the Russians. They had hoped to launch in 2009, but—like the United States with MSL—found they had to delay to 2011.[28] As it turned out, the Russian probe never made it out of Earth orbit and crashed in the ocean.[29]

Mars was *the* destination of choice for many nations, and most acknowledged the virtues of partnership. The robotic program was seen as important in its own right, as well as a prelude to human spaceflight. With all its problems, NASA looked ahead and charted its course for the Red Planet through the robotic program. On November 5, 2009, Charles Bolden, NASA Administrator, and Jean-Jaques Dordain, director general of the ESA, signed a potentially historic agreement to plan together a joint program. Henceforth, if this agreement were adopted and implemented, the United States and Europe would go to Mars in concert.[30]

Mars Planning and Political Discord

As planning got under way, a heated debate erupted between the Obama White House and Congress. It was over the human spaceflight program but affected everything else at NASA, including Mars policy. In February 2010, Obama announced his proposed budget for NASA for the new fiscal year. It provided additional funds but did not start the agency on the $3 billion increment for which the Augustine panel had called. Instead, it terminated Constellation and announced a new program to nurture a commercial industry to ferry crew to the International Space Station, given the looming retirement of the Space Shuttle. It did not give a destination for human spaceflight. Its emphasis was made on a new technology development initiative. There had been no preparation of Congress for the drastic change in policy. NASA, including its Administrator, had largely been left out of the White House decision process.

Congress reacted strongly, immediately, and negatively to the policy. Led by representatives of states with human spaceflight facilities, Congress rejected the Obama plan. In April, Obama sought to assuage Congress by calling for a trip to an asteroid in 2025 and to Mars in the mid-2030s. The Orion space capsule, cancelled along with the rest of Constellation, would be reprieved in the form of a crew-rescue vehicle. A decision on the heavy-lift rocket would be made before the end of his term, President Obama said. The president did not succeed in blunting the rancor. In this environment, little attention was given to robotic Mars policy.

In July 2010, NASA published a report on its current strategic plan for SMD. With respect to MEP, NASA declared that it would seek to launch "successive missions to Mars (roughly every 26 months) to evolve a scientifically integrated architecture of orbiters, landers, and rovers." It said that the existing program organized around projects to "follow the water" was achieving its objectives and announced NASA's new goal: "seeking signs of life." NASA made it clear that it aimed ultimately at "collection and return of samples from Mars." It said it planned to achieve this goal with ESA. NASA proclaimed that the transition from the agency's present to its longer-term future would be led by "the next quantum leap in Mars exploration: the Mars Science Laboratory."[31]

SMD knew precisely what it wanted to do. But NASA as a whole did not, caught as it was between the president and Congress. In October, Congress and the president agreed on a compromise. Part of Constellation came back with new names. Ares I was killed in favor of a commercial industry to be created to service ISS. Rather than waiting, NASA would move ahead as quickly as possible on the Ares V heavy-lift rocket (now called Space Launch System [SLS]). The Orion space capsule would be developed, now named Multi-Purpose Crew Vehicle (MPCV). The asteroid and Mars destinations stayed. Bush's Moon destination was gone.

This compromise did not end the debate, as struggles over funding continued for many months afterward. Also, in November, elections put the Republicans in charge of the House and narrowed the Democrats' majority in the Senate. It was obvious that getting decisions made on NASA policy and budget generally would be extremely difficult.

In early February 2011, President Obama announced his budget for FY 2012. For NASA, there was a freeze—$18.7 billion, the same budget it received in 2010. For the current fiscal year (FY 2011), NASA awaited action by Congress. The power struggle between the Democratic-controlled White House and divided Congress had resulted in no budget from the Hill. NASA operated under a continuing resolution that held its budget static at the previous year's spending. The political and fiscal conflict hindered implementation of existing missions and planning for the future.

The White House projection in the budget for NASA and especially the planetary program was ominous. It showed spending on planetary science increasing by $180 million the next year, to $1.54 billion, and then declining steadily to $1.25 billion by 2016.

NRC's Decadal Survey

The budget flew in the face of scientific recommendations for the planetary program, especially Mars exploration. The NRC, on March 7, released its Decadal Survey for the planetary program. NRC made Mars the top priority among planetary scientists for the ensuing 10 years. It placed the Mars Astrobiology Explorer-Cacher (MAX-C) at the pinnacle of its list of recommended large-scale missions. This would be the flagship for the planetary program in the period 2013–2022. MAX-C, as the first step of a multimission MSR campaign, would collect Martian samples that would be returned to Earth by spacecraft launched in the ensuing decade.[32]

This stretched-out sequence was a function of budget constraint. For the first time, the NRC used an independent analysis of costs in its planetary planning. The cost of MAX-C would be $3.5 billion. That was too high relative to other needs in the planetary program, it said. NRC recommended that NASA reduce the cost to $2.5 billion.

As a second priority, NRC listed an outer planets mission to Jupiter's ice-covered moon Europa. However, this mission's cost was even higher than that listed for MAX-C, at $4.7 billion. Steve Squyres, the chair of the NRC group, spoke of "sticker shock" and the realization of the scientists that expense might preclude any flagship missions, leaving room only for smaller projects.[33]

Commenting on the NRC Decadal Survey at the time of its rollout, Weiler was characteristically blunt: "If we're going to do any more big missions, they're going to be international. The days of $3 billion–$4 billion missions that we do on our own are gone."[34]

Squyres called on the planetary science community—not only the Mars advocates—to back the NRC report and lobby for its funding. Otherwise, there would be few flagships for years to come. James Green, the planetary director, called the Decadal Survey "a guiding light" for NASA as it maneuvered through the short-term political and economic perturbations influencing the program. However, in putting the guiding light into action, the bilateral relationship with ESA would have to be reconfigured, he said. The assumptions about what NASA could back were not likely to be realized under the recent Obama-budget projection.[35]

In late March, Green met with Fabio Favata, head of ESA's Science Planning Office. His purpose was to begin rethinking the bilateral relationship. The existing plan called for ESA (with NASA's help) to launch its ExoMars spacecraft in 2016. This was a combination orbiter-lander, with the orbiter critical to the 2018 mission as a communications system. The 2018 mission consisted of two rovers, one by ESA and one by NASA. The U.S. rover would be akin to the MAX-C, thus initiating MSR. There were different rovers because the two agencies had somewhat distinct requirements, requirements that included work for respective nations and their industries. But now, NASA knew it could not afford an independent rover.[36]

On March 29, Weiler and his ESA counterpart, Southwood, met at JPL. Weiler told Southwood that NASA could not build its own rover and proposed that NASA and ESA combine their interests in a joint rover. Although rumors had circulated that the NASA rover might be cancelled, Southwood was initially "shocked" to hear it directly from Weiler. Southwood, who was soon to retire, saw the bilateral program, aimed at eventual MSR, as his most enduring legacy in leading science at ESA. Was this the breakup of a marriage? he wondered.

But then he thought about the "joint rover" as a solution to ESA's own financial problems. One rover, designed correctly, could incorporate both agencies' needs. What was critical was to get agreement to the change from European nations whose stakes lay in having work for their respective companies and their employees on the joint machine. The same interests applied to the United States, which was also particularly mindful of keeping a Mars scientific group employed at JPL. One rover would presumably be less expensive than two separate rovers.[37]

NASA wanted MAX-C, or something close to it, and the combined rover might make it possible financially to have this machine. NASA would want to include a drill to obtain samples beneath the surface, as well as the ability to rove over a large area to enable soil samples to be cached. The samples would then await a future mission that would collect and return them to Earth. Only in this way—holding to its objectives in a combined rover—would the United States have anything resembling a MAX-C. To be sure, ESA wanted MSR as a long-term objective, but its original 2018 priorities had been somewhat different in terms of payload from those NASA favored.

Weiler and Southwood had a relation that went back years. They trusted one another. They went away from the meeting with a sense of progress in spite of the budget challenge.[38] They established on April 6 a joint engineering working group to investigate how a single spacecraft might be configured.

Shortly thereafter, ESA issued a stop-work order to contractors working on its 2016 mission. For ESA, the two missions, ExoMars 2016 and ExoMars 2018, were mutually dependent and mutually approved in its decision-making process. With the ESA 2018 mission likely to be greatly changed, there was a need to determine its implications for 2016. ESA led the 2016 mission, with NASA contributing instruments and launch services. Who would lead the 2018 mission was to be determined, although the likelihood in April seemed to be NASA. The agency that led would presumably pay the bulk of the costs. In any event, the replanning process got under way immediately. Both agencies were intent on keeping the partnership going. They increasingly saw no alternative in view of budget trends in both political contexts.[39]

MSL's "Cameron" Camera

While planning for a joint flagship mission, NASA also had to implement its existing flagship, MSL. MSL's centerpiece was the rover, and the rover now had a name, Curiosity. NASA's hope was that that name, chosen on the basis of a national competition, would become as familiar as Spirit and Opportunity.

In late March, NASA reluctantly decided that the special 3-D camera being built for MSL would not be ready in time for the November–December MSL launch window. This camera was being developed by Malin Space Science Systems of San Diego. Malin, a longtime Mars enthusiast and contractor, had been joined by James Cameron, whose space science fiction film *Avatar* had been a blockbuster success. Cameron was himself an advocate of Mars exploration and was serving NASA as a public engagement coinvestigator with Malin on MSL.[40]

The camera had been descoped in the Stern period and been reestablished subsequently by Weiler. NASA Administrator Bolden had himself provided additional resources from his own reserve fund to try to speed the development of the camera. For Bolden, Mars was an agency priority, and Mars exploration was one of two science missions to which he gave significant personal attention (the other was the James Webb Space Telescope, a project whose cost overrun was a major headache for the Administrator).[41] But NASA could not make up the lost time from the 2007 descope.[42] This camera (which some called the "Cameron" Camera) provided exceptional imaging capability that would bring the public along as a rider on MSL. There were other cameras on MSL which would help in this respect, just not as dramatically as the Cameron Camera. But the deadline for launch was getting closer, and time precluded waiting for the camera to be fully ready. The total MSL package had to come together.[43]

Bilateral Talks Continue

Throughout April, the joint working group between NASA and ESA labored to find ways to merge the two programs. Southwood retired, as scheduled, and passed his ESA science leadership torch to Alvaro Giménez in May. Giménez regarded the joint effort as a "hot potato," but he agreed with Southwood's analysis of the situation NASA and ESA faced: "As for Mars, the fact is that neither of us can realize the ambitious goals of Mars exploration on our own. We are, I think, linked in this."[44]

NASA was telling ESA in May it would contribute $1.2 billion to the revised 2018 mission, plus an estimated $300 million in launch costs—or $1.5 billion. This was $700 million below what it had projected when Weiler and Southwood had begun discussions and assumed two separate spacecraft. In considering the 2016 and 2018 missions as one ExoMars program, ESA had authorized 1 billion euros (U.S. $1.4 billion). The top official of ESA, Jean-Jacques Dordain, was personally steering the merger through key ESA governing boards.

In selling the revised (but still formative) program to the boards (and nations), Dordain promised to hold to the approved 1 billion euro ceiling and protect the 2018 mission from cost overruns by the 2016 mission. He had ESA take the unusual step of waiving almost all of its usual management fees—a $50 million saving—as a show of commitment to the project and holding the line on costs. Dordain's big problem was that he had to get at least a tentative agreement to the concept of a NASA-ESA merged program from his ESA masters to restart payment to contractors on ExoMars 2016. Because of 2016 launch windows, he felt that work had to recommence July 1.[45]

Dordain wanted a formal assurance from his counterpart, Bolden, that NASA would provide promised funds for the 2018 mission. This assurance would help him persuade his superiors to provide authorization for the July 1, 2016, mission restart. International collaboration, wonderful in theory, was cumbersome in practice. But on May 26–27, Dordain got from ESA's policy body, the Human Spaceflight and Operations Directorate, agreement to the general outlines of the NASA-ESA program. He would now go to the Industrial Policy Committee for financial approval in June. One body provided policy legitimacy, the other money. He needed both affirmations.[46] The verbal promise of U.S. resources helped leverage decisions with ESA to some extent. He awaited a letter from Bolden which he could show the Industrial Policy Committee to close the arrangement.

Spirit and MSL

The rhythm of change within a multimission program—where birth, life, and death of discrete projects occurred in parallel—was seen graphically at this time. For several months, NASA had been working to make contact with one of the two Mars rovers that for seven years had been operating on Mars: Spirit. In May, NASA decided that the task was hopeless. The official date for ending attempts at contact was May 25.[47] Opportunity, however, was still alive and continued.

MSL, meanwhile, almost had a potentially serious setback. It was sitting on a table at JPL, clamped tightly onto a platform. A back shell was attached to MSL. The back shell was designed to protect the car-sized rover as it entered the Martian atmosphere. A crane operator accidentally lifted the back shell and MSL, along with the 2000-pound aluminum table to which MSL was attached. It was an extremely anxious moment for those who witnessed the event, but MSL escaped damage. NASA, JPL, and Mars proponents generally were lucky.[48]

They were not so fortunate when it came to money. On June 8, NASA's inspector general issued a report based on an investigation of MSL. It cautioned that MSL would need additional resources, even at this late date, to be ready to launch on time. Moreover, even with more resources it might have difficulty meeting the launch deadline. There were still technical issues to be resolved. "Project Managers must complete nearly three times the number of critical tasks they originally planned in the few months remaining until launch," the inspector general reported. He also worried that money, time, and technical problems might force NASA to reduce capability to get MSL up. The inspector general charged that NASA was taking MSL down to the wire for the November 25–December 18 window.[49]

NASA said in response that the agency might well have to use remaining reserves of money ($22 million) to launch MSL. That sum, an agency spokesperson said, would be sufficient to deal with the issues raised concerning the $2.5 billion venture. NASA still believed it could resolve whatever technical problems remained by the launch time. It could cope with software development challenges after the launch. The spokesperson said that the software could be uploaded to the rover.[50] Hence, NASA believed that it would make the deadline. The inspector general did not contest the NASA response. The agency thus arranged for MSL's cross-country trip.

The product of almost 10 years and a workforce of 1,000, MSL was at last ready to go from JPL in California to Kennedy Space Center in Florida.[51] On

June 22, an Air Force C-17 transport plane flew MSL to the Florida cape. There was still final assembly to take place, with the next big milestone being in September when the aeroshell would be placed on the car-sized rover.

The Bilateral Program's Uncertainties

As NASA dealt with two missions (Mars Exploration Rovers and MSL) in varying stages along a project cycle, its longer-run hopes for the MAX-C mission rested considerably on ESA and ExoMars decisions. Dordain was having trouble getting consensus from the nations to which he reported. France and Britain expressed reluctance to commit to the 2016 planned launch because of questions about who did what in 2018. Dordain complained he had to have decisions by June 29–30, when the Industrial Policy Committee met. That way he could get contractors moving July 1. "If we are not ready to launch the orbiter in 2016, there is no 2018 mission. If I delay agreeing to ExoMars financing until questions about the rover are settled, industry could later tell me I am responsible for their missing the 2016 launch window. I do not want this." He expected to alleviate some of the concerns expressed about the 2018 mission through the letter from Bolden confirming NASA's intention to jointly develop the 2018 mission with ESA. He expected that letter June 28.[52]

It arrived late in the day on June 29. Bolden wrote that NASA would do its utmost to commit to the 2018 mission by September 15, when it hoped to have more clarity about its budget prospects. At this point, Bolden did not know what resources he would have. That reality meant delay on the 2016 decision until September 29–30 when the Industrial Policy Committee would take up the issue again.

Dordain decided that ESA had sufficient existing authority to fund work on ExoMars 2016 to keep an industrial team working on it at a minimal level for a few months. This would enable the mission to go full speed later to make up time for the 2016 launch date—assuming the United States came through with assurance in September and the Industrial Policy Committee gave its go-ahead. Dordain and the Industrial Policy Committee did not wish to foreclose options. Doing nothing amounted to a decision to kill ExoMars 2016, and without the 2016 mission, the 2018 project could be in jeopardy.[53]

On July 7, Italian Space Agency president Enrico Saggese said that his agency would be willing to sacrifice an entry, descent, and landing module planned for the 2016 flight if that would put the project back on track. Italy's sacrifice mattered. Italy was the biggest contributor to ExoMars 2016, with 33% of the

ESA budget. Saggese said Italy wanted the collaboration to survive and launch ExoMars 2018. Yannick d'Escatha, president of the French Space Agency (with 15% of the spending on the mission), said his agency would also be willing to sacrifice the entry and descent module to reduce the possibility of the 2016 mission exceeding its budget and threatening the 2018 project.

ESA's biggest hurdle was NASA's inability to confirm its role in the 2018 rover mission. Bolden had asked ESA to wait until September, when he could say more about NASA's commitment. But ESA worried that waiting until mid-September before fully approving the 2016 mission would compromise the 2016 launch date. Dordain decided that ESA would wait until October 1 to make a final decision, meaning it would continue to find the money to keep the contractor team going on a skeletal basis with short-term contracts. Missing the 2016 launch date would threaten the 2018 mission since the 2016 mission would provide a telecommunications relay the 2018 rover would need to send information back to Earth.[54]

The MSL Siting Decision

On July 6, NASA announced it had narrowed the number of sites where MSL might land to two. They were both craters: Gale and Eberswalde. In making this decision, NASA dropped two others in the "final four": another crater named Holden and a likely flood channel called Mawrth Vallis. While all four provided evidence of ancient water activity, the two finalists were especially intriguing.

Eberswalde was believed to be an ancient river delta. Gale Crater contained a mountain in its center. If Gale became the final choice, MSL would climb part of the way up this mountain, studying different layers of rock as it went. NASA said it would choose the final site by the end of July. Grotzinger, MSL's lead project scientist, declared at a press conference, "It's like two different flavors of ice cream—do you like the chocolate or vanilla on Mars? So we go back and forth a lot."[55] Weiler told Grotzinger, "John, I want to you to go as if this is the last Mars mission for 50 years. Find the best place to go."[56] Grotzinger got the top MSL managers and scientists together in a proverbial "smoke-filled room." They came down unanimously in favor of Gale and recommended this site to Weiler.[57] Weiler made the choice official, and it was announced on July 22.

Barriers to Bilateralism

In succeeding months and into the fall, the funding prospects on both sides of the Atlantic worsened. Weiler argued with White House budget officials over a NASA Mars budget they were determined to lower. The year before, he had had to endure the paring of the joint missions with ESA from two to one. Now even that one was in jeopardy. As pressure on Weiler and his budget "racheted up," he proposed more modest cuts across the board in NASA science to protect Mars. White House budget officials did not relent. They saw in the bilateral program another multimission big science effort that could become a standing commitment through international connections.[58]

OMB was correct. NASA wanted to move ahead on a program, not a single mission. The fact that the program was bilateral held down costs. But the U.S. expense still would be great, in view of the endeavor's bold goal. Utterly frustrated, Weiler wrote Bolden he had reached the end of the line in his negotiations.[59] On September 30, Weiler retired, thereby concluding an admirable 33-year government career. For Mars supporters inside and outside NASA, he would be greatly missed. It would now be up to others to try to carry on what he and Southwood had begun. That task would be daunting.

The letter from Bolden affirming NASA's intent to support the 2018 Mars mission, which was supposed to come in mid-September, did not arrive. A frustrated Scott Hubbard spoke out against the delay, which was due not to Bolden, but to the White House and OMB. He pointed out that ESA was pledging 1 billion euros to the combined missions; how could the United States not do its part?[60] But the United States was making further decisions to cause angst in Europe. It stated that it could not provide the Atlas V rocket to launch the 2016 mission. That had been part of the original bargain between NASA and ESA. Dordain once more scrambled to find additional money within ESA to keep contractors working until the end of the year. He managed to do so. He earnestly sought to keep open options to maintain the 2016 opportunity. But he now knew he would have to find an alternative rocket, from either Europe or Russia, and additional money from ESA to make up for the U.S. withdrawal. The options for Dordain were narrowing, and there was a distinct possibility ESA would have to abort the 2016 ExoMars launch.

But what about 2018? This was the priority for the United States and also ESA. The 2016 mission was supposed to facilitate it. In the first week of October, Bolden and Dordain met during the International Astronautical Congress in Cape Town, South Africa.[61] Bolden told Dordain that NASA wanted to keep the partnership going, but its budget situation was still too uncertain for him to make commitments.[62]

NASA was living with a continuing resolution most of the year as the president and Congress staggered toward compromises in late 2011. Moreover, Congress had established a "Super Committee" to hammer out extensive budget cuts to reduce the deficit on a long-term basis. The Super Committee failed, however, and draconian cuts across the board would kick in in 2013, unless Congress acted to avert them before then. No one could predict what would ensue. Congress did provide NASA with a budget in November, as well as language that backed "flagship-class missions" that could be implemented with "international partners." However, the Obama administration was silent on a commitment to such a mission. Desperate, Dordain asked Russia to join the U.S.-European alliance to keep the 2016 mission alive. The 2018 mission was still planned, and Bolden was hoping all would be well eventually, but Dordain confessed he was becoming a "doubting Thomas."[63]

MSL Launches

On November 26, MSL, carrying its Curiosity rover, blasted into space. At last, two years late, it was on its way to the Red Planet. Thirteen thousand onlookers watched it soar from Cape Canaveral. It would take eight months for the spacecraft to journey the 352 million miles to Mars. Its goal was to search for evidence that microscopic life might once have lived on Mars—or be capable of living there now. It also contained sensors that would detect radiation affecting the ability of astronauts to land there some day.[64] Administrator Bolden declared, "We are very excited about sending the world's most advanced scientific laboratory to Mars. MSL will tell us critical things we need to know about Mars, and while it advances science, we'll be working on the capabilities for a human mission to the red planet and to other destinations where we've never been."[65]

Everyone connected with the mission was elated, but they also knew the risk of failure. A Russian probe to Phobos, a Mars moon, had launched on November 9 and failed to escape Earth's orbit. The U.S. spacecraft was now on a trajectory to Mars. But the landing, several months hence, would be daunting. And what would the rover find? Time would tell. The future of the Mars program—and maybe NASA—depended greatly on the answer.

NASA Withdraws

Between Thanksgiving and Christmas, Bolden labored to find money for the United States to participate in a serious bilateral program. Negotiating with Dordain and OMB, he could not come up with the money required. He felt he needed $1 billion, and he did not have it, given all the other priorities with influential constituencies which NASA had to fund.

Bolden was talking with Dordain by phone and negotiating in person with OMB on the FY 2013 budget virtually at the same time. It was clear he could not get the money he needed for ExoMars. He called home the NASA technical team working in Paris with European counterparts. Christmas was imminent, and it was pointless to continue the planning effort.

CHAPTER THIRTEEN

Landing on Mars and Looking Ahead

As 2012 got under way, the Mars program was in limbo. The ambitious joint program with ESA was dead, and the National Research Council scientists' goal of Mars Sample Return apparently gone with it. The Mars Science Laboratory landing was scheduled for August, and that mission had stakes not only for the Mars programs but for NASA generally. To have a positive future, Mars advocates had to work hard for recovery on the policy front in Washington. They hoped that the MSL mission, through its Curiosity rover, would give the program a political stimulus it desperately needed, within both the space policy sector and the broader national policy arena. The budget deficit was the overriding priority in Washington. Policy decisions at NASA and between the White House and Congress had put other space programs, including human spaceflight and the James Webb Space Telescope, above Mars exploration in priority. What if MSL, with its never-before-tried sky crane landing system, crashed on Mars? Mars advocates in NASA, the scientific community, and interest groups had only questions and no answers as the year began—an election year that magnified all issues, especially failure in government programs. The NASA Administrator spoke of "Mars Next Decade." What NASA needed were policy decisions assuring there would be a Mars program next decade.

By the end of the year, with MSL successfully landing the Curiosity rover on Mars and Obama's reelection, there was hope among many Mars proponents for moving the Mars program forward. New missions were placed on NASA's agenda. Advocates regrouped and pushed once again for their long-sought MSR.

Reframing Strategy

On February 13, 2012, the FY 2013 budget was released. The president provided a budget of $17.71 billion, a slight decrease from the previous year. The planetary program was cut 20%, with Mars absorbing most of the decrease. The Mars Together effort with ESA ended. The NASA Administrator, at the rollout of the FY 2013 budget, emphasized that NASA was not walking away from Mars exploration missions. He explained that the issue was lack of money for another big science Mars program. "Flagships are expensive; . . . we just could not afford to do another one."[1] But Bolden and his new associate administrator for science, John Grunsfeld, who had come aboard in January, made it clear they were not abandoning Mars. However, they also were reframing the program. The way they did it reflected a different scientific and political strategy.

While Bolden and the 54-year-old Grunsfeld were both technically trained (Bolden in engineering, Grunsfeld in physics), what shaped their approach to space policy most distinctly was their astronaut backgrounds. Both had flown in space, with Grunsfeld's prime experience being in repair of the Hubble Space Telescope. They both saw the divide between the human spaceflight and robotic cultures at NASA as hurtful and unnecessary. In speaking with Grunsfeld, Bolden charged him to bring the programs together in terms of Mars exploration. If he could do that—and it had been tried before and not succeeded—this would help the Mars program by giving it a dual purpose within NASA and also help Bolden sell it to the White House as enabling the president's goal of human spaceflight to Mars in the 2030s.

It was not a strategy that Mars program managers at NASA particularly liked, and it was one that Mars scientists feared. But the two programs shared an interest in "life" questions—"life on" and "life to." The environmental conditions on Mars which would make Martian life possible or present risks to human explorers had to be understood. Life research was the bridge between the two programs. Moreover, MSR required new technologies of descent and exit from Mars that astronauts would require. Bolden and the science director agreed on this strategy. Mars Together with ESA might be replaced with Mars Together within NASA.

Thus, in announcing the new FY 2013 budget on February 13, Bolden stated he had put Grunsfeld in charge of a cross-agency team to determine next steps in the Mars program. The team consisted of Grunsfeld; the associate admin-istrator for human spaceflight, Bill Gerstenmaier; the chief scientist, Waleed Abdalati; and the chief technologist, Mason Peck. What Bolden wanted, he said, was an "integrated strategy to ensure that the next steps for Mars exploration will support science as well as human exploration goals, and potentially take ad-vantage of the 2018–2020 exploration window."[2] Grunsfeld reinforced Bolden's message by stating that Mars missions would track radiation as well as other issues relevant to human exploration.[3] An "integrated strategy" was the new mantra for Mars. NASA would participate in modest ways in Europe's ExoMars program, through certain instruments. But NASA's prime goal would now be a U.S.-run mission in 2018 or 2020.

Bolden followed up with an op-ed in *Space News*. He emphasized his "new strategy that takes into account science objectives, human exploration goals and forward-looking developments in our space technology program." He called for a "synergistic approach" and reminded the space community of his experi-ences. "As a former NASA astronaut who has flown four missions on the space shuttle, including the 1990 flight that deployed the Hubble Space Telescope, I've learned that scientific discovery and human exploration go hand-in-hand." Without human repairs in space, Hubble would have failed to be the "amazing success" it has been. He declared that the "next step" in realizing NASA's vision to explore the unknown was to unravel the questions of "life on Mars." That could best be done, he vowed, by "coordinating NASA's scientific and human exploration programs."[4] By putting Grunsfeld in charge, he was making science the driver in this endeavor. Whether this strategy would work remained to be seen. Grunsfeld embodied the unity of science and human spaceflight: he had flown five missions, three as a Hubble repairman, and he had been NASA chief scientist from 2003 to 2004.

Grunsfeld Moves Ahead

On February 27, Grunsfeld announced at a Mars Exploration Program Analysis Group meeting that he was creating a Mars Program Planning Group. His aim was to have a mission in 2018, one that would be affordable at $700 million. Grunsfeld called 2018 "a sweet spot," a time when the Earth-Mars alignment was especially propitious. Squyres followed him at the meeting and made it clear that he would support such a mission only if it conformed to the NRC Decadal Survey and moved the program toward MSR. Grunsfeld said that he was ap-pointing Orlando Figueroa, former Mars czar and since retired, to head this planning team. Grunsfeld, touting the link with human spaceflight, hinted at the possibility of augmenting the robotic budget through this larger and better-financed directorate. Squyres expressed skepticism that that would happen.[5]

Figueroa, in a subsequent response to reporters, said that whatever NASA did would have to be responsive not only to scientists but also to NASA's budgetary masters. Various outside-NASA scientists expressed concern about the association with human spaceflight, comments echoing similar worries within NASA.[6] Science's goals were not necessarily those of human spaceflight, they pointed out. Human spaceflight cared about safety and operational matters. However, veteran scientist Michael Carr pointed out that life was the link, "whether it's potential Martian life, the effects of Martian dust on humans, or humans' microbial contamination of Mars."[7]

Backlash

The Mars scientific community, led by the Planetary Society, American Geophysical Union, and American Astronomical Society, reacted sharply to the budget cuts. They focused on Congress and the White House to try to restore the money taken from planetary science in general and Mars in particular. The Planetary Society called its campaign "Save Our Science." It generated thousands of signatures on petitions and e-mails of protest. The Society's leader, Bill Nye, personally went from California to Washington to speak to key congressional staffers and lawmakers.[8] Congressman Schiff, representing the district including the Jet Propulsion Laboratory, was active lobbying his colleagues on the House appropriations subcommittee responsible for NASA's budget.

On February 29, congressional hearings were held to consider the president's budget request for NASA. The House appropriations subcommittee, on which Schiff served, was critical. Schiff was a Democrat, but a bipartisan alliance attacked the Mars cuts. Obama's science advisor Holdren, testifying before the subcommittee, endured a fierce grilling. He explained that the White House and senior members of Congress (primarily Senate Democrats) had reached agreement on NASA priorities. These were the heavy-lift Space Launch System rocket and its companion Multi-Purpose Crew Vehicle, the James Webb Space Telescope, and the Commercial Crew Program for independent access to the International Space Station by 2017. With a proposed NASA budget less than the current fiscal year, something had to give, he explained, and it was the planetary program, especially Mars. The lawmakers were not sympathetic.[9]

When Bolden came before the subcommittee, he was similarly lambasted.

Chairman Frank Wolf (R-VA) had a solution for the Mars problem: transfer money from an administration priority, commercial crew, to Mars. Wolf was an avid critic of commercial crew. Bolden resisted such an argument and emphasized that NASA was still aiming for Mars, but in a different way.[10]

The White House was defensive and allowed two major Office of Management and Budget officials involved with NASA oversight to meet with NRC planetary scientists. The administration wanted to make it clear OMB was not against big science flagships, as was being alleged. Paul Shawcross, branch chief for science and space, and Joydip Kundu, who handled NASA's science budget, told a joint meeting of the NRC's Space Studies Board and its Aeronautical and Space Engineering Board that OMB had no bias against flagships. They did not draw the line at $1 billion. OMB looked, rather, at budget coherence.[11] That reassurance did not particularly ameliorate scientists' feelings.

The House subcommittee subsequently provided a modest increment in Mars funding, as did its Senate counterpart. The House panel also required the NRC to certify that the new Mars program would lead to MSR. Otherwise, it said, the money should go for a Europa mission (NRC's second priority for a flagship mission). The congressional moves were part of a larger political struggle between Congress and the president in setting space priorities. But given the political dynamics of the time, there was no way of knowing when Congress would pass a budget and whether additional funds recommended for Mars would survive the larger legislative process.

MSL's Stakes

While worrying about the future Mars program, NASA had to deal with the present challenge. The stakes were immense for MSL. NASA officials concerned with the upcoming MSL landing were increasingly restive. Jim Green, planetary chief at headquarters, admitted that MSL was keeping him awake at night. He was especially nervous about "the seven minutes of terror from the top of the [Mars] atmosphere to landing."[12] Grunsfeld responded to a media inquiry about what would happen to the Mars program if MSL failed. Grunsfeld said that there were "no guarantees," but he thought that support would continue.[13]

NASA wanted to leverage MSL for support, if it were successful, and also to control for damage if it failed. Chris Carberry, the head of a pro-Mars interest group, Explore Mars, Inc., noted the pressure NASA was under, saying that "the stakes have never been higher for a Mars landing." NASA was under intense "budgetary, political, and programmatic pressure," he wrote in an op-ed in *Space*

News. "Every success and failure—no matter how minor—is being scrutinized to an extreme degree." Worse, the landing was taking place "in the heart of the US presidential campaign season, which tends to magnify the impact—positive or negative—of any event."[14]

He said messages had to be prepared in case the mission failed to emphasize the difficulty of what NASA was trying to do. NASA could not spread the messages alone. It needed the help of many advocacy organizations in its support group. The advocacy coalition, in short, had to use MSL Curiosity as an opportunity to build support for NASA programs, especially Mars, in the public and political world.

In July, as the landing date of early August approached, advocacy organizations did indeed prepare. Many planned "celebrations," assuming success. Carberry said that his organization would coordinate a number of "got Curious" landing parties in the United States and abroad. Other organizations were also active, preparing for the event. The Planetary Society planned a "Planetfest" assemblage in Pasadena. Science museums and other organizations scheduled events.[15]

In Atlanta, Science Taxern, an organization that launched science talks for the public, planned five planetary science events leading up to the August landing. It scheduled a party beginning at midnight on the evening when NASA's coverage of the landing commenced. "It's science," said the group's director, Mark Merlin, "but it's also a public celebration of scientific achievement."[16]

NASA was also working to make the most of the event. The agency had to be careful. Under the law, agencies are not permitted to "market" themselves. But they could do "outreach." "We don't try to sell anything," said Robert Jacobs, a NASA spokesperson. "Our job is to clean the windows to give the American public a better view of their space program."[17]

What was NASA doing? For some time, it had been providing information to the media and others about the MSL and its landing. There was a website the public could use to learn more about the project as it developed. In the summer of 2012, NASA's outreach campaign moved into high gear. In June, NASA released "Seven Minutes of Terror," a video that depicted the rover's harrowing ride down to the Martian surface. The video began "with a computer-generated animation of a capsule falling toward the Red Planet, then used stark lighting, thumping music, fancy graphics and dramatic narration" to give the observer an acute sense of the event.[18]

In mid-July, NASA announced a collaboration with Microsoft, under which it had developed a new Xbox "outreach" game called *Mars Rover Landing* designed to give the public a sense of the challenge and adventures of landing in a precise location on the surface. The game was free.[19]

Later in the month, NASA released a video, "Grand Entrance," narrated by William Shatner and Wil Wheaton from *Star Trek*, depicting the spacecraft's entry, descent, and landing. "The goal is to educate the public about Curiosity and build awareness about the landing," NASA said. It also announced that the Toshiba Vision screen in New York's Times Square would provide live coverage of the rover's landing.[20]

Alan Stern, former science director and now frequent critic of NASA, congratulated the agency for making the most of "modern" communication methods. However, he warned that "too much" of this kind of public relations risked "trivializing and making a sideshow of a very expensive and ultimately a very serious endeavor." "It's a fine line," he noted, implying that NASA had better not cross it.[21]

NASA knew that all this ballyhoo could backfire if MSL failed. NASA had to emphasize how difficult was the feat to be attempted. Nye pointed out in an article published July 30 that Europe had tried to land on Mars and failed, while the Soviet Union/Russia had tried 21 times over the decades and failed every time. NASA had recently succeeded, but his message emphasized the difficulty entailed. Nye's other message was that it took a special skill set to succeed and that talent could be lost if the nation did not provide NASA the support the agency needed.[22]

On August 2, as the time for MSL's arrival at Mars drew nearer, the agency initiated a daily round of media events at JPL.[23] On the evening of August 5, MSL approached the Red Planet. In Chicago, the Adler Planetarium held a late-night pajama party so families could follow the landing live. Thousands gathered at the Planetfest in Pasadena. All over the country in institutions and in their homes, people waited to watch. In New York City, crowds gathered in Times Square to view a giant screen that usually only showed ads. NASA began to live-stream the event, and the traffic congested. Up to 23 million people watched one way or another as NASA made pictures available.[24] The time for the real-life "seven minutes of terror" was now.

Seven Minutes of Terror

At 1:25 a.m. (EDT), August 6, 2012, after a journey of eight months and 352 million miles, MSL arrived at Mars. MSL had the most difficult part of its trip ahead. This was the seven-minute period of entry, descent, and landing which the spacecraft would have to endure. On automatic pilot, MSL would need to decrease from 13,000 miles per hour as it met the Mars atmosphere to almost zero to land safely on the surface.

When MSL hit the Mars atmosphere, it began to slow. At seven miles above the surface, it was still going at 900 miles per hour. At this point, MSL unfurled a giant, 51-foot parachute. Then, still falling, it released its heat shield, followed by the firing of retro-rockets. It was now nearing the landing site, but the most uncertain part of the journey remained, even as the spacecraft slowed almost to a stop.[25]

At JPL's mission control, scientists, engineers, administrators, media, and others steeled themselves for the final leg. At one ton of weight, the nuclear-powered, $2.5 billion, car-sized MSL was too heavy for airbags or retro-rockets to land its delicate rover, Curiosity. The rover was five times the weight of Spirit or Opportunity. NASA had invented a sky crane, a wholly novel device for the landing phase. It had been thoroughly tested on Earth, but Earth was no equivalent to what it would face on Mars. The Martian test was what really mattered, and there was much trepidation at NASA about the sky crane's ability to work.

Allen Chen, flight dynamics engineer at JPL, intensely watched the signals from Mars, transmitted by two orbiters, coming to Earth, 14 minutes after the fact. Suddenly, he announced, "Stand by for sky crane." Everyone in mission control (and thousands beyond) sat or stood in complete silence. Less than a minute passed. On Mars, wire cables had emerged from the sky crane, which embraced the MSL rover. Retro-rockets held the combined apparatus two stories above the ground. Then, the sky crane gently lowered the rover the remaining distance to the surface. With its precious cargo positioned and decoupled, the sky crane rocketed safely away. Launched as part of a multifaceted spacecraft from Earth, the Curiosity rover arrived. The landing on Mars took place at 1:32 a.m. (EDT). On Earth, 14 minutes later, an excited Chen said the words for which everyone had waited and hoped: "Touchdown confirmed. We're safe on Mars!"[26]

Success! JPL mission control erupted with yells, hoots, cheers, claps, hugs, and tears. All occurred simultaneously. NASA had gambled and won—so far. With 10 instruments, Curiosity was easily the most technically sophisticated rover ever sent to Mars. Given the costs and already perilous state of NASA funding, failure might have doomed the Mars program for years. Mars advocates understood the stakes. There was universal relief.

Television recorded the jubilant scene. Bobak Ferdowski, a flight engineer who sported a Mohawk hair style with maroon highlights and stars on the side, became an instant Internet celebrity. All the NASA personnel wore similar blue jerseys, but Ferdowski's hair set him apart and transmitted the message that technical people could be "cool," as President Obama later observed.[27]

President Obama made a congratulatory phone call to the team behind the MSL and its Curiosity rover. Calling from Air Force One to JPL, soon after the successful landing, Obama declared, "Due to your dedicated efforts, Curiosity stuck her landing and captured the attention and imagination of millions of people, not just across the country, but people all around the world." JPL director Elachi took the call in the team's mission control area. With him were descent team leader Adam Steltzner; mission managers Peter Theisinger and Richard Cook; project scientist John Grotzinger; and John Grunsfeld, NASA's associate administrator for science.

"You guys should be remarkably proud," Obama said. "Really, what makes us best as a species is this curiosity that we have, and this yearning to discover and know more, and push the boundaries of knowledge. You are perfect examples of that, and we couldn't be more grateful to you." This achievement embodied the American spirit, he declared.

When he had interviewed Bolden for the job of NASA Administrator, Obama had asked him to deliver inspiration to young people. Now, Obama said, "This is the kind of thing that inspires kids across the country. They're telling their moms and dads they want to be part of a Mars mission, maybe even the first person to walk on Mars. And that kind of inspiration is the byproduct of the work of the sort that you guys have done." Obama gave his "personal commitment to protect these critical investments in science and technology."[28]

Throughout the country, NASA won plaudits for the landing from media, politicians, and others. But beyond the immediate bounce in public support, the question remained how sustained it would be. Curiosity had a two-year mission ahead. The future of the Mars program in Washington was still highly uncertain.

Policy Impacts

NASA wasted no time in trying to build on the positive momentum the landing created. On August 20, NASA announced another Mars lander as the next in its midsized Discovery series of planetary exploration missions. This $425 million mission was called InSight, standing for Interior Exploration using Seismic Investigations, Geodesy, and Heat Transport. Led by JPL, it would be launched in 2016. It was a stationary lander to study the interior of Mars via seismic readings and would build on technology used for NASA's 2007 Mars Phoenix.[29]

InSight won against two other non-Mars candidates (a probe to a Saturn moon and one to a comet). Given the excitement surrounding Curiosity and scientific and congressional outcry against the Mars budget cuts, the choice of InSight was not surprising. NASA wanted to send missions to Mars every 26 months. With MAVEN and InSight, NASA was now covered until 2016. The 2018 mission might initiate what Administrator Bolden had called Mars Next Decade. The question was, what would that mission be and how could NASA afford one?

For months, Mars advocates had been pressing Congress and the Obama administration. By September, the Planetary Society had generated 2,000 physical petitions and 17,000 e-mails to Congress, asking for restoration of the cuts to planetary and Mars programs.[30] Although some non-Mars scientists complained of "Mars myopia," there was relative unity in the planetary community, in part because the NRC Decadal Survey had established Mars as the priority, and Squyres had emphasized to scientists the need for a unified front.

The problem for Mars advocates was that Congress and Washington generally were preoccupied with the upcoming presidential election. It was very hard to get anything done in the fall of 2012 if it required congressional action. On Mars, the Curiosity rover began its trip along the surface of the Gale Crater to Mount Sharp. It would take a while to get there. In the meantime, NASA awaited the findings of the Mars Program Planning Group (MPPG) led by Figueroa.

On September 25, Figueroa and Grunsfeld briefed an NRC committee on various options that were being developed thus far by the MPPG. They noted that MPPG was trying to design a program that would be relevant to the missions of space science, human exploration, and technology development. That was the charge Bolden had given Grunsfeld, and Grunsfeld had relayed to Figueroa. Bolden, like other NASA Administrators, saw Mars as a NASA-wide objective.

It was pointed out in the discussion that space science and human spaceflight had historically viewed one another with suspicion. Figueroa likened the relation to that of an elephant and a mouse, with each wary of the other. Although the two programs were on different tracks, Figueroa believed they could come together on MSR. He said MSR remained the top science priority and he could see options for human spaceflight involvement in it. For example, astronauts could collect a sample en route back to Earth from Mars. They could ensure it was safe enough to be brought to the Earth's surface without fear of planetary contamination. Although Figueroa did not specify where astronauts would collect the sample, some observers saw ISS as one possibility.[31]

Both Figueroa and Grunsfeld emphasized that sample return was the best goal to bring the three different directorates together: human spaceflight, science, and technology development. The technologies developed for MSR would benefit human and robotic endeavors. "Sending a mission to go to Mars and return a sample looks a lot like sending a crew to Mars and returning them safely," Grunsfeld pointed out.[32]

The immediate need, Grunsfeld stated, was to make a decision about 2018. If NASA were to launch in 2018, it would have to begin preparations in the next four or five months. The 2018 option was ideal from the standpoint of alignment of Mars and Earth, but was limited by what Grunsfeld called "the $800 million cost bogey." That meant it would be an orbiter. Figueroa said a rover would cost from $1 billion to $1.5 billion.[33]

Speaking at an international conference the next day, Bolden noted that to reach the president's goal of human spaceflight to Mars in the mid-2030s, it would take not only cooperation within NASA but also international cooperation. Bolden came from a human spaceflight emphasis. He later commented that scientists saw MSR as "the Holy Grail" of the robotic program. "The question for many of us is what the timing of accomplishing the Holy Grail is. Do you have to do it before you can send humans? Some would say 'certainly.' But when Neil Armstrong landed on the Moon, we did not have a sample."[34]

There was obviously debate within NASA about emphases and roles in any cooperative endeavor related to Mars, and the NASA Administrator and Science Mission director needed to agree. Sooner or later, issues about who paid for what would have to be worked out. In the fall of 2012, as NASA planned for Mars Next Decade, however, there were many more pressing unanswered questions, and they would have to remain unanswered pending further events on Mars and in the country.

One immediate question was answered in November when Obama was reelected president. For NASA, that election seemed a positive development because a victory for his opponent would have meant one more review of the entire NASA program. The agency desperately needed a measure of stability. There was already enough angst for NASA over the looming sequestration of funds, across agencies, in 2013, unless the president and Congress worked out a deal to avert this calamity. For NASA, sequestration would mean $1.7 billion in reductions.

Meanwhile, on Mars, the extreme excitement that accompanied Curiosity's landing had given way to a muted expectation about what it would discover. The media clung to any word from NASA or the project's lead scientist, Grotzinger, which seemed to have anything to do with Martian life. In late November, Grotzinger said that Curiosity would be making history, and the media speculated that it had found organic molecules—that is, the building blocks of life sought in the mission and which Viking had not found years before. But that possibility was quickly dampened by NASA. There was, to be sure, optimism among Mars exploration advocates that the rover would find something significant for life, sooner or later, but the agency said, in effect, "not yet!"[35]

A Hopeful Decision

One important aspect of the policy continuity arising from Obama's reelection lay with discussions among NASA, OMB, and the Office of Science and Technology Policy about the FY 2014 budget. While that budget, as well as any possible statement about a longer-term Mars Next Decade program, awaited the president's budget announcement in early 2013, there were positive negotiations focused on the near-term issue of a Mars launch in 2018 or 2020.

Grunsfeld, like Weiler, dealt with OMB. OSTP was also involved in negotiations from a policy perspective. Typically, OMB was far more powerful than OSTP unless the president made his perspective clear. In other words, budget usually drove policy. However, Obama had spoken of his desire to personally "protect" the Mars investment. Directly or indirectly, the Obama view meant that policy could drive budget, at least to some extent, at least for this moment.

On December 4, Grunsfeld came to a "town hall" meeting of the American Geophysical Union fall conference in San Francisco. Excitedly, he announced that the White House had approved a $1.5 billion Mars mission for 2020. Grunsfeld said he had wanted to launch in 2018, but became convinced it would be better to wait for 2020 and go for another rover. The science "action is now on the surface" of Mars, he said. The White House had authorized the mission—and, presumably, the early announcement.

How that rover would be designed and what it would do remained to be determined. He said he was setting up a science definition team to help answer these questions. An MSR cache was possible.

Significantly, he praised OMB and OSTP for their assistance in making this new mission possible. Clearly, the successful landing of MSL Curiosity had made a difference in the decision-making process. Grunsfeld called Mars "a special place." The scientific audience was said to be somewhat in a state of "shock." No one had expected this announcement. Ordinarily, it would have come later, if at all, in connection with the FY 2014 budget statement. The White House was sending a message of support for NASA and the Mars community.[36]

This was a time of extreme budgetary and political turmoil. But there was hope and even some optimism for the future of the Mars program.

Conclusion

Mars has always been the compelling prize in exploration for most space enthusiasts. For many involved in the space program, from von Braun's time to today, the dream that has galvanized them has been that of landing human beings on the Red Planet. Fulfilling that dream lies ahead. The reality that has marked the space program profoundly to date has been robotic exploration. Robotic exploration of Mars has been one of the great achievements of the space age. It has not been easy. If Apollo was a one-decade dash, characterized by Kennedy's singular decision to go to the Moon, Mars exploration has been a multidecade marathon, marked by a host of smaller—albeit important—decisions, usually at the NASA level.

Today, the United States and other nations are gradually extending human senses over the millions of miles to Mars, building a permanent infrastructure for exploration with orbiters, landers, rovers, and communication links. The robotic program has been sustained mainly by the quest to determine whether life exists or has ever existed on Mars, and also by the need to send robotic precursors if human beings were ever to go there. There are other rationales that have applied—including international competition and, conversely, international cooperation—but life has been the most influential driver for Mars exploration over the years. In 1976, when it was thought that there was no life on Mars, the program suffered, and it was not until the early 1990s that flight resumed. The

Mars meteorite claims of the mid-1990s were a catalyst for reemphasizing the life rationale in the years that ensued.

This association of Mars with life in the human psyche has made this fabled planet the planetary magnet for the public as well as scientists. It is close enough to visit every 26 months, but still so far away that spaceflight to it, much less landing on it, remains daunting. Over the decades, more flights sent by spacefaring nations have failed than succeeded. The United States has been relatively successful, especially in recent years. But it has also suffered hugely disappointing losses among the victories, losses that have wasted years of effort, cost hundreds of millions of dollars, and adversely affected the space program and those associated with it. With each setback, however, there has been the struggle to respond and the reality of recovery. Mars has been a continuing challenge to the United States and its agency, NASA. For the most part, NASA and the nation have responded to daunting failure with spectacular success. The Mars Science Laboratory landing of 2012 exemplified such remarkable technical achievement.

This study penetrates behind the technical quest for Mars exploration to the agency responsible for it. It looks at the advocates driving the organization politically. The Mars advocacy coalition seeks to make the Red Planet a funding priority for NASA and the nation. For the most part, the influence of this loosely coupled, often changing group of individuals and entities inside and outside NASA has enabled Mars scientists and engineers to achieve their technical goals. It has moved the agency to establish a long-term program and see it carried out successfully. However, when the advocacy coalition has lacked influence in relation to opponents, that fact has inhibited Mars progress. Moreover, even when influential, there have been times when advocates pursued a flawed scientific or political strategy that had a deleterious effect on the evolution of the Mars program.

Mars advocates succeed to the degree they make NASA the institutional embodiment of their quest. NASA is a base for resource acquisition and program action. The advocacy coalition has pushed most effectively for Mars when certain conditions have existed. One is having senior administrators of NASA in the coalition to actively provide passion and power behind the drive for Mars. A second is for the advocates to gain the support of NASA's political masters, thereby neutralizing rivals and helping in dealing with the Office of Management and Budget. A third is maintaining a common front in scientific and political strategy. Absent any of these conditions, the Mars advocacy coalition loses clout.

The Long Journey

Mars is a program in a science directorate in an agency. Not elevated organizationally, it has a high visibility to the political world. That visibility has helped make it a focus of agency attention and controversy over the years. The history of robotic Mars exploration has seen a sequence of overlapping eras. The eras overlap because NASA is usually trying to sell a new program as it is implementing an older one. The history has seen recurring issues. One has been conflict between the priority given Mars and that for other planets. Another has been the tension between those who would explore Mars incrementally and comprehensively and those who favor faster leaps forward and specifically target the search for life. A third is the debate between Mars and space activities other than planetary exploration, such as telescopes or human spaceflight. A fourth is the conflict between NASA and external forces that want to contain space costs generally—sometimes for non-space priorities—and press NASA to cut back expenditures, including those for Mars.

These and other issues have played out in the various eras. They illuminate the politics of Mars. The *first* era involved the pioneering flights of Mariner— the flybys of the 1960s and orbiters of the early 1970s. Mariner took place when the emphasis at NASA was on the Moon. *Second* came the aborted program Voyager, and then Viking, America's initial landings on Mars, in 1976. Viking was an extraordinary success in many ways, but critics saw it as a failure because it did not achieve its avowed goal to find life. There was dispute over the findings, but the scientific consensus was negative as to life, and this perceived failure helped halt momentum in the program. Calls for a mobile Viking (Viking 3) follow-up went nowhere, as did those for Mars Sample Return.

The *third* era was an interregnum, in which advocates of other missions made their claims and Mars proponents struggled to get a hearing. As Mars dimmed on NASA's agenda, the agency's planetary program in general also suffered financially. A relatively few adherents kept the flame of Mars burning, but not brightly. Eventually, Mars Observer launched and approached Mars 17 years after Viking. Although Mars Observer failed as it encountered the Red Planet in 1993, it gave rise to a *fourth* era of Mars exploration. The new era, called Mars Surveyor Program, featured a sequence of two missions that were relatively small and simple and that were launched every 26 months when Mars and Earth were in an optimal alignment. NASA's strategy of "faster, better, cheaper" fit the political times of post–Cold War America. A premature attempt to accelerate

MSR and failure of two Mars probes in succession brought this era to an abrupt close.

The *fifth* era of Mars exploration was the "follow-the-water" Mars Exploration Program. It was more incremental, comprehensive, and realistic about pace and cost and began in 2001 with the Odyssey orbiter.

Those who wished for greater leaps rather than incremental steps made the most ambitious mission of this series, MSL, even more sophisticated than its original planners had recommended. It was the mission that transitioned from following the water to looking for organic carbon compounds and other indicators of life potential. The boldest and most expensive Mars mission since Viking, in many ways MSL was the Viking 3 that never happened in the late 1970s—except that MSL was far more capable than Viking 3 could have been. MSL's cost soared to $2.5 billion as it was deferred to 2011 from its 2007 and then 2009 schedule. It built on everything NASA had learned scientifically and technically up to this point.[1]

What NASA would do after MSL and the smaller project sent in 2013, the Mars Atmosphere and Volatile Evolution Mission, was unclear at the time of writing. What had initially emerged for the *sixth* era had been a bilateral program with Europe. NASA and the European Space Agency had designed a joint program that would begin in 2016 and 2018 and take advantage of succeeding opportunities to build toward an MSR after 2020. Under severe cost-containment pressure from OMB, the United States withdrew from a major role in the 2016 and 2018 missions as originally planned. To help maintain Mars momentum in the wake of the MSL Curiosity landing, NASA instituted a smaller U.S. mission, InSight, for 2016. It also said it would contribute to the European missions via certain instruments. Most significantly, NASA got approval for a $1.5 billion rover in 2020 that would build on MSL's Curiosity. The MSR sequence of missions might or might not be initiated with this later mission. In 2012, NASA Administrator Bolden pledged that NASA would continue Mars exploration and better integrate robotic and human spaceflight requirements in a proposed new programmatic era, which he called Mars Next Decade.[2]

But the initiation and contours of Mars Next Decade were uncertain. There is hope that missions after MSL Curiosity and MAVEN would lead to a coherent program aimed at MSR. In any event, a sixth era presumably will begin in 2016 and build to bolder ventures. That there has been a rocky start to this post-MSL era is not surprising. History shows that Mars exploration has had a long and tortuous journey, consuming decades, with ebbs and flows in momentum. A program of programs, it has not been a steady evolution. It has been marked by punctuation points and key decisions between programs, and sometimes within programs and specific projects.

Mars exploration represents not only a set of missions and hardware but an agreed-upon scientific and political strategy. However, that strategy is a result of conflict and consensus building among interest groups, governmental and nongovernmental. A program constitutes an equilibrium of interests.[3] The equilibrium exemplifies agreements among specialists in a space policy subsystem—bureaucrats, legislators, scientists, others—about a particular course of action. There is relative stability. Events, key individuals, and disagreements within the subsystem, or pressures from larger forces from outside, can disrupt the subsystem and bring about policy change. The task of Mars advocates generally and NASA leaders particularly has been to make the case for Mars. It has been to build and then rebuild consensus within the space sector and relate it to national and international policy as circumstances have necessitated. Situations internal or external to NASA require decision makers to adapt. Change is to be expected. Managing it is an art more than a science.

What and who have been the moving forces behind NASA's journey from Mariner to MSL and beyond? What and who have stood in the way of the Mars proponents? How has their clash of interest influenced the course of Mars exploration? What decisions by NASA have favored one side or the other in the politics of Mars? Where is the program headed?

Advocates

What has energized NASA toward Mars has been a loose coalition of Mars advocates. These proponents—governmental and nongovernmental—have provided the continuing push behind NASA to maintain the quest. What the Mars Underground said of itself—that it was "closely knit but loosely woven"—might be said of Mars advocates generally. They have constituted an inside-outside political coalition, one congealed by shared attitudes rather than overarching structure. There have been scientists, engineers, and managers within NASA who have propelled the Mars program forward. There have also been individuals and institutions outside NASA who have similarly galvanized action in relation to Mars. The Mars coalition is a "special interest" in Washington parlance. NASA has many interests (and constituencies) to satisfy. The robotic Mars program is but one of many agency enterprises, and not the largest. The central strategy of the Mars advocacy coalition has been to make *its* priority a

NASA priority, and to influence NASA to engage academic scientists, industry, the White House, Congress, the media, the American public, and international partners in backing a *sustained* MEP. Especially important has been enlisting (or neutralizing) OMB through broader political support.[4]

The course of Mars exploration has reflected the success and failure of its network of supporters in the yearly competition for priority and funding. Chaikin has written that a cluster of people have had "a passion for Mars."[5] They are the core of the coalition, the activists. In addition, Mars has a long history of being fascinating to a wider audience, and that fact has helped those with a Mars passion to make Mars first among equals in planetary exploration.

The Mars advocacy coalition extends over generations. It has expanded and contracted. Its membership has changed over the decades, and the baton of leadership has been passed on. Some of the prominent early advocates, such as scientists Sagan and Mutch, have had sites on Mars named after them. Others, such as Viking project manager Martin, are virtual legends among many contemporary Mars proponents, particularly engineers and project managers. As the early Mars exploration protagonists have left, others have taken their place. Often, they have been the graduate students of the pioneers, as Squyres was of Sagan and Garvin was of Mutch. Squyres is an example of an outside advocate, while Garvin exemplifies an advocate inside NASA.

Some of the graduate students in the Mars Underground, such as McKay and Stoker, joined NASA as researchers as they advanced professionally. Outside advocates became insiders. Many outside scientists serve on NASA advisory bodies, achieving access and sometimes shaping policy their way. Inside advocates leave NASA and continue their efforts on behalf of Mars from the outside, as seen in the cases of Hubbard and Huntress. Some have been highly visible, such as Zubrin. Others are virtually unknown to the public, as was the situation with Klein at Ames.

Beyond the ad hoc advocacy of individuals, there has been the "institutionalization" of interest. Hubbard and Naderi began a sequence of officials serving, respectively, in Mars director posts at NASA and the Jet Propulsion Laboratory. Certain universities—for example, Cornell, Arizona State, University of Arizona, University of Colorado, Brown, and others—have become continuing focal points for Mars research.[6] JPL is the NASA center that has been the most continually active among NASA centers for big science projects in Mars exploration over the decades. Various companies are closely associated with Mars exploration. Some are huge like Lockheed; others are smaller and more specialized, such as the planetary camera firm of Malin. There are organized interest groups with a Mars emphasis, particularly the Planetary Society. After many years as the Society's executive director, Friedman stepped down, succeeded by Bill Nye, "the science guy," a well-known media commentator. People in positions change, but roles in advocacy continue.

The Planetary Society, based in Pasadena, is associated closely with the robotic program and has JPL roots. The Mars Society is another interest group, particularly oriented toward human spaceflight, but supportive of precursory robotic flight. Core advocates gather allies and attempt to build an ever-widening gyre of support, including politicians, media, and the public. NASA is the target of all pressures. More than an object of pressures, NASA is a force itself. NASA has helped to mobilize Mars advocacy through formation of a Mars program that provides funds to universities, professors, and graduate students. More recently, NASA has sought to build a scientific constituency for astrobiology. Astrobiology (formerly exobiology) was once ridiculed as a science without a known subject. NASA's Ames Research Center helped keep the field going in the late 1970s and 1980s when most scientists abandoned it. But with the revival of life as a credible rationale for Mars exploration in the 1990s, the field has regained respectability and has attracted an interdisciplinary band of able scientists.

Individuals in strategic positions associated with Mars at NASA Headquarters and various centers have provided authoritative leadership to Mars exploration over the years. Some of these individuals have had strong influence in the policy process, and others relatively little. Turning ideas into government programs is hard, especially when resources are extremely limited and competing demands are numerous.

What makes this translation of visions into action so complex is that Mars exploration is big science of a particular kind. It is "distributed" big science. While there have been some billion-dollar missions—for example, Viking and MSL—the program more often has featured a parade of spacecraft of more moderate expense. It has been organized into missions spread across years. Ideally, there is a coherent and integrated sequence of activity, with one mission providing a base of knowledge pointing to what must be done in the one following. For most advocates who provide a "push" toward Mars, there is a major goal that "pulls" them forward. This is the return of samples of rock and soil. At least since Viking, MSR has been the holy grail of the robotic program. Most advocates agree that it is the best way to determine the question of Martian life short of sending scientist-astronauts. However, many Mars activists also see

MSR as a way to develop critical knowledge, technologies, and skills that will enable human spaceflight to Mars. Hence, MSR has a potentially unifying role that makes it a NASA-agency goal, not just a science goal. It relates to the two sides of the life rationale—life on and life to Mars.

It also has symbolic significance. David Southwood, ESA science director who worked with Weiler to initiate a Mars Together program, has declared, "Doing it [MSR] together sends a message. It shows what we can do. It is a big deal. For the robotic program, it is analogous to Armstrong on the Moon."[7] The U.S. withdrawal from a full-partnership role in the program, however, points up the difficulty of accomplishing the goal, whether together or singly.

Opponents

Few individuals or institutions are truly "against" Mars exploration. Opponents are concerned with the issue of priority and whether Mars gets too much versus the outer planets or some other science (or nonscience) option. Throughout the history of Mars exploration, there have been "opponents" who are, in fact, advocates for another priority. Their rhetoric typically calls for "balance." Likewise, there have always been those who oppose federal spending in general, especially for programs they see as nonurgent. Larger, macropolitical forces invariably impinge on decision making in a specific policy sector, such as space. Big science is an inviting target for budget cutters, whether in OMB or Congress.

There have been a number of pressing alternatives within space policy to Mars exploration over the decades, such as Earth's Moon in the 1960s and Jupiter's moon, Europa, in the twenty-first century. Europa also may have life—under its ice. In late 2011, with level NASA funding, Mars came up against huge overruns in the James Webb Space Telescope. This project had extremely influential political support in Congress, more so than Mars. Mars Observer in the 1980s had to wait on the Hubble Space Telescope. Now Mars would wait on Hubble's successor.

There is only so much money for big science (concentrated or distributed). Unless advocates of Mars make their case strongly and well, they will not necessarily get their way. It may be easier to cut a distributed big science program, like Mars, than one that is concentrated in structure, such as the James Webb Space Telescope. A specific mission within a distributed program can be extracted more easily than killing a massive concentrated program, at least one that is well along in implementation.

Politics is about "who gets what, when, and how." Politics applies to planetary science as much as to other fields. Who is to say that Hubble was not deserving of being ahead in line for shuttle launch after the Challenger disaster set it against Mars Observer? Earth observation satellites relate to climate change and, arguably, the long-term survival of the human species. Advocates for this part of the space program have a legitimate case to make. Their advocates have done so. Proponents of human spaceflight continually press NASA—and, indirectly, robotic Mars missions—for resources. In the long run, human spaceflight and Mars exploration are mutually dependent. In the short run, they compete. Figueroa's comment that the relationship between these two major NASA programs is akin to that between an elephant and a mouse is apt. NASA Administrator Bolden, in the wake of the Obama budget proposal for FY 2013, set in motion a planning effort for Mars which more strongly linked robotic and human spaceflight. Administrators have sought such a linkage before and seldom succeeded in forging a true partnership. The human and robotic programs represent two different cultures within NASA.

Unfortunately, there is never enough money for all worthy endeavors. Advocates of alternatives to Mars become opponents of spending on robotic Mars flights even though that is not necessarily their intent. Similarly, flagship missions become barriers to spending on "little science," and there can be divisions within the Mars community. NASA centers vie with one another and with universities and industry. The debate is not about good versus bad, but about various "goods." Over the long haul, Mars exploration has advanced to the extent that it has prevailed over the "opposition," or found ways to reach some measure of accommodation through alliances.

Leadership at NASA

Political pressures from advocates and opponents ultimately affect NASA decisions. NASA is the institutional glue that holds Mars exploration together, sometimes well, sometimes not so well. The advocacy coalition keeps Mars on the NASA and national policy agenda. Opponents within the space sector and outside of it seek displacement of Mars with alternate priorities. Both sides work directly on NASA, and sometimes via end runs to the agency's political masters or the general public. Leaders in NASA respond to events, results, and scientific, bureaucratic, and political pressures, as well as their own predilections. They choose among conflicting options and then work to build internal and external constituencies to effectuate their choices. The end result of the clash of interests can be decisions to establish a new program, to reorient an existing program, or to end a program. One equilibrium in the balance of interests gives way to another.

NASA decisions have to be sold to OMB, the Office of Science and Technology Policy, the president and his political advisors, and Congress. NASA policy at the space subsystem level has to fuse with national policy, largely through the annual budget and appropriations process. Money fuels big science. The process of official decision usually starts within NASA, as the agency sorts out its needs amidst contesting advocacy groups. Within NASA, the decision process requires the associate administrator for science and the Administrator to decide on priorities, the place of Mars among them, and how to build support for those preferences. They are executives with political roles. It is their task to lead. In the words of James Webb, who guided NASA and Apollo in the 1960s, the role of leadership is to integrate "a large number of forces, some countervailing, into a cohesive but essentially unstable whole and keeping it in motion in a desired direction."[8]

A number of individuals in NASA have played these institutional leadership roles with respect to the robotic program over the years, from Mariner to MSL. First were Newell and Glennan. Both downplayed Mars in favor of the Moon, but they empowered Pickering, and the result was Mariner. Newell and Glennan's successor, Webb, maintained Mariner. It was part of the contest between the United States and the Soviet Union. Moreover, Mariner got started at a time when NASA's budget was soaring. There was plenty of support and money for multiple initiatives on several fronts.

In the late 1960s, Newell and Webb looked ahead to post-Apollo NASA. They both wanted to explore the solar system. Webb in particular hoped to use robotic Mars Voyager as a program precursory to human Mars exploration, but did not want to advertise that motivation. He desperately wanted to maintain the Saturn rocket capability and the von Braun team. Newell went along with Webb and saw uses for the Saturn rocket, but many scientists (including a number at JPL) opposed the huge Saturn-driven Voyager. Congress killed the program before it could get started, because critics also saw it (correctly) as a covert precursor to human spaceflight to Mars, and they did not want to go that direction.

Webb moved Newell to another NASA position and told Naugle, his successor, to reshape the Mars program, or it would die. Naugle worked feverishly. He and the NASA Administrator promoted Viking as a replacement. The Saturn rocket and its human-Mars connotation were removed. NASA sold Viking as post-Mariner, not post-Apollo. The search for life became the prime rationale for the robotic program. Naugle solidified scientific support, while Webb built a White House–congressional political base sufficient to get Viking under way. The Mariner political equilibrium ended with the Voyager debacle, and that of Viking gradually ensued.

Webb left, and Paine came on as Administrator. Naugle, a career official, stayed as associate administrator for science, providing continuity. Paine decided that NASA should pursue the most aggressive (and expensive) Viking option Naugle proposed. He was oriented to a human Mars program and saw the precursory potential of Viking. Unfortunately, with Nixon's cutbacks, there was no hope for a human Mars program. In fact, he later had to tell Naugle that Viking could be salvaged only by delaying its launch by two years.

Fletcher succeeded Paine as NASA Administrator. Like Webb and Paine, he saw Mars exploration as not only a science but a NASA priority. He fully backed Viking. More than Webb or Paine, he emphasized the rationale of Viking's quest for exobiological life. He involved himself personally in the Viking project, first with Naugle, then with Naugle's successor, Hinners. In 1976, however, the time came for decisions about what was called Viking 3. President Ford had become a potential target of those wishing to continue Viking, and he awaited a strong push from NASA for a follow-on in his last budget. That push did not come. The scientific consensus was that Viking did not find life. The result was far more ambiguous than a simple "no," but that ambiguity was lost to most observers.

NASA's decision had been to "go for broke," to take an Apollo-like approach to Mars. Apollo's goal had been clear—to beat the Soviet Union to the Moon. In an analogous way, the goal of Viking was to best the Soviet Union in finding life on Mars. Clarifying a goal, making it as simple as possible, can be a way of gaining support. But it is a high-risk strategy, if the goal is not achieved.

Viking's failure to find life after so much concentrated effort, hype, and personal sacrifice on the part of those involved tarnished the allure of Mars. It exhausted and diminished the advocacy coalition. Other non-Mars advocates pressed NASA for "their turn" at priority, specifically for Galileo and Hubble. A follow-on Viking project—seen mainly as a mobile Viking—would be a mission costing $1 billion or more. It would have been so expensive as to preclude other worthy endeavors. NASA could afford only so many big science programs. NASA leaders decided not to press "the Case for Mars," and almost by default Mars exploration moved to the back burner of NASA's agenda.

And there it stayed for years. It took a long time for a new political consensus favorable to Mars to be established. Associate administrators for science and NASA Administrators came and went. They kept Mars exploration alive through "extended missions" studying Viking data, and eventually via Mars Observer, sold as a low-cost mission that would look not for life but for more general geophysical understanding. Life, as a goal, was scientifically unfashionable. And without that special aura, Mars became, de facto, just another planet. Meanwhile, two successive associate administrators for science, Edelson and then Fisk, developed a new global environmental mission for NASA whose significance grew as climate change evolved as an issue. Cost-constrained decision makers chose not to push Mars, whose advocacy coalition had shrunk significantly, while they promoted other projects important in their own right.

The locus of strong advocacy for Mars was outside NASA in the 1980s. External Mars proponents, such as members of the Mars Underground and Carl Sagan, were critical of NASA. Seeking an end run around the agency, Sagan and the Planetary Society used macropolitical rationales, particularly Mars Together with the Soviet Union, to make the Red Planet more salient to the public and politicians. They linked space with international cooperation as a strategy to change NASA priorities. NASA leaders resisted generally when outside advocates sought to alter their priorities. This was particularly the case after the Challenger disaster, when Mars advocates tried to change the shuttle launch schedule in favor of Mars Observer.

Everything changed in respect to Mars when Goldin became Administrator in 1992 and he replaced Fisk with Huntress. Although vastly unlike in personality, Goldin, the political executive, and Huntress, the career official, struck an exceptionally creative alliance and made a huge difference for Mars. The Goldin-Huntress axis was not only extremely Mars oriented but also unusually skilled. Goldin stands out for the passion he had for Mars and ability to work with political forces—Vice President Gore in particular—to further Mars interests. Huntress was crafty as an operator in bureaucracy and with his mercurial boss. He was able to deal well with the science community. Together, Goldin and Huntress used the failure of Mars Observer to trigger a renewal of Mars priority and rebuild what was a weak program. They scheduled missions at every 26-month opportunity. Goldin made Mars *the* flagship of his faster, better, cheaper revolution, thereby enlisting support in the White House and Congress. From the White House perspective, Goldin's efficiency campaign made him a "good soldier."[9] As he was responsive to the White House, it was responsive to him.

Goldin made deft use of the Mars meteorite to rekindle interest in the media and public for the search for life as a rationale. With Huntress on the inside and Sagan as an outside advisor, Goldin worked to revive exobiology, renamed "astrobiology," as a scientific discipline. He sought to accelerate MSR. He linked robotic Mars and human spaceflight more firmly in hopes of enhancing the robotic program's precursory role. When Huntress left and Weiler came on as associate administrator for science, Weiler picked up where Huntress had left off. An experienced and able manager, Weiler also worked in tandem with Goldin.

Mars was emphatically the science and personal priority for Goldin in the 1990s. When the twin Mars failures took place in 1999, Goldin and Weiler retained Mars as a flagship but ended the Mars Surveyor Program. Aided by Hubbard and his team, they made decisions that were more realistic technically and financially. With political support in the White House, OMB played a constructive role in the program redesign, an ally rather than adversary. The "follow-the-water" MEP started a new era for NASA and the Red Planet. When Goldin left NASA, his legacy reflected the flaws of overreach, but it also boasted a Mars program that had been transformed profoundly for the better. Goldin led the advocacy coalition from NASA's summit.

O'Keefe as NASA Administrator maintained the Mars program he inherited, and Spirit and Opportunity helped the agency (and nation) at a time of great psychic need in the post-Columbia period. When the second Bush made his Vision for Space Exploration decision, O'Keefe sought to augment robotic Mars spending and even more strongly link the robotic program with its precursory role for human spaceflight. The dual purpose—life on, life to—was never more explicit, and significantly more funds for Mars were projected. A "Safe on Mars" funding line was planned. "Priority" for Mars as a budgetary strategy was in, "balance" out for O'Keefe. A backlash from advocates of other space science programs came quickly and intensely against what they saw as too extreme a Mars emphasis.

Griffin came on as Administrator, listened to non-Mars advocates, and "rebalanced" the science program away from Mars. He was most determined to launch NASA's human return to the Moon—as prelude to Mars—but he lacked a presidential funding commitment required for Moon-Mars. With far too much on NASA's plate, and the shuttle costing more than projected, he decided to cut science to help fund human spaceflight, and Mars was not excluded from the pain. The "Safe on Mars" funding element went away, along with much else that was not near-term.[10] Griffin hoped his 2007 choice as associate administrator for science, Stern, would help him design a Mars exploration program that was scientifically sound, politically acceptable, and affordable. He agreed when Stern wished to again accelerate MSR. But the way Stern attempted to get to MSR proved extremely controversial. Whatever might have been said for the scientific MSR goal, the Mars program strategy proposed by Stern did not get the support of the scientific community it had to have to be viable. Then, Stern and Griffin clashed over MSL. When Stern directed a cut to the iconic Spirit and Opportunity rovers, without consulting Griffin, the NASA Administrator overruled him publicly, and Stern was forced to resign.

It was up to Weiler to repair the damage, as he returned to rechart the Mars program. He sought to restore the political equilibrium undergirding Mars exploration which he found had been disrupted. He started by adroitly getting the NASA Administrator on his side. The first big decision he and Griffin made regarding the Red Planet was to delay MSL by two years. Behind the decision was Weiler's understanding and finesse in working the NASA setting. In achieving the same decision Stern had sought—delay—Weiler carefully gathered support within NASA and JPL. This decision added to already substantial MSL costs, but made eventual success more likely. Griffin made it abundantly clear he regarded Mars exploration as a top priority for the Science Mission Directorate—*and* NASA. He decided that MSL would get the money it needed to succeed, even if there had to be cuts to other worthy programs as a consequence. But Spirit and Opportunity were not to be touched!

Leaders matter. They make difficult decisions that have large consequences. They engineer choices within the space policy sector and relate those choices to the broader national and international policy world. Many others can advocate, advise, lobby, and complain. But officials in the key positions of associate administrator for science and NASA Administrator have formal authority to decide, and making choices is never easy when there is not enough money for all that needs to be done. The essence of science policy lies with decisions about priority. Spirit and Opportunity would never have succeeded had not Goldin and Weiler found the money to make them happen. And Weiler had to do that more than once during the rovers' development. Also, in government, how decisions are made or sold can be as important as the decisions themselves.

The Bush administration gave way to that of Obama, and Bolden became NASA leader. Weiler, who remained as associate administrator for science until 2011, planned for the next era of Mars exploration—a Mars Together program with Europe and possibly other nations. As before, the robotic program had a prime science goal—to find evidence of present or past life. The means for achieving this purpose remained MSR. This means is also itself an interim goal, a vital enabling one. There was virtual unanimity among JPL, the Mars science community, NASA decision makers, and Mars enthusiasts generally about MSR. The challenge, as always, was to find the money to realize this objective.

Weiler, the NASA decision maker, became an advocate to OMB and White House staff. He hoped that international cooperation would provide a helpful political rationale for its achievement, symbolizing that in austere times nations could collaborate on grand and worthwhile challenges, while sharing the risks and costs. He aimed at a new political equilibrium or consensus, based on a "Mars Together" rationale. He could not persuade NASA's budgetary overseers, and he resigned at least in part as an act of protest.

Notwithstanding NASA's withdrawal from the planned European partnership, the desire for joint missions remained. NASA Administrator Bolden and his new associate administrator for science, Grunsfeld, did not give up on collaboration. Nor would their successors likely do so. There are realities about bold ambitions and an austere funding environment which shape what leaders do. What NASA needs is help in getting resources to match scientific vision. That can come through alliance with domestic groups as well as international partners. It can also come through exciting discovery. Long-term programs need periodic catalysts. Mars had one with the Mars meteorite in the mid-1990s. MSL's Curiosity could produce a stimulus through exciting findings. Successes reinvigorate a lengthy program; failures bog it down. Discoveries or dramatic events can elevate Mars from sectoral policy to national policy. They can help attract political leaders, as the meteorite did Clinton and Gore. But discoveries or events become catalysts for funding only when astute Mars advocates and their allies make good use of them, engage the media, win the public, and maneuver skillfully in the political/policy process. The larger the advocacy coalition, and the better it is led, the more powerful the push for Mars.

The Journey Ahead

Caltech's Grotzinger has called the recent period of Mars exploration a "golden era." He has marveled how missions have built systematically on one another. The coordination and integration of missions have, he wrote, "brought us ever closer to fathoming the broad range of environmental processes that have transformed the surface of Mars, beginning over four billion years ago." Mars exploration is going from following the water to searching for the building blocks of life.[11] Beyond that is the investigation of Mars samples for past and present life itself. Optimism among scientists about Martian life has returned as a prime motivator of national and international planning.

The achievements in science and technology would not have been possible without organization and politics. NASA has pulled the components of a distributed multimission big science program together and obtained resources for implementation. Orbiters and landers have been linked, and they pointed the way for the best places for rovers to go to search for traces of life. The program has not gone as consistently or smoothly as Mars advocates would have liked. The journey has been anything but steady. Science may provide a "guiding light," but politics influences how fast and how well government and the researchers it supports can follow it.[12] The political process can result in a pause in activity, as well as acceleration.

The history of robotic Mars exploration has been one of progress, setback, and renewed dedicated effort.[13] It is filled with human drama that is at times heroic and at other times tragic. The future of Mars exploration will likely emulate the rhythm of the past. It will advance, hit barriers, and then advance again. Over the long haul, Mars exploration moves forward. What gives the robotic program direction is that it has relative consensus on a clear technical goal akin to the Apollo lunar landing. For scientists, engineers, and NASA Administrators, it is called MSR.

That goal ties together individual missions distributed over time. It is a siren call for most Mars specialists. It has been the compelling goal for decades—and many of the most important conflicts around Mars policy have entailed issues of when and how to reach the MSR objective. For virtually everyone, specialists and general public alike, the goal is also a means to answer deeper and broader questions about life which underlie Mars. A culmination for the robotic program, it is seen by NASA and its Mars constituency also as a potentially big step toward human spaceflight.

Why Explore?

Mars exploration takes an intrinsic need and turns it into governmental action. The cost is billions of dollars over time. Humans explore because of science, but also because of an innate drive to know what is on the next frontier. We want to extend the human presence outward. How do we fit into the universe? Are we alone? Mars embodies scientific and larger human needs. Mars is attainable. It may have answers to age-old questions that are among the most profound that humans ask. Robots are there today and will continue to forge a trail. They may find life or evidence of past life. Whether or not they do, they will address fundamental issues. The quest itself lifts the spirit. Robots go first as pioneers. Ultimately, men and women will bring life to the Red Planet. Mars calls because we want to know about ourselves.

Notes

Introduction

1. Jeffery Kluger, "Live from Mars," *Time*, Aug. 20, 2012, 20–25; "This Is What NASA Should Be Doing," *Aviation Week and Space Technology*, Aug. 13, 2012, 58; "Bullseye," *Space News*, Aug. 13, 2012, 26.

2. The White House, "National Space Policy," Sept. 19, 1996, www.fas.org/spp/military/docops/national/nstc-8.htm.

3. John Grotzinger, "Beyond Water on Mars," *Nature Geoscience*, Apr. 2009, 1.

4. *World Book Encyclopedia*, vol. 13 (Chicago: World Book, 1984), s.v. "Mars," 180–182b.

5. Grotzinger, "Beyond Water on Mars."

6. Marcia Smith, "NASA Officials Cheer $17.7 Billion Request for FY2013," Feb. 14, 2012, www.spacepolicyonline.com/news/nasa-officials-cheer-17-7-billion-request-for-FY2013.

7. Charles Bolden, "NASA Recharting Its Path to Mars," *Space News*, Feb. 20, 2012, 17.

8. NASA spoke of the second decade of the twenty-first century as constituting a "Robust Multi-Year Mars Program" that would combine existing missions and propose new missions for the decade, culminating in the 2020 rover. "NASA Announces Robust Multi-Year Mars Program; New Rover to Close Out Decade of New Missions," NASA News Release, Dec. 4, 2012, www.nasa.gov/home/hqnews/2012/dec/HQ_12–420_Mars_2020.html.

9. Paul Sabatier and Hank Jenkins-Smith, "The Advocacy Coalition Framework: An Assessment," in Paul Sabatier, ed., *Theories of the Policy Process* (Boulder, CO: Westview, 1999), 117–166.

10. James Webb, *Space Age Management: The Large-Scale Approach* (New York: McGraw-Hill, 1969), 135–136.

11. James L. True, Bryan Jones, and Frank R. Baumgartner, "Punctuated-Equilibrium Theory: Explaining Stability and Change in American Policymaking," in Sabatier, *Theories of the Policy Process*.

12. Leonard David, "NASA Group Eyes 2011 for Mars Mission," *Space News*, Aug. 28–Sept. 1, 1998, 3.

13. The term "exploring machines" is used in Oran Nicks, *Far Traveler: The Exploring Machines* (Washington, DC: NASA, SP-480, 1985).

CHAPTER ONE: The Call of Mars

1. Cited by Malcolm Walter, *The Search for Life on Mars* (Cambridge, MA: Perseus, 1999), 97.

2. Bruce Murray, *Journey into Space: The First Three Decades of Space Exploration* (New York: Norton, 1988), 32.

3. Amy Paige Snyder, "NASA and Planetary Exploration," in John Logsdon, ed., *Exploring the Unknown*, vol. 5 (Washington, DC: NASA, 2001), 263–266.

4. Roger Launius and Howard McCurdy, *Robots in Space* (Baltimore, MD: Johns Hopkins University Press, 2008), 64–65.

5. David Portree, *Humans to Mars* (Washington, DC: NASA, 2001), 1–3.

6. Steven Dick, "50 Years of NASA History," in Rhonda Carpenter and Ana Lopez, eds., *NASA: 50 Years of Exploration and Discovery* (Tampa, FL: Faircount Media Group, 2008), 31.

7. Eileen Galloway, "Sputnik and the Creation of NASA: A Personal Perspective," in Carpenter and Lopez, *NASA*, 48.

8. Roger Launius, "Leaders, Visionaries, and Designers," in Carpenter and Lopez, *NASA*, 258.

9. Homer Newell, *Beyond the Atmosphere: Early Years of Space Science* (Washington, DC: NASA, 1980), 100.

10. Steven Dick and James Strick, *The Living Universe: NASA and the Development of Astrobiology* (New Brunswick, NJ: Rutgers University Press, 2004), 18.

11. Homer Newell, interview by Edward C. Ezell, Oral History, May 25, 1977, NASA History Office Files.

12. Edward Goldstein, "NASA's Planet Quest," in Faircount Media Group, *NASA*, 129.

13. Newell, *Beyond the Atmosphere*, 261.

14. Ibid., 262. See also Douglas Mudgway, *William Pickering: America's Deep Space Pioneer* (Washington, DC: NASA, 2007).

15. Newell, *Beyond the Atmosphere*, 260–266.

CHAPTER TWO: Beginning the Quest

1. A. J. S. Rayl, "NASA Engineers and Scientists: Transforming Dreams into Reality," in Rhonda Carpenter and Ana Lopez, eds., *NASA: 50 Years of Exploration and Discovery* (Tampa, FL: Faircount Media Group, 2008), 271.

2. Ibid., 274.

3. Edward Ezell and Linda Ezell, *On Mars: Exploration of the Red Planet, 1958–1978* (Washington, DC: NASA, 1984), 34.

4. Ibid.

5. Asif Siddiqi, *Challenge to Apollo: The Soviet Union and the Space Race* (Washington, DC: NASA, 2000), 337.

6. Robert Reeves, *The Superpower Space Race* (New York: Plenum, 1994), 314.

7. Ibid., 318.

8. W. Henry Lambright, *Powering Apollo: James E. Webb of NASA* (Baltimore: Johns Hopkins University Press, 1995).

9. John F. Kennedy, "Message to Congress, May 25, 1961," *Public Papers of the Presidents of the United States*, Jan. 20–Dec. 31, 1961 (Washington, DC: GPO, 1962), 404.

10. Homer Newell, *Beyond the Atmosphere: Early Years of Space Science* (Washington, DC: NASA, 1980), 113.

11. John Logsdon, "Ten Presidents and NASA," in Carpenter and Lopez, *NASA*, 229.

12. Siddiqi, *Challenge to Apollo*, 337.

13. Murray, *Journey into Space*, 38.

14. Clayton Koppes, *JPL and the American Space Program: A History of the Jet Propulsion Laboratory* (New Haven, CT: Yale University Press, 1982), 167.

15. Douglas Mudgway, *William H. Pickering: America's Deep Space Pioneer* (Washington, DC: NASA, 2008), 151–153.

16. "Initial Scientific Interpretation of Mariner IV Photography," Press Conference, NASA Headquarters, July 29, 1965, Webb, Pickering, et al., Mariner Mission Results Info. JPL files, Box 551 Folder 11.

17. Murray, *Journey into Space*, 44.

18. William Poundstone, *Carl Sagan: A Life in the Cosmos* (New York: Henry Holt, 1999), 87.

19. "The Dead Planet," *New York Times*, July 30, 1965, cited in Ezell and Ezell, *On Mars*, 81.

20. Philip Abelson, "The Martian Environment," *Science*, Feb. 12, 1965, 683; see Ezell and Ezell, *On Mars*, 81.

21. Andrew Chaikin, *A Passion for Mars: Intrepid Explorers of the Red Planet* (New York: Abrams, 2008), 61.

22. Keay Davidson, *Carl Sagan: A Life* (New York: John Wiley, 1999), 180.

23. Ezell and Ezell, *On Mars*, 81.

24. Mudgway, *William H. Pickering*, 153.

CHAPTER THREE: Leaping Forward

1. NASA, *Preliminary History of NASA: 1963–69* (Washington, DC: NASA, 1969), 4:36.

2. Edward Ezell and Linda Ezell, *On Mars: Exploration of the Red Planet* (Washington, DC: NASA, 1984), 127.

3. Bruce Murray, *Journey into Space: The First Three Decades of Space Exploration* (New York: Norton, 1989), 49.

4. Bruce Murray and Merton Davies, "A Comparison of US and Soviet Efforts to Explore Mars," *Science*, Feb. 25, 1966, 945–954.

5. Oran Nicks, *Far Travelers: The Exploring Machines* (Washington, DC: NASA, 1985), chap. 12, 3.

6. Clayton Koppes, *JPL and the American Space Program: A History of the Jet Propulsion Laboratory* (New Haven, CT: Yale University Press, 1982), 191.

7. Ibid.

8. Michael Neufeld, *Von Braun: Dreamer of Space, Engineer of War* (New York: Knopf, 2007), 417.

9. W. Henry Lambright, *Powering Apollo: James E. Webb of NASA* (Baltimore, MD: Johns Hopkins University Press, 1995).

10. NASA, *Preliminary History*, 2:19.

11. William Burrows, *Exploring Space: Voyages in the Solar System and Beyond* (New York: Random House, 1990), 197–198.

12. NASA, *Preliminary History*, 4:37.

13. John Naugle, interview by author, June 21, 2009.

14. Ezell and Ezell, *On Mars*, 135–136.

15. Ibid., 135; NASA, *Preliminary History*, 2:21.

16. John Naugle, interview by author, June 21, 2009; John Naugle, correspondence, June 19, 2009.

17. John Naugle, interview by author, June 21, 2009; John Naugle, correspondence, June 19, 2009.

18. Ezell and Ezell, *On Mars*, 125.

19. John Naugle, interview by author, June 21, 2009; John Naugle, correspondence, June 19, 2009; Ezell and Ezell, *On Mars*, 125, 140.

20. Lambright, *Powering Apollo*.

21. John Naugle, interview by author, June 21, 2009; John Naugle, correspondence, June 19, 2009.

22. Ezell and Ezell, *On Mars*, 151–152.

23. Burrows, *Exploring Space*, 207; John Naugle, interview by author, June 21, 2009; John Naugle, correspondence, June 19, 2009.

24. John Naugle, interview by author, June 21, 2009; John Naugle, correspondence, June 19, 2009.

25. Gerald Soffen, Oral History, Feb. 28, 1980, NASA History Office Files.

26. Ezell and Ezell, *On Mars*, 152–153.

27. Andrew Chaikin, *A Passion for Mars: Intrepid Explorers of the Red Planet* (New York: Abrams, 2008), 37.

28. Ezell and Ezell, *On Mars*, 186.

29. Chaikin, *Passion for Mars*, 55.

30. John Naugle, interview by author, June 21, 2009; John Naugle, correspondence, June 19, 2009.

31. John Logsdon, "The Evolution of US Space Policy and Plans," in Logsdon, ed., *Exploring the Unknown*, vol. 1 (Washington, DC: NASA, 1995), 385. Chaikin, *Passion for Mars*, 49–50.

32. Ezell and Ezell, *On Mars*, 188–189; John Naugle, interview by author, June 21, 2009; John Naugle, correspondence, June 19, 2009.

33. Chaikin, *Passion for Mars*, 50.

34. Ezell and Ezell, *On Mars*, 191.

35. Logsdon, "Evolution," 385.

36. Chaikin, *Passion for Mars*, 52–53.

37. Roger Launius, "A Western Mormon in Washington, D.C.: James C. Fletcher, NASA, and the Final Frontier," *Pacific Historical Review* 64, no. 2 (May 1995): 217–241.

38. John Naugle, interview by author, June 21, 2009; John Naugle, correspondence, June 19, 2009.

39. Robert Reeves, *The Superpower Space Race: An Explosive Rivalry through the Solar System* (New York: Plenum, 1994), 372–373.

40. Chaikin, *Passion for Mars*, 63–64; Richard Pothier, "Life on Mars? Some Scientists Believe So," *Miami Herald*, Nov. 14, 1971, NASA History Office Files.

41. Carl Sagan, *The Demon-Haunted World: Science as a Candle in the Dark* (New York: Ballantine, 1996), 213.

42. Reeves, *Superpower Space Race*, 376.
43. Chaikin, *Passion for Mars*, 65.
44. Ibid., 65–70.
45. Michael Carr, interview by author, June 17, 2011.
46. Chaikin, *Passion for Mars*,73.
47. Logsdon, "Evolution," 386.
48. Ibid., 387.
49. Ibid., 388.
50. Ibid.

CHAPTER FOUR: Searching for Life

1. James Fletcher, interview by Roger Launius, Oral History, Sept. 19, 1991, NASA History Office Files.
2. Paul Jankowski, "Voyage of the American Viking," *Du Pont Context Magazine*, no. 2 (1975), NASA History Office Files.
3. Marvin Miles, "Life on Mars? Scientists Unsure," *Los Angeles Times*, Mar. 5, 1972, NASA History Office Files.
4. George Low to John Naugle, memorandum, May 8, 1972, NASA History Office Files; John Naugle to James Fletcher, memorandum, Aug. 4, 1972, NASA History Office Files.
5. Beverly Crindorff, "Viking Manager Drives Team," June 20, 1975, NASA History Office Files.
6. John Naugle, interview by author, June 21, 2009; John Naugle, correspondence, June 19, 2009.
7. Edward Ezell and Linda Ezell, *On Mars: Exploration of the Red Planet, 1958–1978* (Washington, DC: NASA, 1984), 251.
8. Thomas O'Toole, "Unmanned Spacecraft Off to Mars," *Washington Post*, Aug. 21, 1975, A1.
9. Ezell and Ezell, *On Mars*, 232–234.
10. Ibid., 235.
11. Thomas O'Toole, "Mars Spacecraft Cost Soars," *Washington Post*, May 19, 1994, A3.
12. Ibid.
13. William Shumann, "Viking Cost Boost Eludes Space Outlook," *Aviation Week and Space Technology*, Dec. 6, 1974, 46–48.
14. O'Toole, "Unmanned Spacecraft."
15. Ezell and Ezell, *On Mars*, 252.
16. A. Thomas Young, Oral History, Nov. 18, 1997, NASA History Office Files.
17. Ezell and Ezell, *On Mars*, 304–305.
18. Carl Sagan to James Martin, Feb. 3, 1973, NASA History Office Files.
19. James Fletcher to John Naugle, memorandum, Feb. 22, 1973, NASA History Office Files.
20. John Naugle to James Fletcher, memorandum, Mar. 7, 1973, NASA History Office Files.
21. Ezell and Ezell, *On Mars*, 313.

22. Harold Schmeck Jr., "2 Mars Sites Chosen for Unmanned Landings in '76," *New York Times*, May 7, 1973, NASA History Office Files.

23. Ezell and Ezell, *On Mars*, 268–269; "Viking Cost Upped to $930 Million," *Defense Space Daily*, Nov. 22, 1974, 121.

24. John Naugle, interview by author, June 21, 2009; John Naugle, correspondence, June 19, 2009.

25. "Appropriations Committee Issues Warning on Viking," *Defense Space Daily*, Aug. 7, 1974, 198.

26. Noel Hinners, interview by author, Jan. 28, 2009.

27. Ezell and Ezell, *On Mars*, 313.

28. Shumman, "Viking Cost Boost."

29. Deputy Administrator (Low) to Administrator (Fletcher), memorandum, Sept. 12, 1974, NASA History Office Files.

30. Fletcher to Dixie Lee Ray, Nov. 23, 1973, NASA History Office Files.

31. James Fletcher to Edgar Cortright, Aug. 7, 1974, NASA History Office Files.

32. Noel Hinners to James Fletcher, memorandum, Nov. 4, 1974, NASA History Office Files.

33. Deputy Administrator (Low) to Associate Administrator for Space Science (Hinners), memorandum, Nov. 11, 1974, NASA History Office Files; James Fletcher to Dr. Low, memorandum, Nov. 6, 1974, NASA History Office Files; Associate Administrator for Space Science (Hinners) to Administrator (Fletcher), memorandum, Nov. 4, 1974, NASA History Office Files.

34. George Low to Noel Hinners, memorandum, Nov. 11, 1974, NASA History Office Files.

35. Victor McElheny, "'76 Mars Landing Gains with Launching of Titan," *New York Times*, Dec. 18, 1974, 15.

36. Robert Reeves, *The Superpower Space Race: An Explosive Rivalry through the Solar System* (New York: Plenum, 1994), 380–384.

37. Viking Dep. Program Manager, Orbiter, to Viking Program Manager, memorandum, July 18, 1974, NASA History Office Files.

38. Noel Hinners, interview by author, Jan. 28, 2009.

39. Carl Sagan to Noel Hinners, Mar. 14, 1975, NASA History Office Files.

40. Dan Partner, "Mars Launch Gear Readied," *Denver Post*, Apr. 10, 1975, NASA History Office Files.

41. Robert Kraemer to James Martin, July 2, 1975, NASA History Office Files.

42. Jonathan Spivak, "Two Unmanned Spacecraft Will Seek Signs of Life on Mars after Year-Long Flight," *Wall Street Journal*, Aug. 11, 1975, 28.

43. Noel Hinners, interview by author, Jan. 28, 2009.

44. O'Toole, "Unmanned Spacecraft."

45. John Naugle to Director, Lunar and Planetary Science, memorandum, Dec. 22, 1975, NASA History Office Files.

46. "Viking to Change to Smaller Orbit," *Virginia-Pilot*, June 21, 1976, NASA History Office Files.

47. "Clearest Mars Photos Show Hazardous Terrain," *Pasadena Star-News*, June 24, 1976, NASA History Office Files.

48. George Alexander, "Viking's Landing on Mars Delayed; Safer Site Sought," *Los Angeles Times*, June 28, 1976, NASA History Office Files.

49. "Mars Landing Off for July 4," *Washington Post*, June 28, 1976, NASA History Office Files.

50. Noel Hinners, interview by author, Jan. 29, 2009.

51. "July 17 Landing on Mars Set," *Virginia Pilot*, July 2, 1976, NASA History Office Files.

52. David Sallsbury, "Space Cowboys on the Martian Frontier," *Technology Review*, Dec. 1976, 8.

53. George Alexander, "Mars Less Foreboding in Latest Photos," *Los Angeles Times*, June 25, 1976, NASA History Office Files.

54. David Sallsbury, "Finding the Best Site for Delayed Mars Landing," *Christian Science Monitor*, June 28, 1976, NASA History Office Files.

55. Sallsbury, "Space Cowboys."

56. Jonathan Spivak, "Viking Craft Is Set to Land on Mars Tuesday in Attempt to Detect Any Life That May Exist," *Wall Street Journal*, July 16, 1976, 1.

57. John Naugle, interview by author, June 21, 2009; John Naugle, correspondence, June 19, 2009.

58. Diane Aimsworth, "The Spirit of Viking," *JPL Universe*, July 12, 1996, 3.

59. Thomas O'Toole, "Viking Lands Safely, Transmits Detailed Pictures of Mars," *Washington Post*, July 21, 1976, 1.

60. John Naugle, interview by author, June 21, 2009; John Naugle, correspondence, June 19, 2009.

61. "The News from Mars," *New York Times*, July 21, 1976, 30.

62. John Noble Wilford, "Viking on Mars," *New York Times*, July 21, 1976, 1.

63. Cristine Russell, "If Life Does Exist on Mars, It May Be Merely Microbial," *Washington Star*, June 28, 1976, NASA History Office Files.

64. "Viking Will Tell Whether Life on Mars," *Washington Star*, July 18, 1976, NASA History Office Files.

65. Noel Hinners, interview by author, Jan. 28, 2009.

66. Joel N. Shurkin, "Viking Goes Hunting for Some Martians," *Philadelphia Inquirer*, June 27, 1976, NASA History Office Files.

67. "Space Cut Is Feared If Viking Doesn't Discover Life on Mars," *New York Times*, July 4, 1976, 15.

68. Andrew Chaikin, *A Passion for Mars: Intrepid Explorers of the Red Planet* (New York: Abrams, 2008), 167.

69. Ibid., 165.

70. "Viking 1 Finds Life or Strange Chemistry," *Defense/Space Business Daily*, Aug. 3, 1976, 169–170.

71. Chaikin, *Passion for Mars*, 188.

72. Donald Fink, "Mars Scrutiny Enters New Phase," *Aviation Week and Space Technology*, Aug. 6, 1976, 14.

73. Ezell and Ezell, *On Mars*, 408.

74. Chaikin, *Passion for Mars*, 166.

75. James Fletcher to President Gerald Ford, Sept. 2, 1976, NASA History Office Files.

76. Walter Sullivan, "Second Mars Landing Set for Tonight," *New York Times*, Sept. 3, 1976, NASA History Office Files.

77. "Mars-Life Proof Lies with Second Ship," *Baltimore Sun*, Sept. 7, 1976, NASA History Office Files.

78. "Viking 2 Digs Trench, Puts Soil into Lab," *Washington Post*, Sept. 26, 1976, NASA History Office Files.

79. "Lack of Life on Mars Puts Viking 3 in Doubt," *Glendale News-Press*, Oct. 4, 1976, NASA History Office Files.

80. John Noble Wilford, "Viking 2 Tests Finds No Organic Matter," *New York Times*, Oct. 1, 1976, 13.

81. David Des Marais, interview by author, Oct. 8, 2009.

82. Gentry Lee, interview by author, July 22, 2011.

83. Chaikin, *Passion for Mars*, 168.

84. Ibid.; "Viking News Conference," transcript (Washington, DC: NASA, Nov. 9, 1976), NASA History Office Files.

85. Chaikin, *Passion for Mars*, 168.

CHAPTER FIVE: Struggling to Restart

1. Amy Paige Snyder, "NASA and Planetary Exploration," in John Logsdon, ed., *Exploring the Unknown*, vol. 5 (Washington, DC: NASA, 2001), 288.

2. President Ford, Jim Martin, and John Naugle, transcript of phone conversation, Sept. 4, 1976, NASA History Office Files.

3. "Martin Predicts Mobile Viking for 1981 Launch Program," *Newport News Daily Press*, Sept. 5, 1976, NASA History Office Files.

4. Carl Sagan, *Cosmos* (New York: Ballantine, 1980), 107–108.

5. David Sallsbury, "Stage May Be Set for Viking 3 Planning," *Christian Science Monitor*, Sept. 7, 1976, NASA History Office Files.

6. "NASA Looks at Mars 'Rover' Mission as Viking Follow-On," *Defense/Space Daily*, Sept. 16, 1976, 88.

7. John Noble Wilford, "Large Amounts of Water Discovered by Viking 2," *Washington Star*, Sept. 23, 1976, A-3.

8. Donald Fink, "Viking Successes Spur Rover Mission," *Aviation Week and Space Technology*, Sept. 27, 1976, 40–42.

9. William Pickering to Elmer Groo, Associate Administrator for Center Operations, Oct. 21, 1975, JPL Files, Box 214, Folder 140.

10. Geoffrey Briggs, interview by author, July 15, 2009.

11. "Lack of Life on Mars Puts Viking 3 in Doubt," Oct. 4, 1976, NASA History Office Files.

12. John Naugle, interview by author, June 21, 2009.

13. "Lack of Life."

14. Oran Nicks, *Far Travelers: The Exploring Machines* (Washington, DC: NASA SP-480, 1985), 14.

15. "National Aeronautics and Space Administration's Fiscal 1978 Budget Currently Contains No Specific Viking Follow-On," *Aviation Week and Space Technology*, Nov. 19, 1976, NASA History Office Files.

16. "Viking (1): End of First Phase of 70's Space Spectacular," *Science*, Nov. 19, 1976, NASA History Office Files.

17. "NASA Weighs 1981 vs. 1984 for Viking Follow-on," *Aerospace Daily*, Dec. 7, 1976, 176–178.

18. Ibid.

19. Noel Hinners, interview by author, Jan. 28, 2009.

20. "Ford to Ask Acceleration of Earthquake Research in the Forthcoming Budget," *New York Times*, Dec. 17, 1976, A-21.

21. Bill Delany, "Manager of Viking Project Pays Final Visit to Langley," *Hampton Daily Press*, Dec. 18, 1976, NASA History Office Files.

22. Jimmy Carter, "To the Members of the Viking Team," Apr. 1, 1977, NASA History Office Files.

23. Robert Frosch, "Introduction Letter," *Journal of Geophysical Research*, June 23, 1977, NASA History Office Files.

24. John Noble Wilford, "Mars Probe Showed No Sure Sign of Life," *New York Times*, Sept. 18, 1977, NASA History Office Files.

25. Andrew Chaikin, *A Passion for Mars: Intrepid Explorers of the Red Planet* (New York: Abrams, 2008), 168.

26. Bruce Murray, *Journey into Space: The First Thirty Years of Space Exploration* (New York: Norton, 1989), 244. He used the term "gray mice" for scientific missions of interest mainly to scientists.

27. John Logsdon, "Ten Presidents and NASA," in Rhonda Carpenter and Ana Lopez, eds., *NASA: 50 Years of Exploration and Discovery* (Washington, DC: NASA, 2008), 235.

28. Noel Hinners, interview by author, Jan. 28, 2009.

29. Gentry Lee, interview by author, July 22, 2011.

30. Craig Covault, "Jupiter Mission Approval Saves Planetary Capability, 300 Jobs," *Aviation Week and Space Technology*, July 25, 1977, 21–22.

31. Logsdon, "Ten Presidents," 235.

32. William Poundstone, *Carl Sagan: A Life in the Cosmos* (New York: Henry Holt, 1999), 253.

33. Bruce Murray to Noel Hinners, Feb. 20, 1979, regarding decisions on Mars Program, Feb. 9, JPL Files, Box 223, Folder 31.

34. Jeffrey Lenorowitz, "Funding Reallocation Forces Cutback in Unmanned Mars Programs," *Aviation Week and Space Technology*, Mar. 19, 1979, 44–48.

35. Steve Squyres, interview by author, June 1, 2010.

36. Terrance McGarry, "NASA Planning Dramatic Return to Planet Mars," *Houston Chronicle*, Aug. 28, 1980, NASA History Office Files.

37. A. J. S. Rayl, "The Third Time's a Charm: The Saga of Mars Climate Sounder," *Planetary Report*, May/June 2006, 9.

38. Murray, *Journey into Space*, 257.

39. Carol Stoker, interview by author, July 15, 2009.

40. Chris McKay, interview by author, Oct. 13, 2009.

41. Chaikin, *Passion for Mars*, 133–134; Robert Zubrin, *The Case for Mars* (New York: Touchstone, 1996), 70–74.

42. Chris McKay, interview by author, Oct. 13, 2009.

43. Chaikin, *Passion for Mars*, 137.

44. Ibid., 136.

45. Craig Covault, "Biology Stress for Mars 1984 Debated," *Aviation Week and Space Technology*, July 25, 1977, 60.

46. Louis D. Friedman, "30 Years of the Planetary Society," *Planetary Report*, March/April 2010, 4.

47. Murray, *Journey into Space*, 348.

48. Ibid., 350.

49. Harold Klein, interview by Steven Dick, Oral History, May 14, 1997, NASA History Office Files; see also Steven Dick and James Strick, *The Living Universe: NASA and the Development of Astrobiology* (New Brunswick, NJ: Rutgers University Press, 2004).

50. David Des Marais, interview by author, Oct. 8, 2009.

51. Ibid.

52. Michael Meltzer, *When Biospheres Collide: A History of NASA's Planetary Protection Program* (Washington, DC: NASA, 2011), 458.

53. David Des Marais, interview by author, Oct. 8, 2009.

54. William Burrows, *Exploring Space: Voyages in the Solar System and Beyond* (New York: Random House, 1990), 377.

55. Chaikin, *Passion for Mars*, 106.

56. Murray, *Journey into Space*, 262.

57. Snyder, "NASA and Planetary Exploration," 289.

58. Howard McCurdy, *The Space Station Decision: Incremental Politics and Technical Choice* (Baltimore: Johns Hopkins University Press, 1990).

59. Murray, *Journey into Space*, 216; Joseph Boyce, interview by author, Sept. 17, 2008.

60. "Galileo Cut," *Aviation Week and Space Technology*, Nov. 16, 1981, 16; Bruce Murray to Jay Keyworth, Dec. 8, 1981, and Dec. 15, 1981, JPL Files, Box 198, Folder 480.

61. Murray, *Journey into Space*, 279.

62. John Logsdon, "The Survival Crisis of the U.S. Solar System Exploration Program" (unpublished, June 1989), cited in Snyder, "NASA and Planetary Exploration," 290.

63. John Naugle, interview by author, June 21, 2009.

64. Murray, *Journey into Space*, 284.

65. Geoffrey Briggs, interview by author, July 15, 2009.

66. M. Mitchell Waldrop, "Planetary Science: Up from the Ashes," *Science*, Nov. 12, 1982, 665–666; David Morrison and Noel Hinners, "A Program for Planetary Exploration," *Science*, May 6, 1983, 561–567, esp. 566n7; Solar System Exploration Committee, *Planetary Exploration through Year 2000*, Executive Summary (Washington, DC: GPO, 1983).

67. Waldrop, "Planetary Science"; Morrison and Hinners, "Program for Planetary Exploration"; Solar System Exploration Committee, *Planetary Exploration through Year 2000*; Snyder, "NASA and Planetary Exploration," 292.

68. Rayl, "Third Time's a Charm," 91.

69. Waldrop, "Planetary Science," 665.

70. Chaikin, *Passion for Mars*, 123.

71. George Keyworth, interview by John Logsdon, Oral History, Mar. 5, 1986, NASA History Office Files.

72. Richard Norwind, "Is a Trip to Mars in the Budget?," *Los Angeles Herald Examiner*, Feb. 2, 1984, NASA History Office Files.

73. David Des Marais, interview by author, Oct. 8, 2009.

74. Richard Norwind, "JPL Scientists Bid a Fond Farewell to Viking Lander 1," *Los Angeles Herald Examiner*, May 6, 1983, NASA History Office Files.

CHAPTER SIX: Moving Up the Agenda

1. Geoffrey Briggs, interview by author, July 15, 2009.

2. Bruce Murray, *Journey into Space: The First Three Decades of Space Exploration* (New York: Norton, 1989), 320–321.

3. Ibid., 321.

4. Ibid., 315.

5. Ibid., 332.

6. Ibid., 332.

7. Ibid., 332–333.

8. James Fletcher to Rep. Edward Boland, Jan. 2, 1987, NASA History Office Files.

9. Thomas Frieling, "The Devastating Delay," *Space World*, July 1987, 27–30.

10. Ibid., 30.

11. Murray, *Journey into Space*, 333.

12. M. Mitchell Waldrop, "Boland, NASA at Odds over Launch of Mars Observer," *Science*, Feb. 13, 1987, 743.

13. "NASA Postpones a Mars Mission for Two Years," *New York Times*, Mar. 15, 1987, 25.

14. Dale Myers to Cong. Bill Nelson, Apr. 20, 1987, NASA History Office Files.

15. Dale Myers to Bruce Murray, Apr. 21, 1987, NASA History Office Files.

16. Murray, *Journey into Space*, 334.

17. Shirley Green to Lew Allen, Apr. 20, 1987, NASA History Office Files.

18. William Burrows, *Exploring Space: Voyages in the Solar System and Beyond* (New York: Random House, 1990), 380, 386.

19. Murray, *Journey into Space*, 335.

20. Geoffrey Briggs, interview by author, July 15, 2009.

21. Roger Launius, "A Western Mormon in Washington, DC: James C. Fletcher, NASA, and the Final Frontier," *Pacific Historical Review* 64, no. 2 (May 1995): 217–241; see also W. Henry Lambright, *NASA and the Environment: The Case of Ozone Depletion* (Washington, DC: NASA, 2005).

22. Len Fisk, interview by author, Oct. 19, 2009.

23. "NASA Aides Set Sights on Flight to Explore Mars," *Los Angeles Times*, Apr. 1, 1987, 3; "Houston Conference Sets Tone for Mars Adventure," *JPL Universe*, Apr. 3, 1987, 1; "NASA Briefs Contractors on Mars Rover/Sample Return Mission," *Aviation Week and Space Technology*, Apr. 6, 1987, 22–23.

24. Sally Ride, *Leadership and America's Future in Space — a Report to the Administrator* (Washington, DC: NASA, 1987), 32.

25. Ibid., 5–6.

26. Thor Hogan, *Mars Wars: The Rise and Fall of the Space Exploration Initiative* (Washington, DC: NASA, 2007), 40–41.

27. Murray, *Journey into Space*, 336.

28. Ibid.

29. Louis Friedman, "World Watch," *Planetary Report*, March/April 1988.

30. Ibid.

31. Eliot Marshall, "Mars Mania and NASA," *Science*, Apr. 1, 1988, 18.

32. David Portree, *Humans to Mars: Fifty Years of Mission Planning* (Washington, DC: NASA, 2001), 77.

33. Amy Paige Snyder, "NASA and Planetary Exploration," in John Logsdon, ed., *Exploring the Unknown*, vol. 5 (Washington, DC: NASA, 2001), 292.

34. Andrew Chaikin, *A Passion for Mars: Intrepid Explorers of the Red Planet* (New York: Abrams, 2008), 189–191.

35. Noel Hinners, interview by author, Jan. 28, 2009.

36. Robert Pepin, Chairman, COMPLEX, to Geoffrey Briggs, July 12, 1988 (Washington, DC: NAS-NRC, 1988).

37. Charles Polk, *Mars Observer Project History* (Pasadena, CA: JPL, 1990), 80.

38. Marshall, "Mars Mania"; Murray, *Journey into Space*, 337–338; Geoffrey Briggs to William Purdy Jr., July 25, 1988, NASA History Office Files.

39. Kathy Sawyer, "Soviets Launch Probe of Mars' Moon," *Washington Post*, July 8, 1988, A1.

40. Burrows, *Exploring Space*, 372.

41. Ibid., 373.

42. Ibid., 374–375.

43. Ibid., 370.

44. John Logsdon, "Ten Presidents and NASA," in Rhonda Carpenter and Ana Lopez, eds., *NASA: 50 Years of Exploration and Discovery* (Tampa, FL: Faircount Media Group, 2008), 236.

45. Len Fisk, interview by author, Oct. 19, 2009.

46. Hogan, *Mars Wars*, 179–81.

47. Ibid., 40–41, 86–87.

48. Colin Norman, "Bush Budget Highlights R&D," *Science*, Feb. 2, 1990, 517–519.

49. Hogan, *Mars Wars*, 109.

50. Ibid., 110.

51. Ibid., 111.

52. Scott Hubbard, interview by author, July 14, 2009.

53. Ibid.

54. Joseph Boyce, interview by author, Sept. 17, 2008.

55. Douglas Isbell, "Penetrators, Price Cut in NASA Mars Plan," *Space News*, May 28–June 3, 1990, 3.

56. Donna Shirley, with Danelle Morton, *Managing Martians* (New York: Broadway Books, 1998), 123, 125, 129.

57. Colin Norman, "Science Budget: Growth amid Red Ink," *Science*, Feb. 8, 1991, 251.

58. DPIVIROTTO to GSHUBBARD, et al., memoranda, "Notes and Strategy Session," "Comments on Mike's Paper," and "An Adaptive Strategy for Mars Exploration," Mar. 6, 1991, NASA History Office Files.

59. Len Fisk, correspondence, Feb. 24, 2010.

60. David Hamilton, "Future Budget Squeezes Bode Ill for NASA Space Science," *Science*, Sept. 6, 1991, 1083.

61. Lennard Fisk to Ed Stone, Director JPL, regarding Mars Environmental Survey, Jan. 28, 1991, JPL Files, Box 259, Folder 185.

62. Scott Hubbard, interview by author, July 14, 2009.

63. Lennard Fisk to Ed Stone, Director JPL, regarding MESUR Phase A Studies, Nov. 8, 1991, JPL Files, Box 259, Folder 185.

64. Len Fisk, interview by author, Oct. 9, 2009.

65. Wes Huntress, interview by author, Sept. 18, 2008.

66. Ibid.

CHAPTER SEVEN: Prioritizing Mars

1. Colin Norman, "Science Budget: Selective Growth," *Science*, Feb. 7, 1992, 674.

2. W. Henry Lambright, *Transforming Government: Dan Goldin and the Remaking of NASA* (Washington, DC: IBM, 2001).

3. Stephanie Roy, "The Origin of the Smaller, Faster, Cheaper Approach in NASA's Solar System Exploration Program," *Space Policy* 14 (1998): 153–171.

4. Eliot Marshall, "Making Less Do More at NASA," *Science*, Oct. 2, 1992, 20.

5. Ibid., 21.

6. Roy, "Origin of the Smaller, Faster, Cheaper Approach," 166.

7. Wesley Huntress, interview by Rebecca Wright, Jan. 9, 2003, NASA History Office Files.

8. Steven Squyres to Carl Pilcher, Apr. 8, 1992, NASA History Office Files; Michael Carr to Wesley Huntress, Apr. 20, 1992, NASA History Office Files; Jonathan Lunine to Wesley Huntress, May 22, 1992, NASA History Office Files.

9. Carl Pilcher to Steven Squyres, Apr. 8, 1992, NASA History Office Files.

10. Donna Shirley, with Danelle Morton, *Managing Martians* (New York: Broadway Books, 1998).

11. Wesley Huntress, interview by author, Sept. 18, 2008.

12. Norm Haynes, interview by author, July 13, 2009.

13. William Poundstone, *Carl Sagan: A Life in the Cosmos* (New York: Henry Holt, 1999), 352.

14. Michael Carr, interview by author, Aug. 17, 2011.

15. John Noble Wilford, "After 17 Years, NASA Prepares for a Return Trip to Mars," *New York Times*, Sept. 22, 1992, C1; Robert Cowen, "US Mars Observer Heads for the Red Planet," *Christian Science Monitor*, Oct. 28, 1992, 14.

16. Wesley Huntress, interview by author, Sept. 18, 2008.

17. Wesley Huntress, interview by Rebecca Wright, Jan. 9, 2003, NASA History Office Files.

18. Don Goldin, interview by author, May 23, 2011.

19. Wesley Huntress, "The New Era in Mars Exploration," *Space Times*, Nov.–Dec. 1992, 8–10.

20. W. Henry Lambright, "Leadership and Large-Scale Technology: The Case of the International Space Station," *Space Policy* 21 (2005): 195–203; W. Henry Lambright, "Leading Change at NASA: The Case of Dan Goldin," *Space Policy* 23 (2007): 33–43.

21. Louis Friedman, interview by author, May 23, 2010.

22. John Travis, "Mars Observer's Costly Solitude," *Science*, Sept. 3, 1993, 1267.

23. Kathy Sawyer, "Silence at Red Planet Reverberates on Earth," *Washington Post*, Nov. 13, 1993, 1; Andrew Chaikin, *A Passion for Mars: Intrepid Explorers of the Red Planet* (New York: Abrams, 2008), 191–192.

24. Travis, "Mars Observer's Costly Solitude," 1284.

25. James Garvin, interview by author, Sept. 12, 2008.

26. Kathy Sawyer, "Panel Suspects Rupture in Propulsion System Doomed Mars Observer," *Washington Post*, Jan. 6, 1994, A3.

27. Len Fisk, interview by author, Oct. 9, 2009.

28. David Coia, "Panel to Probe Loss of Mars Observer," *Washington Times*, Aug. 27, 1993, A3; "NASA Names Team to Study Return Trip to Mars," *NASA News*, Sept. 1, 1993, NASA History Office Files.

29. Wesley Huntress, interview by author, Sept. 18, 2008.

30. Robert Nelson, "Big Science vs. Small Science in Our Space Quest," *Houston Chronicle*, Sept. 1, 1993, 25.

31. Sharon Begley, "The Stars of Mars," *Newsweek*, July 21, 1997, 26.

32. Wesley Huntress, interview by author, Sept. 18, 2008.

33. Amy Paige Snyder, "NASA and Planetary Exploration," in John Logsdon, ed., *Exploring the Unknown*, vol. 5 (Washington, DC: NASA, 2001), 297.

34. Wesley Huntress, interview by author, Sept. 18, 2008.

35. Len Fisk, interview by author, Oct. 19, 2009; Roy, "Origin of the Smaller, Faster, Cheaper Approach," 160; Snyder, "NASA and Planetary Exploration," 292.

36. Kathy Sawyer, "Money Worries Add to Agency Problems," *Washington Post*, Feb. 10, 1994, A25.

37. Acting Dep. Admin. to Administrator, "Senior Management Meeting," memorandum, Mar. 24, 1994, NASA History Office Files.

38. "NASA Streamlining Took 'Seven or Eight Months' off Mars Surveyor Award," *Aerospace Daily*, July 12, 1994, 53.

39. Wesley Huntress, interview by author, Sept. 18, 2008.

40. Liz Tucci, "Administrator Blames Bureaucracy for NASA Waste," *Space News*, Apr. 18–24, 1994, 10.

41. "Landing Site Is Chosen for Next US Mission to Mars," *New York Times*, Sept. 11, 1994; Ben Iannotta, "JPL's Mars Pathfinder Program Taking Shape," *Space News*, Dec. 19–25, 1994, 9.

42. Wesley Huntress, interview by author, Sept. 18, 2008.

43. Warren Ferster, "NASA Floats '98 Mars Mission," *Space News*, Oct. 31– Nov. 6, 1994, 4.

44. Marcia Dunn, "Controversial NASA Chief Says, 'I Y'am What I Y'am,'" *Wheeling West Virginia News Register*, July 17, 1994, NASA History Office Files.

45. Linda Billings, "Mars on Hold," *NASA Magazine*, Winter 1994, 26–29.

46. Iannotta, "JPL's Mars Pathfinder Program," 9.

47. Jeffrey Mervis, "Clinton Holds Line on R&D," *Science*, Feb. 10, 1995, 780–782.

48. James Asker, "Goldin Halts NASA Hiring, Vows Wind Tunnel Fight," *Aviation Week and Space Technology*, Feb. 20, 1995, 29.

49. David Morrison, "Low-Rent Space," *National Journal*, Apr. 29, 1995, 1028.

50. "NASA Sets Up New Headquarters Office to Cut Operations Costs," *Aerospace Daily*, June 30, 1995, 511.

51. Theresa Foley, "Mr. Goldin Goes to Washington," *Air and Space*, April/May 1995, 36–34.

52. Kathy Sawyer, "NASA, Seeking Savings, Puts Robots on a Weight-Loss Plan," *Washington Post*, May 15, 1995, A3.

53. "An Exobiological Strategy for Mars Exploration," unpublished report for NASA, Jan. 1995.

54. Sawyer, "NASA, Seeking Savings," A3.

55. Ibid.

56. Morrison, "Low-Rent Space," 1029.

57. "Space Shots," Dan Goldin speaking at a Steps to Mars 2 Conference, July 15, 1995, *Space News*, July 24–30, 1995, 24.

58. Ben Iannotta, "Russian Instrument May Ride on Mars Lander," *Space News*, Nov. 13–26, 1995, 6.

CHAPTER EIGHT: Accelerating Mars Sample Return

1. Andrew Lawler, "Research Knows No Season as Budget Cycle Goes Awry," *Science*, Feb. 2, 1996, 589.

2. Andrew Lawler and Jeffrey Mervis, "Battle Lines Drawn for 1997 R&D Budget," *Science*, Mar. 22, 1996, 1658; "Crunch Ahead for Space Science," *Science*, Mar. 22, 1996, 1660; Andrew Lawler, "A Slippery Slope for Science," *Science*, Mar. 29, 1996, 1796–1797.

3. William Broad, "Probes Will Carry New Hope for Detecting Life on Mars," *New York Times*, Feb. 20, 1996, C1.

4. Alan Boss, *Looking for Earths: The Race to Find New Solar Systems* (New York: Wiley, 1998).

5. Mike Meyer, interview by author, Mar. 4, 2009.

6. Sen. Barbara Mikulski to Daniel Goldin, June 12, 1996, NASA History Office Files.

7. Andrew Chaikin, *A Passion for Mars: Intrepid Explorers of the Red Planet* (New York: Abrams, 2008), 151.

8. Robert Zubrin, with Richard Wagner, *The Case for Mars* (New York: Touchstone, 1997).

9. Ben Iannotta, "Robotic Missions Set Up Human Sojourns to Mars," *Space News*, July 8–14, 1996, 8.

10. Wesley Huntress, interview by author, Sept. 18, 2008.

11. Ibid.

12. Ibid.

13. Kathy Sawyer, *The Rock from Mars* (New York: Random House, 2006), 125.

14. Ibid., 127.

15. Ibid., 128.

16. Ibid., 129.

17. Ibid., 114, 132.

18. Jack Gibbons, interview by author, June 17, 2010.

19. Sawyer, *Rock from Mars*, 133.

20. David S. McKay et al., "Search for Past Life on Mars: Possible Relic Biogenic Activity in Martian Meteorite ALH84001," *Science*, Aug. 16, 1996, 924–930.

21. Sawyer, *Rock from Mars*, 157.

22. Chaikin, *Passion for Mars*, 158.

23. "Life on Mars," *Washington Post*, Aug. 10, 1996, A18.

24. "Remarks by the President upon Departing," *White House*, Aug. 7, 1996, NASA History Office Files.

25. Kathy Sawyer, "NASA Releases Images of Mars Life Evidence," *Washington Post*, Aug. 8, 1996, A3.

26. John Noble Wilford, "The Attraction of Mars," Proceedings of Nov. 22, 1998, Symposium, *Life in the Universe: What Can the Martian Fossil Tell Us?* (Washington, DC: George Washington University, Planetary Society, National Space Society, 1996), NASA History Office Files.

27. "Life on Mars," *New York Times*, Aug. 8, 1996, A26; "Life on Mars," *Washington Post*, Aug. 10, 1996, A18.

28. Malcolm Browne, "Planetary Experts Say Mars Life Is Still Speculative," *New York Times*, Aug. 8, 1996, D20; David Colton, "Discovery Would Equal Finding the New World," *USA Today*, Aug. 8, 1996, 1.

29. Andrea Petersen, "Star-Struck Meteorite Collectors Are Pushing Rock Prices Sky High," *Wall Street Journal*, Aug. 8, 1996, B2.

30. Wesley Huntress to Dan Goldin, memorandum, Aug. 14, 1996, D. Goldin Papers, National Archives, Record 71517, 71518, Box 63.

31. Dan McCleese, interview by author, Mar. 30, 2011.

32. Ibid.

33. Ibid.; Kathy Sawyer, "Evidence of Ancient Life Alters Mars Missions," *Washington Post*, Aug. 19, 1996, A3; Leonard David, "NASA Group Eyes 2011 for Mars Mission," *Space News*, Aug. 28–Sept. 1, 1998, 3.

34. Sawyer, "Evidence of Ancient Life," A3.

35. Dan McCleese, interview by author, Mar. 3, 2011.

36. The White House, "National Space Policy," Sept. 19, 1996, www.fas.org/spp/military/docops/national/nstc-8.htm.

37. W. Henry Lambright, "Leading Change at NASA: The Case of Dan Goldin," *Space Policy* 23, no. 1 (2007): 37.

38. Report, "The Search for Origins: Findings of a Space Science Workshop," Oct. 28–30, 1996, NASA History Office Files.

39. Andrew Lawler, "Building between the Big Bang and Biology," *Science*, Nov. 8, 1996, 912.

40. John Noble Wilford, "Craft Is Launched to Explore Mars," *New York Times*, Nov. 8, 1996, A1.

41. "Mars Probe Heads for Mapping," *Washington Times*, Nov. 8, 1996, A3.

42. Dan Goldin, "We Will Go to Mars," Proceedings of Nov. 22, 1996, Symposium, *Life in the Universe, What Can the Martian Fossil Tell Us?* (Washington, DC: George Washington University, Planetary Society, National Space Society, 1996), NASA History Office Files.

43. Andrew Lawler, "Partisan Battle Mars Markup of Joint Authorization Bill," *Science*, May 3, 1996, 640.

44. Dan Goldin to T. J. Glauthier, OMB, Nov. 19, 1996, NASA History Office Files; Dan Goldin to Franklin Raines, OMB, Nov. 19, 1996, NASA History Office Files.

45. Lawler, "Building between the Big Bang and Biology," 912.
46. Paul Hoversten, "First Step to Mars: Getting There," *USA Today*, Dec. 2, 1996, 3A.
47. Ibid.
48. William Broad, "Even in Death, Carl Sagan's Influence Is Still Cosmic," *New York Times*, Dec. 1, 1998, www.nytimes.com/1998/12/01/science/even-in-death-carl-sagan-s-influence-is-still-cosmic.html?scp=1&sq=even%20in%20death%20carl%20sagan&st=cse.
49. Sawyer, *Rock from Mars*, 189.
50. Steve Dick, notes, "Vice President Gore and Mars," Dec. 12, 1996; Larry Klaes, notes, "Science Panel Briefing Gore about ET Life," Dec. 12, 1996; Vice President's Statement at the Space Science Symposium, Dec. 11, 1996, NASA History Office Files; Andrew Lawler, "Origins Researchers Win Gore's Ears, Not Pocketbook," *Science*, Dec. 20, 1996, 2003.
51. Lawler, "Origins Researchers," 2003.
52. President Bill Clinton, State of the Union Address, Feb. 4, 1997, http://clinton2.nasa.gov/WH/SOU97/, NASA History Office Files.
53. Andrew Lawler, "Smoother Road for R&D Spending," *Science*, Feb. 14, 1997, 916–919.
54. Wesley Huntress, interview by author, Sept. 18, 2008.
55. Lawler, "Smoother Road," 919.
56. Ibid., 917.
57. Ibid.
58. David S. F. Portree, *Humans to Mars: Fifty Years of Mission Planning, 1950–2000* (Washington, DC: NASA, 2001), 95.
59. Dan Goldin to Mina, Dec. 12, 1996, NASA History Office Files.
60. Wesley Huntress, interview by author, Sept. 18, 2008.
61. Jeff Faust, "Comments by Dan Goldin," in "Planetary Scientists Mark 'Amazing Year' at Conf." (Sept. 1997), www.coseti.org/dangold/l.htm, originally in Sept. 1997 issue of *Space News*, NASA History Office Files.
62. Todd Halvorson, "Pathfinder One Month Later: A Celestial Success Story," *Florida Today Space Online*, Aug. 2, 1997, www.flatoday.com/space/explore/stories/1997b/080297d.htm, NASA History Office Files.
63. Wesley Huntress, interview by Rebecca Wright, Jan. 9, 2003, NASA History Office Files.
64. Halvorson, "Pathfinder One Month Later."
65. "Mars Mission '100 Percent Success,' NASA says," *Florida Today*, Aug. 9, 1997, 5.
66. Sharon Begley, "Greetings from Mars," *Newsweek*, July 14, 1997, 25.
67. Halvorson, "Pathfinder One Month Later."
68. Ibid.
69. Ibid.
70. The problem of calculating dollars across decades is revealed by the fact that Donna Shirley had used a $4 billion figure. See her comment in chapter 7.
71. "Pathfinder Named for Late Carl Sagan," *CNN Sci-Tech*, July 6, 1997, www.cnn.com/tech.9701/06/pathfinder.sagan.ap/index.h, NASA History Office Files.
72. Bruce Smith, "Mars Mapping Mission Delayed, Shortened," *Aviation Week and Space Technology*, Nov. 17, 1997, 36.

73. Andrew Lawler, "Science Catches Clinton's Eye," *Science*, Feb. 6, 1998, 796.

74. Wesley Huntress, interview by author, Sept. 18, 2008.

75. Joseph Boyce, interview by author, Sept. 17, 2008.

76. Ed Weiler, interview by author, Mar. 11, 2003.

77. Kathy Sawyer, "New Research Shakes Theory of Life on Mars," *Washington Post*, Jan. 16, 1998, A3.

78. Paul Recer, "Life on Mars," *Wire Service*: AP-N (AP US & World), Aug. 5, 1998, NASA History Office Files.

79. Steven Dick and James Strick, *The Living Universe: NASA and the Development of Astrobiology* (New Brunswick, NJ: Rutgers University Press, 2004), 210–211.

80. Ibid., 214.

81. G. Scott Hubbard, "Astrobiology: Its Origins and Development," in Rhonda Carpenter and Ana Lopez, eds., *NASA: Fifty Years of Exploration and Discovery* (Tampa, FL: Faircount Media Group, 2008), 156–163.

82. Daniel Goldin, "Go for Launch," Commencement Address, University of Arizona, May 16, 1998, NASA History Office Files; Daniel Goldin, Space address at "Alan Shepard Memorial Service," Houston, TX, Aug. 1, 1998, ftp://ftp.hq.nasa.gov/pub/pao/Goldin/dg_shepard.html, NASA History Office Files.

83. Carol Stoker, interview by author, July 15, 2009.

84. Sandra Blakeslee, "Society Organizes to Make a Case for Mars," *New York Times*, Aug. 18, 1998, F3; Robert Zubrin, "Mars: Time for Political Action," *Space News*, Nov. 25, 1998, 15.

85. Diane Ainsworth, "New Mars Plan Targets Sample Return," *Jet Propulsion Laboratory Universe*, Nov. 13, 1998, 1, 3.

86. Norm Haynes, interview by author, July 13, 2009.

87. Claude Canizares and Ronald Greeley to Carl Pilcher, Nov. 11, 1998, with attachments.

88. Kathy Sawyer, "With Mars Images, NASA Says: Face It, It's a Mesa," *Washington Post*, Apr. 7, 1998, A1.

89. Zubrin, "Mars: Time for Political Action."

90. President Bill Clinton, interview by Walter Cronkite, CNN, Oct. 29, 1998, www.whitehouse.gov.wh/new/html/1998/029–20023.html, NASA History Office Files.

CHAPTER NINE: Overreaching, Rethinking

1. William O'Neil, "Mars Sample Return Overview," July 26, 1999, NASA History Office Files.

2. David Malakoff, "2000 Budget Plays Favorites," *Science*, Feb. 5, 1999, 778–780.

3. Dan Goldin, interview by author, May 23, 2010.

4. Steve Squyres, interview by author, June 1, 2010.

5. James Garvin, interview by author, Sept. 12, 2008.

6. Wesley Huntress, "Grand Challenges for Space Exploration," Second Annual Carl Sagan Memorial Lecture, *Space Times*, May–June 1999, 4–13.

7. Charles Elachi and Louis Friedman, "Building toward Mars: A Vision for the Future," *Planetary Report*, Mar./Apr. 1999, 14–18.

8. Kathy Sawyer, *The Rock from Mars* (New York: Random House, 2006), 199.

9. Ibid.

10. Ibid. The $750 million figure comes from an interview with Scott Hubbard, July 14, 2009. The $500 million figure was cited in an interview with Ed Weiler, Apr. 16, 2010. Both men pointed out that those numbers were exceptionally wrong.

11. Andrew Chaikin, "Why Did Mars Polar Lander Fail? A Conversation with Donna Shirley," Space.com, Jan. 24, 2000, www.space.com/people/shirley-Q&A_000121.html, NASA History Office Files.

12. "Mars Exploration Becomes Global Enterprise," *Florida Today Space Online*, June 16, 1999, www.flatoday.com:80/space/today/061699c.htm, NASA History Office Files.

13. "Get a Life," *Aviation Week and Space Technology*, July 19, 1999, 23.

14. Peter Kendall, "NASA Chief Dazes Some Astronomers: Idea Floated for Tiny Spaceships," *Chicago Tribune*, June 4, 1999, 3.

15. K. C. Cole, "Science File: Questions and Answers with NASA Chief Dan Goldin," *Los Angeles Times*, July 1, 1999, B-2.

16. Kathy Sawyer, "Staff, Training Faulted in Loss of Mars Probe," *Washington Post*, Nov. 11, 1999, 1; Robert Lee Hotz, "String of Missteps Doomed Orbiter; JPL Found at Fault," *Los Angeles Times*, Nov. 11, 1999, NASA History Office Files.

17. Sawyer, "Staff, Training Faulted."

18. Eliot Marshall, "A System Fails at Mars, a Spacecraft Is Lost," *Science*, Nov. 19, 1999, 1457–1459.

19. Matthew Fordahl, "A Lot Is Riding on Success of Mars Lander Missions; Shoestring Budget Philosophy Will Be Put to Test," *New Orleans Times-Picayune*, Nov. 26, 1999, A23.

20. Hotz, "String of Missteps."

21. Kathy Sawyer, "Polar Lander Speeds toward Tricky Mars Mission," *Washington Post*, Nov. 29, 1999, A11.

22. Kathy Sawyer, "Polar Lander Fails to Signal Mars Arrival," *Washington Post*, Dec. 4, 1999, 1.

23. "NASA Ponders Lessons-Learned Exercise as Hopes for Mars Lander Fade," *Aerospace Daily*, Dec. 7, 1999, NASA History Office Files.

24. Noel Hinners, interview by author, Jan. 28, 2009.

25. Andrew Bridges, "Mars Polar Lander Investigation Panel Embarks on 2-Month Task," Space.com, Jan. 14, 2000, www.space.com/, NASA History Office Files.

26. Ed Weiler, interview by author, Apr. 16, 2010.

27. Chaikin, "Why Did Mars Polar Lander Fail?"

28. David Page, "Reflections on the Loss of the Mars Polar Lander," *Planetary Report*, Jan./Feb. 2000, 19.

29. Andrew Lawler, "Reports Will Urge Overhaul and Delays to NASA's Mars Missions," *Science*, Mar. 10, 2000, 1722–1723.

30. Ibid.

31. Scott Hubbard, *Exploring Mars: Chronicles from a Decade of Discovery* (Tucson: University of Arizona Press, 2012), 18.

32. Scott Hubbard, interview by author, July 7, 2005; Scott Hubbard, interview by author, July 14, 2009.

33. Ibid.

34. Andrew Lawler, "Clinton Seeks 'Major Lift' in U.S. Research Programs," *Science*, Jan. 28, 2000, 558–559.

35. President Bill Clinton to the Honorable Dan Goldin, Mar. 27, 2000, NASA History Office Files.

36. Keith Cowing, "NASA Reveals Probable Cause of Mars Polar Lander and Deep Space-2 Mission Failures," SpaceRef.com, Mar. 28, 2000, www.spaceref.com/news/viewnews.html?id-105, NASA History Office Files.

37. Warren Leary, "NASA Chief Faulted for Failure of Mars Lander," *International Herald Tribune*, Mar. 30, 2000, http:/nt.excite.com/ntd?uid=311698db3829bc11:page=show:topic=space-science:sb=date, NASA History Office Files.

38. Brian Berger, "Pressure, Tight Budgets Doomed Mars Missions," *Space News*, Apr. 10, 2000, 1.

39. Noel Hinners, interview by author, Jan. 28, 2009.

40. Steven Siceloff and Frank Oliveri, "Goldin Waffles on Blame for Mars Failures," *Florida Today Space Online*, June 23, 2001, www.flatoday.com/news/space/station/stories/2001a/Jun/spa0623oa.html, NASA History Office Files.

41. Dan Goldin, interview by author, May 23, 2010.

42. Matthew Fordahl, "NASA Chief Takes Blame," ABCNews.com, Mar. 30, 2000, http://nt.excite/ntd.dcg?uid=311169bdb3829bc11:page=show:topic=space+science:sb=date, NASA History Office Files.

43. "Midcourse Correction," *JPL Universe*, Mar. 31, 2000, NASA History Office Files.

44. "Commerce Committee Senators Request Mars Polar Lander Documents," SpaceRef.com, Mar. 28, 2000, www.spaceref.com/news/viewpr.html?pid=1282.

45. "Deep Space on the Cheap," *Washington Post*, Mar. 30, 2000, A20.

46. "NASA Faces Capitol Hill Heat but No Fire over Its Mars Program," *Aerospace Daily*, June 21, 2000, NASA History Office Files.

47. "Planetary Society Calls for Cool Heads in NASA Mars Debate," *Florida Today Space Online*, Mar. 28, 2000, www.flatoday.com/space/explore/stories/2000a/032800marsplanet.htm, NASA History Office Files.

48. Louis Friedman, interview by author, May 23, 2010.

49. Hubbard, *Exploring Mars*, 38.

50. Firouz Naderi, interview by author, Nov. 12, 2009.

51. Mal Peterson to Dan Goldin, memorandum, Mar. 20, 2000, Box 77956, Goldin Papers, National Archives.

52. James Garvin, interview by author, Sept. 12, 2008.

53. Scott Hubbard, interview by author, July 14, 2009.

54. Ed Weiler, interview by author, Apr. 16, 2010.

55. Ibid.

56. Firouz Naderi, interview by author, Nov. 12, 2009.

57. Scott Hubbard, interview by author, July 14, 2009.

58. Ibid.

59. Ibid.; Scott Hubbard, interview by author, July 7, 2005.

60. Hubbard, *Exploring Mars*, 54, 90–97.

61. Ed Weiler, interview by author, Apr. 16, 2010.

62. Scott Hubbard, interview by author, July 14, 2009.

63. Hubbard, *Exploring Mars*, 61.
64. Ibid., 66.
65. Scott Hubbard, interview by author, July 14, 2009.
66. Hubbard, *Exploring Mars*, 80.
67. "NASA Sets Press Conference Today as News of Mars Water Leaks Out," *Aerospace Daily*, June 22, 2000, NASA History Office Files.
68. Firouz Naderi, interview by author, Nov. 12, 2009.
69. Hubbard, *Exploring Mars*, 83.
70. Scott Hubbard, interview by author, July 14, 2009; Scott Hubbard, interview by author, July 7, 2005.
71. Firouz Naderi, interview by author, Nov. 12, 2009.
72. Scott Hubbard, interview by author, July 7, 2005.
73. John Noble Wilford, "NASA to Send Large Rover to Explore Surface of Mars," *New York Times*, July 28, 2000, http://partners.nytimes.com/library/national/science/072800sci-mars-rover.html.
74. Scott Hubbard, interview by author, July 7, 2005.
75. Hubbard, *Exploring Mars*, 86.
76. Ibid.
77. Steve Squyres, *Roving Mars: Spirit and Opportunity and the Exploration of the Red Planet* (New York: Hyperion, 2005), 91.
78. Ibid., 91–92; Scott Hubbard, interview by author, July 7, 2005; Charles Elachi, interview by author, Nov. 11, 2009.
79. "NASA Makes Sending Two Rovers to Mars in '03 an 'Agency Priority,'" *Aerospace Daily*, Aug. 11, 2000, NASA History Office Files.
80. Hubbard, *Exploring Mars*, 87–88.
81. Dan Goldin, interview by author, May 23, 2010; Dan McCleese, interview by author, Mar. 30, 2011.
82. Firouz Naderi, interview by author, Nov. 12, 2009.
83. Hubbard, *Exploring Mars*, 108.
84. Scott Hubbard, interview by author, July 7, 2005.
85. Hubbard, *Exploring Mars*, 35–36.
86. Ibid., 114.
87. "NASA Plans Mars Exploration for Next Two Decades," *Spaceflight Now*, Oct. 27, 2000, NASA History Office Files.
88. Ibid.; Paul Recer, "NASA Announces Plan for Robotic Exploration of Mars," *Associated Press Newswires*, Oct. 26, 2000, NASA History Office Files.
89. "NASA Retools Its Mars Campaign," *MSNBC Technology Space News*, Oct. 26, 2000, www.msnbc.com/news/481378.asp, NASA History Office Files.
90. Ibid.

CHAPTER TEN: Adopting "Follow the Water"

1. W. Henry Lambright, *Executive Response to Changing Fortune: Sean O'Keefe as NASA Administrator* (Washington, DC: IBM, 2005), 14; see also Frank Sietzen Jr. and Keith Cowing, *New Moon Rising* (Burlington, ON: Collector's Guide Publishing, 2004), 44.
2. James Garvin, interview by author, June 29, 2011.

3. Scott Hubbard, *Exploring Mars: Chronicles from a Decade of Discovery* (Tucson: University of Arizona Press, 2012), 140.

4. David Malakoff, "For All But NIH, the Devil Is in the Details," *Science*, Apr. 13, 2001, 182–183.

5. "NASA Mars Chief: Planned Budget Boost Helps Revamped Mars Exploration," *Aerospace Daily*, Apr. 11, 2001, NASA History Office Files.

6. Hubbard, *Exploring Mars*, 141–145.

7. "'Mars Czar' Hubbard Steps Down; Interim Replacement Named," Space.com, Apr. 19, 2001, www.space.com/news/spaceagencies/hubbard_resigns_010419.html, NASA History Office Files.

8. Fuk Li, interview by author, Mar. 3, 2011.

9. James Garvin, interview by author, June 29, 2011.

10. Andrew Lawler, "Goldin Quits Top Space Agency Post, but His Legacy Lingers," *Science*, Oct. 26, 2001, 758–759.

11. Andrew Lawler, "Pluto and Pork Win Out at NASA," *Science*, Nov. 16, 2001, 1430.

12. Lambright, *Executive Response*.

13. David Malakoff, "War Effort Shapes US Budget, With Some Program Casualties," *Science*, Feb. 8, 2002, 952–954.

14. Andrew Lawler, "Planetary Science's Defining Moment," *Science*, Jan. 4, 2002, 32–37.

15. Ibid.

16. Address by the Honorable Sean O'Keefe, NASA Administrator, Apr. 12, 2002, Maxwell School of Citizenship and Public Affairs, entitled "Pioneering the Future," www.maxwell.syr.edu/news/okeefe_041202.html, NASA History Office Files.

17. Richard Kerr, "Safety versus Science on Next Trips to Mars," *Science*, May 10, 2002, 1006–1008.

18. Ed Weiler, interview by author, Apr. 16, 2010.

19. Brian Berger, "NASA Faces Higher Cost, Deadline for Mars Rover Mission," *Space News*, Sept. 30, 2002, 8.

20. Richard Kerr, "NASA's New Road to Faster, Cheaper, Better Exploration," *Science*, Nov. 15, 2002, 1320–1322.

21. James Garvin, interview by author, June 29, 2011; Firouz Naderi, "The Mars Program I Recall," *JPL Universe*, Jan. 2009, 1, 2, 8.

22. Firouz Naderi, interview by author, Nov. 12, 2009.

23. Steve Squyres, interview by author, June 6, 2010.

24. "NASA Rovers Slated to Examine Two Intriguing Sites on Mars," *NASA News Release*, Apr. 11, 2003, NASA History Office Files.

25. William Harwood, "NASA Hopes to Tap Mystery of Water on Mars," *Washington Post*, May 12, 2003, A7.

26. William Harwood, "NASA Launches First of Two Rovers," *Washington Post*, June 11, 2003, A14.

27. Lambright, *Executive Response*.

28. Ibid.

29. Ibid.

30. William Harwood, "Mars Lander Remains Silent," *Washington Post*, Dec. 27, 2003, A2.

31. Kathy Sawyer, "Bracing for 'Six Minutes from Hell,'" *Washington Post*, Jan. 4, 2004, A3.

32. Firouz Naderi, interview by author, Nov. 12, 2009.

33. Kathy Sawyer, "A Triumphant Landing on Mars," *Washington Post*, Jan. 5, 2004, 1.

34. Charles Elachi, interview by author, Nov. 11, 2009.

35. Leonard David, "Spirit Landing Opens New Chapter of Mars Exploration," *Space News*, Jan. 12, 2004, 16.

36. Andrew Lawler, "Scientists Add Up Gains, Losses in Bush's New Vision for NASA," *Science*, Jan. 23, 2004, 444–445; Mike Allen and Eric Pianin, "Bush Outlines Space Agenda," *Washington Post*, Jan. 15, 2004, A1.

37. Len Fisk, interview by author, Oct. 9, 2009.

38. Ibid.

39. Andrew Lawler, "How Much Space for Science?," *Science*, Jan. 30, 2004, 610–612; Lambright, *Executive Response*.

40. Kathy Sawyer, "2nd Rover Opens Eyes to Wonder of Mars," *Washington Post*, Jan. 26, 2004, A1; Brian Berger, "Spirit's Success Attributed to Lessons from Past Mistakes," *Space News*, Jan. 12, 2004, 3.

41. David Malakoff, "2005 Budget Makes Flat a Virtue," *Science*, Feb. 6, 2004, 748–750; Charles Siefe, "Highlights from the Budget," *Science*, Feb. 6, 2004, 749.

42. Len Fisk, interview by author, Oct. 9, 2009.

43. Jefferson Morris, "'Mars Testbed' Missions to Begin in 2011," *Aviation Week's Aerospace Daily*, Feb. 23, 2004, 4; Leonard David, "Rovers Aiding NASA in Planning, Prioritizing Future Missions," *Space News*, Mar. 8, 2004, 6.

44. "Opportunity Rover Finds Evidence 'Salty Sea' Existed on Mars," *Aviation Week's Aerospace Daily*, Mar. 24, 2004, NASA History Office Files.

45. "Opportunity Finds Evidence of Liquid Water on Mars, NASA Says," *Aviation Week's Aerospace Daily*, Mar. 3, 2004, NASA History Office Files; David, "Rovers Aiding NASA"; "Opportunity Rover Finds Evidence 'Salty Sea.'"

46. Andrew Lawler, "Scientists Fear Collateral Damage from NASA's Revised Vision," *Science*, Mar. 26, 2004, 1952–1953.

47. Len Fisk, interview by author, Oct. 9, 2009.

48. Andrew Lawler, "NASA Officials Consider Major Reorganization," *Science*, Apr. 30, 2004, 663.

49. Andrew Lawler, "Rising Costs of Shuttle and Hubble Could Break NASA Budget," *Science*, Sept. 24, 2004, 1882–1883.

50. Jeffrey Mervis, "Bush Victory Leaves Scars—and Concerns about Funding," *Science*, Nov. 12, 2004, 1110–1113.

51. Doug McCuistion, interview by author, Mar. 5, 2009.

52. Andrew Lawler and Jeffrey Mervis, "Science Agencies Caught in Post Election Spending Squeeze," *Science*, Dec. 3, 2004, 1662–1663.

53. Andrew Lawler, "O'Keefe to Go, but Hubble Remains a Battleground," *Science*, Dec. 17, 2004, 2018–2019.

54. Chris Shank, interview by author, Dec. 3, 2008.

55. Jeffrey Mervis, "Caught in the Squeeze," *Science*, Feb. 11, 2005, 832–834.

56. James Garvin, interview by author, June 29, 2011.

57. Ibid.; Lambright, *Executive Response*.

58. Donald Kennedy, "Confusion at the Space Agency," *Science*, Mar. 11, 2005, 1533.

CHAPTER ELEVEN: Implementing amidst Conflict

1. W. Henry Lambright, *Launching a New Mission: Michael Griffin and NASA's Return to the Moon* (Washington, DC: IBM, 2009).

2. Andrew Lawler, "Balancing the Right Stuff," *Science*, Apr. 22, 2005, 484–487.

3. Ibid.; Lambright, *Launching a New Mission*.

4. Lawler, "Balancing the Right Stuff"; Andrew Lawler, "Nominee Wins Quick Praise for His Technical Expertise," *Science*, Mar. 18, 2005, 1709.

5. Andrew Lawler, "Griffin Names Winners and Losers in Cost Squeeze," *Science*, May 20, 2005, 1100.

6. James Garvin, interview by author, June 29, 2011; Doug McCuistion, interview by author, Mar. 5, 2009.

7. Brian Berger, "NASA Mars Telecom Orbiter Axed as Space Agency Priorities Shift," *Space News*, July 25, 2005, 1, 4.

8. Tariq Malik, "New NASA Orbiter to Pave Way for Future Mars Missions," *Space News*, Aug. 8, 2005, 16; Andrew Lawler, "Mars Bound," *Science*, Aug. 19, 2005, 1105.

9. James Garvin, interview by author, June 29, 2011.

10. Andrew Lawler, "Budget Woes Greet NASA Science Chief," *Science*, Aug. 19, 2005, 1165.

11. "Mars Science Lab Can Wait," *Space News*, Aug. 8, 2005, 18.

12. Louis Friedman, "Mars Cannot Wait," *Space News*, Aug. 22, 2005, 18.

13. Leonard David, "Mars Rover Hits Milestone; Spectacular Viewpoint," *Space News*, Aug. 29, 2005, 20.

14. Marcia Dunn, "NASA Scans Hurricane Damage to Shuttle Ops," *Washington Post*, Sept. 8, 2005, www.washingtonpost.com/, NASA History Office Files.

15. Warren Leary, "NASA Planning Return to Moon within 13 Years," *New York Times*, Sept. 20, 2005, 1.

16. Ibid.

17. Ibid.; Marcia Dunn, "NASA Planning Moon Launch for 2018," *Washington Post*, Sept. 9, 2005, http://washingtonpost.com/wp/wp-dyn/content/article/2005/09/19/AR2005091900705_p, NASA History Office Files.

18. Lambright, *Launching a New Mission*, 11.

19. Carol Gramling and Andrew Lawler, "You Make the Call, NASA Chief Tells Scientists," *Science*, Oct. 21, 2005, 425.

20. Brian Berger, "Griffin Takes NASA Back to the Future in Exploration," *Space News*, Dec. 9, 2005, 10.

21. Eli Kintisch, "U.S. Science Budgets Emerge," *Science*, Nov. 11, 2005, 957.

22. Jeffrey Mervis, "NIH Shrinks, NSF Crawls as Congress Finishes Spending Bills," *Science*, Jan. 6, 2006, 28.

23. Donald Kennedy, "NASA: Back to Eating Seed Corn," *Science*, Nov. 25, 2005, 1245.

24. Lambright, *Launching a New Mission*, 16–17; Michael Griffin, interview by author, Feb. 29, 2008.

25. Lambright, *Launching a New Mission*, 16–17.

26. Eli Kintisch and Jeffrey Mervis, "A Budget with Big Winners and Losers," *Science*, Feb. 10, 2006, 702–704.

27. Ibid.

28. Andrew Lawler, "A Space Race to the Bottom Line," *Science*, Mar. 17, 2006, 1540–1543.

29. Rocco Mancinelli, "Destroying Astrobiology Would Be a Disaster," *Space News*, Mar. 20, 2006, 19.

30. Louis Friedman, "World Watch: Washington DC," *Planetary Report*, Mar./Apr. 2006, 17.

31. Lawler, "Space Race to the Bottom Line," 1542.

32. Ibid.

33. Ibid., 1543.

34. Ibid., 1540.

35. Ibid., 1543.

36. Guy Gugliotta, "NASA Craft Reaches Mars," *Washington Post*, Mar. 11, 2006, A5.

37. Tariq Malik and Leonard David, "MRO Enters Mars Orbit, Will Search for Future Landing Sites," *Space News*, Mar. 20, 2006, 14.

38. Brian Berger, "Draft Plan Scales Back NASA's Mars Agenda," *Space News*, Apr. 3, 2006, 1, 14.

39. Andrew Lawler, "Crisis Deepens as Scientists Fail to Rejigger Space Research," *Science*, May 12, 2006, 824–825.

40. Ibid.

41. Ibid., 824.

42. Richard Cook, interview by author, June 6, 2011.

43. Dan McCleese, interview by author, Mar. 30, 2011.

44. Rob Manning, interview by author, Aug. 8, 2011.

45. Gentry Lee, interview by author, Aug. 8, 2011.

46. Richard Cook, interview by author, June 6, 2011.

47. Gentry Lee, interview by author, Aug. 8, 2011.

48. Rob Manning, interview by author, Aug. 8, 2011.

49. Gentry Lee, interview by author, Aug. 8, 2011.

50. Richard Kerr, "In Search of the Red Planet's Sweet Spot," *Science*, June 16, 2006, 1588–1590.

51. Leonard David, "Science Panel Pushes for Mars Surface Network," *Space News*, July 10, 2006, 10; Andrew Lawler, "Long-Term Mars Exploration under Threat, Panel Warns," *Science*, July 14, 2006, 157.

52. Andrew Lawler, "NASA Chief Blasts Science Advisers, Widening Split with Researchers," *Science*, Aug. 25, 2006, 1032.

53. Craig Covault, "Victoria's Secret," *Aviation Week and Space Technology*, Oct. 2, 2006, 25–27.

54. Craig Covault, "Spying on Mars," *Aviation Week and Space Technology*, Oct. 23, 2006, 24–26.

55. Leonard David, "Mars Global Surveyor Stays Silent, Is Feared Lost," *Space News*, Nov. 27, 2006, 13.

56. Rick Weiss, "Mars Photos May Indicate the Recent Flow of Water," *Washington Post*, Dec. 7, 2006, A2.

57. Richard Kerr, "Mars Orbiter's Swan Song: The Red Planet's A-Changin'," *Science*, Dec. 8, 2006, 1528–1529.

58. Jeffrey Mervis, "NIH Trims Award Size as Spending Crunch Looms," *Science*, Dec. 22, 2006, 1862.

59. Jeffrey Mervis, "Research Rises—and Falls—in the President's Spending Plan," *Science*, Feb. 9, 2007, 750, 753; Jeffrey Mervis, "Science Adviser Says That Pruning Is the Key to a Healthy Budget," *Science*, Feb. 16, 2007, 927.

60. Andrew Lawler, "Astrobiology Fights for Its Life," *Science*, Jan. 19, 2007, 318–321.

61. Brian Berger, "Mars Phoenix Needs More Funding to Launch on Time," *Space News*, Jan. 15, 2007, 13.

62. Andrew Lawler, "Stern but Kind at NASA," *Science*, Feb. 2, 2007, 585.

63. Lambright, *Launching a New Mission*, 22.

64. Andrew Lawler, "Taking a Stern Look at NASA Science," *Science*, Mar. 16, 2007, 1484.

65. Lambright, *Launching a New Mission*, 22.

66. Andrew Lawler, "Expert Panel Faults Expert Panels," *Science*, May 4, 2007, 675.

67. "Mars Rover Spirit Unearths Surprise Evidence of Wetter Past," NASA News Release 07–118, May 21, 2007.

68. S. Alan Stern and Gerry Griffin, "U.S. Needs Near-Term Results in Human Space Exploration," *Space News*, Sept. 26, 2011.

69. Michael Meyer, interview by author, Mar. 4, 2009.

70. Alan Stern, interview by author, July 20, 2011.

71. Frank Morring Jr., "NASA May Move Up Mars Sample Return Mission," *Aviation Week Aerospace Daily and Defense Report* 223, no. 4 (July 6, 2007).

72. Leonard David, "Mars Sample Return Proposal Stirs Excitement, Controversy," *Space News*, July 23, 2007, 19.

73. Ibid.

74. Robert Zubrin, "Don't Wreck the Mars Program," *Space News*, July 23, 2007, 17.

75. David, "Mars Sample Return Proposal."

76. Morring, "NASA May Move Up."

77. Charles Petit, "Northern Exposure," *Air and Space*, Aug. 2007, 50–55.

78. Frank Morring Jr., "NASA Trims Mars Science Lab Rover to Cover $75 Million Shortfall," *Aviation Week Aerospace Daily and Defense Report* 223, no. 53 (Sept. 14, 2007); Brian Berger, "NASA Science Chief Says Cuts Should Keep MSL on Schedule," *Space News*, Sept. 24, 2007, 9.

79. Berger, "NASA Science Chief."

80. Brian Berger, "NASA Restores Two Instruments aboard Mars Science Laboratory," *Space News*, Nov. 2, 2007, 4; Frank Morring Jr., "Back on Board," *Aviation Week and Space Technology*, Nov. 19, 2007, 17.

81. Richard Kerr, "Majority Rules in Finding a Path for the Next Mars Rover," *Science*, Nov. 9, 2007, 908–909.

82. Jeffrey Mervis, "Promising Year Ends Badly After Fiscal Showdown Squeezes Science," *Science*, Jan. 4, 2008, 18–19.

83. "Missed Opportunity," *Aviation Week and Space Technology*, Jan. 7, 2008, 14.

84. Andrew Lawler, "Scientists Hope to Adjust the President's Vision of Space," *Science*, Feb. 1, 2008, 565–566.

85. Jeffrey Mervis, "A Science Budget of Choices and Chances," *Science*, Feb. 8, 2008, 714–715.

86. Lawler, "Scientists Hope to Adjust"; Andrew Lawler, "War of the Worlds?," *Science*, Feb. 29, 2008, 1174–1176.

87. Ed Weiler, interview by author, Sept. 19, 2008. This was what Weiler heard from others. He personally never heard Stern say that.

88. Jefferson Morris, "NASA's Mars Budget Dips $1.1B Compared to FY '08 Runout," *Aviation Week Aerospace Daily and Defense Report* 225, no. 88 (Feb. 4, 2008); Robert Braun, "The Future of Our Mars Exploration Program," *Space News*, Mar. 3, 2008, 19, 21.

89. Lawler, "War of the Worlds?," 1176.

90. Ibid.

91. Chris McKay, interview by author, Oct. 13, 2009.

92. Lawler, "War of the Worlds?," 1175–1176.

93. Alan Stern, interview by author, July 20, 2011.

94. Orlando Figueroa, interview by author, Feb. 26, 2009; James Garvin, interview by author, June 29, 2011; John Grotzinger, interview by author, June 27, 2011.

95. Alan Stern, interview by author, July 20, 2011.

96. James Green, interview by author, Aug. 9, 2011; Alan Stern, interview by author, July 20, 2011.

97. Alan Stern, interview by author, July 20, 2011.

98. Director of Planetary Science Division to Jet Propulsion Laboratory, Program Manager, Mars Exploration Program, Mar. 19, 2008. Memo provided to author by James Green, NASA.

99. Leonard David, "Stern's Resignation Underscores NASA's Science Budget Challenges," *Space News*, Mar. 31, 2008, 1, 4.

100. Mark Kaufman, "Mars Rovers Survive NASA's Budget Crunch," *Washington Post*, Mar. 26, 2008, A17.

101. Alan Stern, interview by author, July 20, 2011. Griffin told the media that he was reluctant to see Stern go. See Brian Berger, "NASA Taps Experienced Hand to Replace Science Chief," *Space News*, Mar. 31, 2008, 5.

102. Kaufman, "Mars Rovers Survive," A17; David, "Stern Resignation Underscores," 1, 4.

103. Alan Stern, interview by author, July 20, 2011.

104. Berger, "NASA Taps Experienced Hand"; Andrew Lawler, "NASA's Stern Quits over Mars Exploration Plans," *Science*, Apr. 4, 2008, 31.

105. Berger, "NASA Taps Experienced Hand."

106. Lawler, "NASA's Stern Quits."

107. Ed Weiler, interview by author, Apr. 16, 2010.

108. Guy Webster and Mark Whalen, "Phoenix Stands Tall," *JPL Universe*, June 2008, 1.

109. Craig Covault, "Phoenix Delivers," *Aviation Week and Space Technology*, June 9, 2008, 34–35; Clara Moskowitz, "Phoenix Scientists Find Evidence of Ice on Mars," *Space News*, June 23, 2008, 17; Marc Kaufman, "Mars Soil 'Friendly' to Life, Tests Show," *Washington Post*, June 27, 2008, A6; Craig Covault, "Phoenix Scores Again," *Aviation Week and Space Technology*, Aug. 4, 2008, 35; David Fahrenthold, "Existence of Water on Mars Confirmed," *Washington Post*, Aug. 1, 2008, A2.

110. Marc Kaufman, "Mars Finding Doesn't Rule Out Life," *Washington Post*, Aug. 6, 2008, A2; Craig Covault, "Martian Chronicles Revisited," *Aviation Week and Space Technology*, Aug. 11, 2008, 30–32.

111. "Delayed Scout," *Aviation Week and Space Technology*, Sept. 22, 2008, 20.

112. Andrew Lawler, "Rising Costs Could Delay NASA's Next Mission to Mars and Future Launches," *Science*, Sept. 26, 2008, 1754.

113. Peter B. de Selding, "Study: Mars Sample Return Would Take 10 Years, Cost $5 Billion-Plus," *Space News*, July 14, 2008, 1, 11.

114. Lawler, "Rising Costs Could Delay."

115. "NASA Sticks with 2009 for Mars Science Lab Launch," *Space News*, Oct. 13, 2008, 3.

116. "Window Closing," *Aviation Week and Space Technology*, Oct. 20, 2008, 21.

117. Alan Stern, "Viewing NASA's Mars Budget with Resignation," *Science*, Oct. 31, 2008, 672.

118. Alan Stern, "NASA's Black Hole Budgets," *New York Times*, Nov. 24, 2008, www .nytimes.com/2008/11/24/opinion/24stern.html?pagewanted=all.

119. James Garvin, "The Price of Exploration," *Science*, Nov. 28, 2008, 1324.

120. John F. Mustard to Doug McCuistion and Michael Meyer, NASA, Aug. 15, 2008. Letter provided author by James Green, NASA.

121. "NASA Drops Sample Canister from 2009 Mars Science Lab," *Space News*, Nov. 24, 2008, 3.

122. Ed Weiler, interview by author, Apr. 16, 2010.

123. Richard Cook, interview by author, June 6, 2011.

124. Rob Manning, interview by author, Aug. 8, 2011.

125. Doug McCuistion, interview by author, Mar. 5, 2009.

126. Ed Weiler, interview by author, Apr. 16, 2010.

127. Andrew Lawler, "Delays in Mars Mission Will Ripple across Space Science," *Science*, Dec. 12, 2008, 1618.

128. Rachel Courtland, "Over-budget Mars Rover Mission Delayed until 2011," *New Scientist*, Dec. 4, 2008, www.newscientist.com/, NASA History Office Files; Clive Simpson and Tim Furniss, "NASA Delays Mars Science Laboratory to 2011," *Spaceflight*, Feb. 2009, 44, NASA History Office Files.

129. Lawler, "Delays in Mars Mission."

CHAPTER TWELVE: Attempting Alliance

1. Andrew Lawler, "NASA, ESA Choose King of Planets for Flagship Mission in 2020," *Science*, Feb. 27, 2009, 1154.

2. Firouz Naderi, interview by author, Nov. 12, 2009; Firouz Naderi, "The Mars Program I Recall," *JPL Universe*, Jan. 2009, 1, 2, 8.

3. Richard Cook, interview by author, June 6, 2011.

4. Richard Cook, "Managing Complexity," unpublished paper.

5. Gentry Lee, interview by author, Aug. 8, 2011.

6. Andrew Lawler, "Can a Shotgun Wedding Help NASA and ESA Explore the Red Planet?," *Science*, Mar. 27, 2009, 1666–1667; Peter B. de Selding, "NASA Poised to Take Key Role in European ExoMars Rover Mission," *Space News*, June 22, 2009, 1, 17.

7. Lawler, "Can a Shotgun Wedding Help."

8. Leonard David, "Mars Rover Scientist Squyres to Chair Planetary Decadal Survey," *Space News*, Mar. 30, 2009, 22.

9. "Hubbard to Lead NASA Mars Program Review," *Space News*, Apr. 20, 2009, 9.

10. Andrew Lawler, "Trouble on the Final Frontier," *Science*, Apr. 3, 2009, 34–35.

11. "JPL Makes More Changes to MSL Management Team," *Space News*, Mar. 23, 2009, 9.

12. "Hubbard to Lead Mars Program Review."

13. Louis Friedman, "World Watch," *Planetary Report*, May/June 2009, 19.

14. Jeffrey Mervis, "Stimulus Spending Looms Large as Obama Charts a Course for Science," *Science*, May 15, 2009, 864–866; "NASA Asks Augustine to Point the Way," *Science*, May 22, 2009, 999.

15. David Southwood, interview by author, Apr. 1, 2011.

16. Edward Weiler, interview by author, Apr. 16, 2010.

17. Peter B. de Selding, "Europe-US Mars Agreement Silent on ExoMars Mission," *Space News*, July 13, 2009, 4.

18. Peter B. de Selding, "Italy Skeptical of US-European Mars Collaboration," *Space News*, July 27, 2009, 6.

19. Michael A. Taverna and Frank Morring Jr., "Exploring Together," *Aviation Week and Space Technology*, June 29, 2009, 54.

20. "Joint US-European Exploration of Mars," *AAAS Policy Alert*, July 22, 2009, 2; Marcia Smith, "Planetary Science Decadal Survey Steering Committee Meeting Summary," SpacePolicyOnline.com, July 10, 2009.

21. Amy Klamper, "White House Urges Restraint in Space Science Plans," *Space News*, July 13, 2009, 14.

22. Frank Morring Jr., "On Duty," *Aviation Week and Space Technology*, July 27, 2009, 36–37.

23. Laura Delgado, "The Augustine Committee Report—Review of US Human Space Flight Plans," SpacePolicyOnline.com Fact Sheet, Sept. 14, 2009; "Review of US Human Space Flight Plans Committee," *Summary Report*, Sept. 8, 2009, NASA History Office Files.

24. Norman Augustine et al., *Seeking a Human Spaceflight Program Worthy of a Great Nation, Executive Summary* (Washington, DC: NASA, 2009), 15.

25. Debra Werner, "Augustine Defends Panel's Findings to U.S. Lawmakers," *Space News*, Sept. 21, 2009, 6.

26. Augustine et al., *Seeking a Human Spaceflight Program*, 9–17.

27. Peter B. de Selding and Amy Klamper, "U.S., Europe Making Progress on Joint Mars Exploration Plan," *Space News*, Oct. 5, 2009, 11.

28. "Russia Delays Phobos-Grunt Mars Mission until 2011," *Space News*, Sept. 28, 2009, 9.

29. "Russia's Failed Mars Probe Crashes into Pacific," *New York Times,* Jan. 15, 2012, www.nytimes.com/.

30. Charles Bolden Jr. and Jean-Jacques Dordain, "Statement of Intent for Potential Joint Robotic Exploration of Mars," Nov. 5, 2009. Document provided by NASA to author.

31. NASA, *2010 Science Plan for NASA's Science Mission Directorate* (Washington, DC: NASA, 2010), 52.

32. Amy Svitak, "NASA Budget Outlook Relegates Flagship Probes to Back Burner," *Space News,* Mar. 2011, 4; Richard Kerr, "Price Tags for Planet Missions Force NASA to Lower Its Sites," *Science,* Mar. 11, 2011, 1254–1255.

33. Kerr, "Price Tags for Planet Missions." The "sticker shock" comment appears on p. 1254.

34. Svitak, "NASA Budget Outlook," 4.

35. Marcia Smith, "Planetary Scientists Need to Make Their Case to Congress," SpacePolicyOnline.com, Mar. 7, 2011, www.spacepolicyonline.com/.

36. Amy Svitak, "NASA Money Woes Batter Planetary Flagship Project," *Space News,* Mar. 21, 2011, 14.

37. Peter B. de Selding, "ESA Halts ExoMars Orbiter Work to Rethink Red Planet Plans with NASA," *Space News,* Apr. 25, 2011, 1, 6.

38. David Southwood, interview by author, Apr. 1, 2011.

39. De Selding, "ESA Halts ExoMars Orbiter Work."

40. Leonard David, "NASA Nixes 3-D Camera for Mars Science Laboratory," *Space News,* Apr. 4, 2011, 15.

41. Charles Bolden, interview by author, Aug. 18, 2011.

42. James Garvin, interview by author, June 29, 2011.

43. David, "NASA Nixes 3-D Camera."

44. "Profile: Alvaro Giménez," *Space News,* May 9, 2011, 18.

45. Peter B. de Selding, "France, Britain Reluctant to Recommit to Revised but 'Risky' ExoMars Mission," *Space News,* June 27, 2011, 1, 17.

46. Ibid.

47. "Troubled Spacecraft," *Aviation Week and Space Technology,* May 30, 2011, 14.

48. Frank Morring Jr., "Close Call," *Aviation Week and Space Technology,* May 30, 2011, 31.

49. Marcia Smith, "NASA IG: MSL May Not Be Ready by November, Power Supply Problems Will Reduce Mission Capabilities," SpacePolicyOnline.com, June 8, 2011, www.spacepolicyonline.com/; Office of Inspector General, *NASA's Management of the Mars Science Laboratory Project* (Washington, DC: NASA, June 8, 2011). The quote is on p. iv of the IG report.

50. Dan Leone, "Mars Science Lab Needs $44 M More to Fly, NASA Audit Finds," *Space News,* June 13, 2011, 12.

51. Eryn Brown, "Rival Rovers," *Syracuse Post-Standard,* June 20, 2011, C-3.

52. De Selding, "France, Britain Reluctant," 1.

53. Ibid.

54. Peter B. de Selding, "Italy OK with Canceling ExoMars Demonstration Lander," *Space News,* July 11, 2011, 7.

55. Mike Wall, "Hunt for MSL's Landing Site Down to Two Craters," *Space News,* July 11, 2011, 10.

56. John Grotzinger, interview by author, June 27, 2011.

57. Richard Kerr, "How an Alluring Geological Enigma Won the Mars Rover Sweepstakes," *Science*, July 29, 2011, 508–509.

58. Marcia Smith, "NASA Starts Planning for Smaller Mars Mission in 2018," SpacePolicyOnline.com, Feb. 27, 2012, www.spacepolicyonline.com/.

59. "Weiler Quit NASA over Cuts to Mars Program," *Science*, Feb. 17, 2012, 780–781.

60. "Martian Mystery," *Aviation Week and Space Technology*, Oct. 3, 2011, 13.

61. Peter B. de Selding, "NASA Cannot Launch 2016 ExoMars Orbiter," *SpaceNews*, Oct. 3, 2011, 4, 13.

62. Peter B. de Selding, "ESA Formally Invites Roscosmos to Join ExoMars Mission as Full-Fledged Partner," *Space News*, Oct. 17, 2011, 1, 4.

63. Ibid.

64. "NASA Launches Sophisticated Rover on Journey to Mars," *New York Times*, Nov. 27, 2011, 24.

65. "NASA Launches Most Capable and Robust Rover to Mars," *NASA News Release*, Nov. 26, 2011.

CHAPTER THIRTEEN: Landing on Mars and Looking Ahead

1. Jeffrey Merus, "Science Spared Brunt of Ax in Budget Request," *Science*, Feb. 17, 2012, 784.

2. Dan Leone, "NASA Rebooting Mars Program in Hunt for Cheaper Missions," *Space News*, Feb. 20, 2012, 7.

3. Ibid.

4. Charles Bolden, "NASA Recharting Its Path to Mars," *Space News*, Feb. 20, 2012, 17.

5. Marcia Smith, "NASA Starts Planning for Smaller Mars Missions in 2018," SpacePolicyOnline.com, Feb. 27, 2012, www.spacepolicyonline.com/.

6. Dan Leone, "Former Mars Czar Tapped to Lead NASA's Mars Reboot," *Space News*, Mar. 5, 2012, 7.

7. Richard Kerr, "Planetary Science Is Busting Budgets," *Science*, July 27, 2012, 404.

8. Casey Drier, "The Battle Continues," *Planetary Report*, Sept. 2012, 20.

9. Dan Leone, "U.S. House Appropriators Push Back on Obama's Planned Mars Cuts," *Space News*, Mar. 5, 2012, 6.

10. "Rep. Wolf Eyes Commercial Crew Budget for Mars Relief," *Space News*, Mar. 26, 2012, 7.

11. Marcia Smith, "No Prohibition against Science Flagship Missions, OMB Tells NRC," SpacePolicyOnline.com, Apr. 5, 2012, www.spacepolicyonline.com/.

12. Marcia Smith, "What Keeps You Awake at Night?," SpacePolicyOnline.com, Apr. 15, 2012, www.spacepolicyonline.com/news/what-keeps-you-awake-at-night.

13. Marcia Smith, "NASA Seeks Input for Planning for New Mars Mission," SpacePolicyOnline.com, Apr. 13, 2012, www.spacepolicyonline.com/news/nasa-seeks -input-for-new-mars-mission.

14. Chris Carberry, "The Power of Curiosity," *Space News*, July 23, 2012, 19.

15. Ibid.

16. Kenneth Chang, "Simulated Space 'Terror' Offers NASA an Online Following," *New York Times*, July 11, 2012, A14.

17. Ibid.

18. Ibid.

19. Linda Billings, "MSL: Public Spectacle or Learning Opportunity?," *Space News*, Oct. 1, 2012, 19.

20. Ibid.

21. Amina Khan, "Mars Rover Curiosity's Other Mission: PR," *Los Angeles Times*, Dec. 1, 2012, http://articles.latimes.com/2012/dec/01/science/la-sci-nasa-jpl-curiosity-pr-20121201.

22. Bill Nye, "Planetary Science—Because It's Hard," *Space News*, July 30, 2012, 19.

23. Billings, "MSL."

24. Jeffrey Kluger, "Live from Mars," *Time*, Aug. 20, 2012, 22.

25. Ibid.

26. Ibid.

27. "NASA/JPL Mars Curiosity Rover Operations Team," *Space News*, Aug. 27, 2012, 11; Alan Boyle, "Obama Tells Mars Rover Team: Let Me Know If You See Martians," NBCnews.com, Aug. 13, 2012, http://cosmiclog.nbcnews.com/.

28. Ibid.

29. Dan Leone, "NASA Picks Mars Lander as Next Discovery Mission," *Space News*, Aug. 27, 2012, 17.

30. Drier, "Battle Continues."

31. Marcia Smith, "Mars Program Planning Group Identifies Options for Future Mars Exploration—Update," SpacePolicyOnline.com, Sept. 25, 2012, www.spacepolicyonline.com/.

32. Frank Morring Jr., "Working Together," *Aviation Week and Space Technology*, Oct. 1, 2012, 36–37.

33. Dan Leone, "Mars Planning Group Endorses Sample Return," *Space News*, Oct. 1, 2012, 1, 4.

34. Morring, "Working Together."

35. Marcia Smith, "JPL Clarifies Curiosity's Lack of Historic Discovery So Far," SpacePolicyOnline.com, Nov. 29, 2012, www.spacepolicyonline.com/new/jpl-clarifies-curiositys-lack-of-historic-discover.

36. Marcia Smith, "NASA Announces New $1.5 Billion Mars Rover for Launch in 2020," SpacePolicyOnline.com, Dec. 4, 2012, www.spacepolicyonline.com/news/nasa-announces-new-1-5-billion-mars-rover-for-launch-in-2020.

Conclusion

1. Chris Scolese, interview by author, Dec. 4, 2008.

2. Charles Bolden, "NASA Recharting Its Path to Mars," *Space News*, Feb. 20, 2012, 17.

3. James L. True, Bryan Jones, and Frank R. Baumgartner, "Punctuated-Equilibrium Theory: Explaining Stability and Change in American Policymaking," in Paul Sabatier, ed., *Theories of the Policy Process* (Boulder, CO: Westview, 1999).

4. Scott Hubbard, *Exploring Mars: Chronicles from a Decade of Discovery* (Tucson, AZ: University of Arizona Press, 2012), 99.

5. Andrew Chaikin, *A Passion for Mars: Intrepid Explorers of the Red Planet* (New York: Abrams, 2008).

6. Charles Elachi, interview by author, Nov. 11, 2009.

7. David Southwood, interview by author, Apr. 1, 2011.

8. James E. Webb, *Space Age Management: The Large-Scale Approach* (New York: McGraw-Hill, 1989), 89.

9. Jack Gibbons, interview by author, June 17, 2010.

10. Doug McCuistion, interview by author, Apr. 16, 2010.

11. John Grotzinger, "Beyond Water on Mars," *Nature Geoscience*, Apr. 1, 2009, 1.

12. NASA planetary director Jim Green has spoken of science's "guiding light," and NASA Administrator Dan Goldin commented on politics as a determinant of pace and effectiveness.

13. Samuel Florman, "Technology and the Tragic View," in Albert Teich, ed., *Technology and the Future* (Boston, MA: Wadsworth, 2009), 36–44.

Index

About the Author

W. Henry Lambright is a professor of public administration and international affairs as well as of political science at the Maxwell School of Citizenship and Public Affairs at Syracuse University.

Frequently quoted in the media, Dr. Lambright has performed research for the National Science Foundation, NASA, the Department of Energy, the Department of Defense, the State Department, IBM, and other organizations. He is the author or editor of seven previous books, including *Powering Apollo: James E. Webb of NASA*; *Presidential Management of Science and Technology: The Johnson Presidency*; *Governing Science and Technology*; and *Space Policy in the Twenty-First Century*. In addition, he has written more than 300 articles, papers, and reports. He is a Fellow of both the American Association for the Advancement of Science and the National Academy of Public Administration.

His doctorate is from Columbia University, where he also received his master's degree. Dr. Lambright received his undergraduate degree from Johns Hopkins University.